Springer Optimization and Its Applications

Volume 147

Aims and Scope
Optimization has been expanding in all directions at an astonishing rate during the last few decades. New algorithmic and theoretical techniques have been developed, the diffusion into other disciplines has proceeded at a rapid pace, and our knowledge of all aspects of the field has grown even more profound. At the same time, one of the most striking trends in optimization is the constantly increasing emphasis on the interdisciplinary nature of the field. Optimization has been a basic tool in all areas of applied mathematics, engineering, medicine, economics and other sciences.

The series *Springer Optimization and Its Applications* publishes undergraduate and graduate textbooks, monographs and state-of-the-art expository works that focus on algorithms for solving optimization problems and also study applications involving such problems. Some of the topics covered include nonlinear optimization (convex and nonconvex), network flow problems, stochastic optimization, optimal control, discrete optimization, multi-objective programming, description of software packages, approximation techniques and heuristic approaches.

More information about this series at http://www.springer.com/series/7393

Ding-Zhu Du • Panos M. Pardalos • Zhao Zhang
Editors

Nonlinear Combinatorial Optimization

 Springer

Editors
Ding-Zhu Du
Department of Computer Science
The University of Texas at Dallas
Richardson, TX, USA

Panos M. Pardalos (iD)
Department of Industrial & Systems
Engineering
University of Florida
Gainesville, FL, USA

Zhao Zhang
Department of Computer Science
Zhejiang Normal University
Jinhua, Zhejiang, China

ISSN 1931-6828   ISSN 1931-6836   (electronic)
Springer Optimization and Its Applications
ISBN 978-3-030-16196-5   ISBN 978-3-030-16194-1   (eBook)
https://doi.org/10.1007/978-3-030-16194-1

Mathematics Subject Classification (2010): 90C27

This Springer imprint is published by the registered company Springer Nature Switzerland AG.
The registered company address is: Gewerbestrasse 11, 6330 Cham, Switzerland

Preface

With the recent developments of computer technology, more and more optimization problems appear to have discrete structure with nonlinear objective function and/or nonlinear constraints. They form a new research area, *nonlinear combinatorial optimization*. Let us mention a few examples.

In wireless sensor networks, the energy efficiency is an important issue and, usually, the energy is a nonlinear function with respect to communication radius. The energy-efficient topological control problem is to minimize the total energy consumption over *all* spanning tree in homogeneous sensor systems, or over all strongly connected subnetworks in heterogeneous sensor systems, and hence it is a nonlinear combinatorial optimization problem.

In cloud computing, the processing time of a job depends on the energy assigned to the job and is actually a nonlinear function with respect to the energy. Therefore, the job scheduling problem with the total energy constraint is also a nonlinear combinatorial optimization problem with nonlinear constraint.

In social networks, the expectation of influence, i.e., the expectation of the number of infected nodes, is a nonlinear function with respect to the seed set and hence its maximization is a nonlinear optimization problem.

In machine learning, a lot of problems can be formulated into submodular optimizations, which gives a big motivation to study submodular optimization with machine learning approaches.

There are various types of nonlinear combinatorial optimization problems, such as submodular cover and submodular knapsack problems, submodular optimization and nonsubmodular optimization, offline and online nonlinear combinatorial optimization, set function optimization and integer lattice optimization, and the discrete DC (i.e., the difference of two convex functions) programming. There also are many new approaches produced when two areas, nonlinear continuous optimization and discrete optimization, meet in this interdisciplinary domain.

While the nonlinearity is merged into combinatorial optimization, the nonlinear optimization method is getting involved into those problems. For example, discrete Newton method has been successfully used for solving the inverse problem of combinatorial optimization, convex relaxation and algorithms for continuous convex program play important roles in dealing with submodular optimization, and discrete convex analysis plays an important role in solving the DC programming. On the other hand, some methodologies are extended from linear to nonlinear, such as the Graver basis, which is an important tool to study the linear integer programming, and now is involved in solving the nonlinear integer programming.

This book is a collection of extraordinary chapters written by invited leading experts in the area of nonlinear combinatorial optimization. The subjects cover theoretical developments, such as discrete Newton methods, the Graver basis, submodular optimization, and set function optimization, and various applications, such as topological control in wireless networks, influence maximization, friending, rumor blocking in social networks, and multi-document extractive summarization in machine learning. All chapters in this book provide a clear and authoritative picture of what nonlinear combinatorial optimization is and the direction in which research is going on. Thus, we hope that the book would serve as a useful reference for university students, professors, and researchers with interest in this area, possibly from applied mathematics, computer science, industrial and system engineering, and management science.

Richardson, TX, USA   Ding-Zhu Du
Gainesville, FL, USA   Panos M. Pardalos
Jinhua, China   Zhao Zhang

Contents

A Role of Minimum Spanning Tree

Zhao Zhang and Xiaohui Huang

Abstract In a wireless sensor network, a topology control problem aims to adjust power of sensors so that the topology supported by the power has some desirable property and the total power is as small as possible. In this chapter, we shall present studies on the topology control problem with the properties of containing a spanning tree, a strongly connected spanning digraph, and a broadcast tree. Minimum spanning tree plays an important role in all these studies, serving as a linearization method for these nonlinear problems.

1 A Property of Minimum Spanning Tree

For a connected graph $G = (V, E)$ with edge lengths $\{w(e)\}_{e \in E}$, a *minimum-length spanning tree of G* (MST) is a spanning tree T of G whose length $w(T) = \sum_{e \in E(T)} w(e)$ is minimum. In this section, the following property of minimum spanning tree is proved.

Lemma 1 *Let T and T^* be an arbitrary spanning tree and a minimum-length spanning tree of G, respectively. Then there is a one-to-one onto mapping $\sigma : E(T) \to E(T^*)$ such that for every $e \in E(T)$, $w(e) \geq w(\sigma(e))$.*

Proof It is known that an MST can be found by the following greedy algorithm.

Let $\widetilde{T}$ be the MST found by Algorithm 1. Suppose $|E(\widetilde{T})| = t$, $E(\widetilde{T}) = \{e_1^*, \ldots, e_t^*\}$, and $w(e_1^*) \leq \ldots w(e_t^*)$. Order edges in an arbitrary spanning tree T as $e_1, \ldots, e_t$ such that $w(e_1) \leq \ldots \leq w(e_t)$. We claim that $w(e_i^*) \leq w(e_i)$ for $i = 1, \ldots, t$. Suppose this is not true, let k be the first index with $w(e_k^*) > w(e_k)$. Denote $I = \{e_1^*, \ldots, e_{k-1}^*\}$ and $J = \{e_1, \ldots, e_k\}$. If for every $e_j \in J \setminus I$, $I \cup \{e_j\}$ has a cycle, then I is a spanning forest of graph $G[I \cup J]$ (the subgraph of G induced by edge set $I \cup J$), and thus any spanning forest of $G[I \cup J]$ has $k - 1$

Z. Zhang (✉) · X. Huang
Zhejiang Normal University, Jinhua, Zhejiang, China
e-mail: hxhzz@sina.com

© Springer Nature Switzerland AG 2019
D.-Z. Du et al. (eds.), *Nonlinear Combinatorial Optimization*,
Springer Optimization and Its Applications 147,
https://doi.org/10.1007/978-3-030-16194-1_1

Algorithm 1 Greedy Algorithm for MST

Input: A connected graph $G = (V, E)$ and a non-negative length function w on E.
Output: A minimum-length spanning tree T of G.

1: $T \leftarrow \emptyset$.
2: Sort edges as $\{e_1, \ldots, e_n\}$ such that $w(e_1) \leq \ldots \leq w(e_n)$.
3: **for** $i = 1$ to n, **do**
4:      If $T \cup \{e_i\}$ is acyclic, then $T = T \cup \{e_i\}$.
5: **end for**
6: Output T.

---

edges. However, J is an acyclic subgraph of $G[I \cup J]$ which contains k edges, a contradiction. So, there exists an edge $e_j \in J \setminus I$ such that $I \cup \{e_j\}$ is acyclic. But then, the greedy algorithm should choose e_j instead of e_k in the k-th iteration, because $w(e_j) \leq w(e_k) < w(e_k^*)$, again a contradiction.

For any MST T^* with $E(T^*) = \{f_1^*, \ldots, f_t^*\}$ and $w(f_1^*) \leq \ldots \leq w(f_t^*)$. By the above argument, $w(e_i^*) \leq w(f_i^*)$ for $i = 1, \ldots, t$. Since $\sum_{i=1}^{t} w(f_i^*) = \sum_{i=1}^{t} w(e_i^*)$, we must have $w(f_i^*) = w(e_i^*)$ for $i = 1, \ldots, t$. Hence the mapping σ determined by $\sigma(e_i) = f_i^*$ $(i = 1, \ldots, t)$ satisfies the requirement of the lemma.

This lemma indicates that a spanning tree is minimum for edge weight $\{w(e)\}_{e \in E}$ if and only if it is minimum for edge weight $\{w(e)^\alpha\}_{e \in E}$ for any constant $\alpha > 0$. This property allows us to transfer a nonlinear problem to a linear problem through minimum spanning tree. In the next several sections, we give some examples. An interesting common aspect in those examples is that the minimum spanning tree plays an important role in designing good approximation algorithms for them.

2   Symmetric Topological Control

The energy of wireless devices is often supplied with batteries, which means that the energy supply is usually limited. Due to this fact, the energy efficiency becomes an important issue in the study of wireless networks.

The communication range of a wireless station (node) is closely related with its energy consumption and antenna type. For an omnidirectional antenna at a station s, the signal power received at a location t is decreasing as the distance $d(s, t)$ is increasing. Suppose at s, the signal power is $p(s)$. Then at location t, it is $\frac{p(s)}{d(s,t)^\alpha}$, where α is a constant usually between 2 and 5 [33]. Suppose c is the quality threshold for the signal power, that is, in order to receive the signal correctly, the signal power has to be at least c. This means that for location t to receive the signal correctly, it must have $\frac{p(s)}{d(s,t)^\alpha} \geq c$, i.e.,

$$p(s) \geq c \cdot d(s, t)^\alpha.$$

Thus, the communication range at each station is a disk with radius r where r is determined by the power p through formula

$$p = cr^{\alpha}.$$

Often, the transmission quality threshold c is normalized to 1.

When the power p at each node is adjustable, one often studies the problem of minimizing the total energy consumption under certain network connection constraints in order to meet the requirement for certain duty. There exist many combinatorial optimization problems of this type in the literature. The following is a general mathematical formulation.

Given a set of nodes V and a distance table $\{d(u, v)\}_{(u,v) \in V \times V}$, denote $w(u, v) = d(u, v)^{\alpha}$. If $d(u, v) = d(v, u)$ holds for every pair of nodes u and v, the problem is said to have *symmetric power requirement*. Otherwise, the problem has *asymmetric power requirement*. In an *asymmetric topological control* problem, the power for the presence of arc (u, v) is that the power assigned to node u satisfies $p(u) \geq w(u, v)$. In a *symmetric topological control* problem, the power for the presence of edge (u, v) is that $p(u) \geq w(u, v)$ and $p(v) \geq w(v, u)$. The problem is to find a power assignment $p : V \rightarrow R^{+}$ such that the graph with vertex set V and arc set $\{(u, v) \mid p(u) \geq w(u, v)\}$ (in the asymmetric topological control problem) or edge set $\{(u, v) \mid p(u) \geq w(u, v), p(v) \geq w(v, u)\}$ (in the symmetric topological control problem) satisfies certain properties and the total power $\sum_{u \in V} p(v)$ is minimized. There is an equivalent statement of this problem. For a directed graph $G = (V, E)$ with arc weight w, the minimum power for the existence of G is

$$P(G) = \sum_{u \in V} p_G(u), \tag{1}$$

where

$$p_G(u) = \max\{w(u, v) : (u, v) \in E\} \tag{2}$$

is the minimum power that can be assigned to u in order to guarantee the presence of all those arcs $\{(u, v) : (u, v) \in E(G)\}$. When G is clear under the context, subscript G is omitted. An undirected graph can be viewed as a directed graph with each edge replaced by two opposite arcs. The problem is to find a directed graph (or an undirected graph) G satisfying certain properties such that $P(G)$ is minimized.

In order that nodes in a wireless network can communicate with each other, the network is required to contain a spanning tree (in symmetric topological control problem) or a strongly connected spanning subgraph (in asymmetric topological control problem). Considering fault-tolerance, one may require that the network has higher connectivity. Considering transmission delay, one may require that the network has bounded diameter.

Fig. 1 An in-arborescence

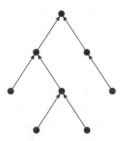

This section considers symmetric topological control problem under symmetric power requirement, which can be stated as follows:

MIN-POWER SYMMETRIC CONNECTIVITY: Given a wireless network with power-adjustable nodes and symmetric power requirement, find a power assignment to minimize the total power and to keep the network connected symmetrically.

As the above argument shows, this problem is equivalent to finding a *minimum power spanning tree*, that is, a spanning tree T with $P(T)$ being minimized.

The NP-hardness of this problem was showed by Blough et al. [3]. To study its approximation solutions, consider an in-arborescence A. An in-arborescence is a directed rooted-spanning tree in which every node except for the root has out-degree exactly one (see Figure 1). Let T be the spanning tree obtained from A by ignoring the directions on all arcs. Notice that the weight of each arc is added exactly once in $P(A)$, and the weight of its corresponding edge is added exactly once in $w(T)$. Hence we obtain the following lemma.

Lemma 2 (In-Arborescence Lemma) *For any in-arborescence A, let T be the spanning tree obtained from A by removing the directions of all arcs. Then*

$$P(A) = w(T).$$

Corollary 3 *For any spanning tree T, $P(T) \geq w(T)$.*

Proof Choose an arbitrary node as the root of T, and orient every edge towards the root. Then one obtains an in-arborescence A. This corollary follows from the In-Arborescence Lemma by observing that $p_T(u) \geq p_A(u)$ for every node $u \in V$.

On the other hand,

$$P(T) = \sum_{u \in V(T)} \max_{(u,v) \in E(T)} w(u, v) \leq \sum_{u \in V(T)} \sum_{v:(u,v) \in E(T)} w(u, v) = 2 \cdot w(T). \quad (3)$$

Suppose T_{mst} is a minimum-weight spanning tree with edge weight $w(\cdot)$ and T^{opt} is an optimal spanning tree for MIN-POWER SYMMETRIC CONNECTIVITY (called *minimum power spanning tree*). Then by (3) and Lemma 1,

$$P(T_{mst}) \leq 2 \cdot w(T_{mst}) \leq 2 \cdot w(T^{opt}) \leq 2 \cdot P(T^{opt}).$$

Fig. 2 A tight example for Theorem 1 with $2n$ points on a line

This means that T_{mst} is a 2-approximation for MIN-POWER SYMMETRIC CONNEC-TIVITY. Since we have seen in the last section that a minimum-length spanning tree is also a minimum-weight spanning tree, we have the following.

Theorem 1 *The minimum-length spanning tree is a polynomial-time 2-approximation for* MIN-POWER SYMMETRIC CONNECTIVITY.

The significance of using a minimum-length spanning tree instead of a minimum-weight spanning tree is that a minimum-length spanning tree in Euclidean plane can be computed in $O(n \log n)$ time, while computing a minimum-weight spanning tree directly may need $O(m \log m)$ time, where $m = O(n^2)$ is the number of edges.

The following example given in [8] shows that using a minimum-length spanning tree as an approximation for MIN-POWER SYMMETRIC CONNECTIVITY, performance ratio 2 is tight.

Let $v_1, v_2, \ldots, v_{2n}$ be $2n$ points lying along a line in Figure 2 with the distance as follows:

$$d(v_1, v_2) = d(v_3, v_4) = \cdots = d(v_{2n-1}, v_{2n}) = 1,$$
$$d(v_2, v_3) = d(v_4, v_5) = \cdots = d(v_{2n-2}, v_{2n-1}) = \varepsilon,$$

where ε is a sufficiently small positive number. The minimum power for keeping symmetric connectivity is

$$nc(1 + \varepsilon)^\alpha + (n - 1)c\varepsilon^\alpha + c \to (n + 1)c \text{ as } c \to 0$$

which is achieved when $v_1, v_3, \ldots, v_{2n-1}$ are assigned with power $c(1 + \varepsilon)^\alpha$, $v_2, v_4, \ldots, v_{2n-2}$ are assigned with power $c\varepsilon^\alpha$, and v_{2n} is assigned with power $c1^\alpha$. The minimum-length spanning tree is the path $(v_1, v_2, \ldots, v_{2n})$ with power-cost

$$2nc1^\alpha = 2nc.$$

Therefore, the performance ratio is

$$\frac{2nc}{nc(1 + \varepsilon)^\alpha + (n - 1)c\varepsilon^\alpha + c} \to 2$$

as $\varepsilon \to 0$ and $n \to \infty$.

Is there a polynomial-time approximation with performance ratio less than 2 for MIN-POWER SYMMETRIC CONNECTIVITY? The answer is yes. To see it, we introduce a type of decomposition of a tree.

A *k-restricted decomposition* of a tree T is a partition of T into a set of edge-disjoint subtrees each with at most k nodes. Consider a k-restricted decomposition of T, which is denoted as $Q = \{T_1, T_2, .., T_q\}$. The power-cost of Q is defined by

$$P(Q) = \sum_{i=1}^{q} P(T_i).$$

Clearly, the power-cost of Q differs from the power-cost of T at the joints of subtrees. That is, extra power will be created at a node which belongs to more than one subtree. It can be expected that when k is sufficiently large, the power-cost of Q will approach the power-cost of T. The following result due to Calinescu et al. confirms this intuition.

Define

$$\rho_k = \sup_{T} \min_{Q} \frac{P(Q)}{P(T)},$$

where T is over all trees and Q is over all k-restricted decompositions of T.

Theorem 2 (Calinescu et al. [10]) *For any $k \geq 3$, $\rho_k \leq 1 + 1/\lfloor \log_2 k \rfloor$.*

Proof First, transform the edge-weighted tree T into a node-weighted rooted binary tree B in the following way (see Figure 3). Let $e_0 = u_0 v_0$ be an edge of T with the maximum weight. Then $T - e_0$ has two subtrees rooted at u_0 and v_0, respectively. The *parent-, children-,* and *sibling-relations* among edges are comprehended in the natural way with respect to these two rooted subtrees. For example, in Figure 3a, e_2 is the parent-edge of e_4 and e_5, and e_1, e_2, and e_3 are the sibling-edges of each other. Edge e_0 is viewed as the parent-edge of those edges which are incident with u_0 and v_0. For each edge $e \neq e_0$, if e has the maximum weight among its siblings, then define $next(e)$ to be the parent-edge of e. Otherwise, define $next(e)$ to be its next heavier sibling-edge, where "next" is with respect to the sorting of siblings in increasing weights. For example, in Figure 3a, $next(e_1) = e_2$, $next(e_2) = e_3$, $next(e_3) = e_0$, etc. Let B be the tree with node set $V(B) = E(T)$ and edge set $E(B) = \{(e, next(e)): e \in E(T)\}$ (see Figure 3b). The weight on each node of B is defined to be the weight on its corresponding edge in T (we still use w to denote the node-weight function on B). So, the node weight of B is the same as the edge weight of T. Observe that B is a binary tree rooted at e_0. In fact, every node $e \in V(B)$ has at most two children in B, one corresponds to its heaviest child-edge in T and the other corresponds to its previous sibling in T, where "previous" is also with respect to the sorting of siblings in increasing weights.

The decomposition of T is constructed as follows. For simplicity of statement, denote $K = \lfloor \log_2 k \rfloor$. For $i = 0, \ldots, K - 1$, let $L_i = \{e \in V(B): dist_B(e, e_0) \equiv i \mod K\}$, where $dist_B(e, e_0)$ is the number of edges on the (e, e_0)-path along B. By the pigeonhole principle, there exists an index i_0 with

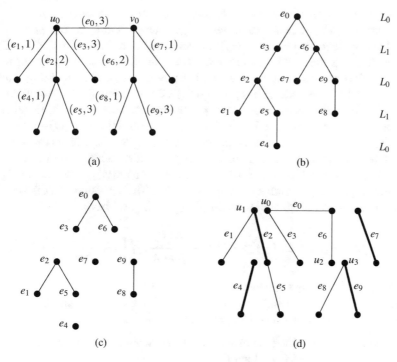

Fig. 3 (a) Tree T. The numbers in the brackets indicate the weights of corresponding edges. (b) Tree B for $K = 2$. The right labels indicate layers $L_0, \ldots, L_{K-1}$. (c) Node-disjoint decomposition $\{B_0, \ldots, B_q\}$ of B for $i_0 = 0$. (d) Edge-disjoint decomposition $Q = \{T_0, \ldots, T_q\}$ of T. Each blackened edge corresponds to a root of some B_i, and this edge has the maximum weight among those edges incident with the root of T_i according to our construction rule

$$w(L_{i_0}) \leq \frac{w(B)}{K} = \frac{w(T)}{\lfloor \log_2 k \rfloor}. \tag{4}$$

Removing parent-edges of those nodes in L_{i_0}, B is divided into a set of node-disjoint subtrees $\{B_0, \ldots, B_q\}$, where B_0 is rooted at e_0 (see Figure 3c). Observe that each B_i corresponds to a tree T_i of T with $|V(T_i)| = |V(B_i)| + 1$ (see Figure 3d). Let $Q = \{T_0, \ldots, T_q\}$.

Since each B_i is a binary tree with depth at most $K - 1$, we have $|V(B_i)| \leq 2^K - 1 \leq k - 1$. Hence $|V(T_i)| \leq k$. For $i = 1, \ldots, q$, let f_i be the root of B_i. Observe that

$$P(Q) = P(T) + \sum_{i=1}^{q} \min\{w(f_i), w(next(f_i))\} \leq P(T) + \sum_{i=1}^{q} w(f_i). \tag{5}$$

To see the above equality, consider Figure 3a, d. For one example, node u_1 is split away from u_0. In T, node u_0 has power $w(e_0)$. After the splitting, node u_0 has power $w(e_0)$ and node u_1 has power $w(e_2)$. Power $w(e_2)$ is a surplus in $P(Q) - P(T)$. Notice that e_2 is the root of a subtree of B. For another example, consider u_2 and u_3. In T, they are the same node which has power $w(e_9)$. After splitting, node u_2 has power $w(e_6)$ and node u_3 has power $w(e_9)$. Power $w(e_6)$ is a surplus in $P(Q) - P(T)$. Notice that $e_6 = next(e_9)$ and e_9 is the root of a subtree of B. In general, if we denote by g_i the root of T_i for $i = 1, \ldots, q$, then surpluses in $P(Q) - P(T)$ occur at $\{g_i\}_{i=1}^q$. According to our construction rule, the edge of T_i which corresponds to f_i has the maximum weight among those edges incident with the root of T_i. Hence the node of T from which g_i is split away has power $\max\{w(f_i), w(next(f_i))\}$, and thus the surplus at g_i is $\min\{w(f_i), w(next(f_i))\}$. Then, relation (5) follows.

Combining (5) with the observation $\{f_i\}_{i=1}^q \subseteq L_{i_0}$, we have

$$P(Q) \leq P(T) + w(L_{i_0}) \leq P(T) + \frac{w(T)}{\lfloor \log_2 k \rfloor} \leq \left(1 + \frac{1}{\lfloor \log_2 k \rfloor}\right) P(T),$$

where the last inequality follows from Corollary 3.

Hence Q is a k-restricted decomposition of T whose power-cost approximates $P(T)$ within a factor of $1 + 1/\lfloor \log_2 k \rfloor$. The theorem is proved.

Li et al. [27] determined the exact value of ρ_k, which equals to the inverse of the k-Steiner tree ratio determined in [4].

Theorem 3 (Li et al. [27]) *For any $k \geq 3$,*

$$\rho_k = \frac{(r+1)2^r + s}{r2^r + s},$$

where $k = 2^r + s$, $0 \leq s < 2^r$.

The proof for the above theorem starts from the same transformation of T into B, and makes use of a more complicated labeling procedure which is similar to the proof for the k-Steiner tree ratio in [4]. In particular, $\rho_3 = 5/3$, which was also proved by Althaus et al. in [1].

Calinescu et al. [10] indicated that the k-restricted decomposition plays a same role in the study of MIN-POWER SYMMETRIC CONNECTIVITY as the k-restricted Steiner tree in the NETWORK STEINER MINIMUM TREE problem [21, 32, 34, 39]. In fact, by Theorem 2, a minimum power spanning tree can be approximated by a k-*restricted minimum power spanning tree*, that is, a set of edge-disjoint subtrees $Q = \{T_1, \ldots, T_q\}$ with minimum power-cost $P(Q)$, the union of which is a spanning tree and each subtree has at most k nodes.

Some of the algorithms for k-RESTRICTED STEINER MINIMUM TREE can be adapted to find a k-restricted minimum power spanning tree. For example, Zelikovsky [38] presented a relative greedy algorithm for k-RESTRICTED STEINER MINIMUM TREE, achieving performance ratio $1 + \ln 2 \approx 1.69$. Prömel and

Steger [32] gave a random algorithm for 3-RESTRICTED STEINER MINIMUM TREE, which has performance ratio $1 + \varepsilon$ with probability at least $1/2$. This was achieved by transforming the 3-RESTRICTED STEINER MINIMUM TREE problem to a MINIMUM SPANNING TREE IN 3-UNIFORM HYPERGRAPHS problem (the latter has a randomized fully polynomial-time scheme [32]). In these two algorithms, considering power has no big difference from considering cost, and thus they can be adapted to produce $(1 + \ln 2 + \varepsilon)$-approximation (when k is sufficiently large) and $(5/3 + \varepsilon)$-approximation (recall that $\rho_3 = 5/3$) for MIN-POWER SYMMETRIC CONNECTIVITY, respectively.

Although some new techniques [23, 26] have been introduced to study MIN-POWER SYMMETRIC CONNECTIVITY, $(5/3 + \varepsilon)$ is still currently the best known performance ratio for this problem. Not every technique for NETWORK STEINER MINIMUM TREE can be easily applied to MIN-POWER SYMMETRIC CONNECTIVITY. In fact, polynomial-time approximation for NETWORK STEINER MINIMUM TREE has been improved by Byrka et al. [5] to $(\ln 4 + \varepsilon)$, which is smaller than 1.39 for sufficiently small ε. However, it is an open problem whether the new technique used in [5] can be applied to MIN-POWER SYMMETRIC CONNECTIVITY.

Problem 1 Find an approximation algorithm for MIN-POWER SYMMETRIC CONNECTIVITY with a performance ratio better than $5/3 + \varepsilon$.

3 Asymmetric Topological Control

It is interesting to notice that the minimum spanning tree is also a polynomial-time 2-approximation for the asymmetric topological control problem under symmetric power requirement, which can be stated as follows.

> MIN-POWER STRONG CONNECTIVITY: Given a wireless network with power-adjustable nodes and symmetric power requirement, find a power assignment to minimize the total power and to keep the network strongly connected.

This problem was initially studied by Chen and Huang [16]. They showed that the minimum-length spanning tree gives a 2-approximation. Kirousis et al. [24] showed the NP-hardness of the problem in 3-dimensional Euclidean space with $\alpha = 2$. Clementi et al. [17, 19] showed that the NP-hardness remains in 2-dimensional Euclidean plane.

Recall that in an asymmetric topological control problem, an arc (u, v) exists if and only if the communication range of node u covers node v, i.e., $p(u) \geq w(u, v)$. Also recall that the minimum power for the existence of a directed graph $H = (V, E)$ is $P(H) = \sum_{u \in V(H)} p_H(u)$, where $p_H(u) = \max_{(u,v) \in E(H)} w(u, v)$. If H is strongly connected, then H contains an in-arborescence A. In fact, such A can be obtained by a breadth-first search from an arbitrarily chosen root in $\overleftarrow{H}$, where $\overleftarrow{H}$ is the directed graph obtained from H by reversing every arc of H. Since $P(H) \geq P(A)$, by In-Arborescence Lemma, we obtain the following.

Lemma 4 *For any strongly connected directed graph H and any node r of H, there is an in-arborescence A with root r such that A is contained in H and*

$$P(H) \geq w(A),$$

where $w(A) = \sum_{(u,v)\in E(A)} w(u, v)$ is the total weight of in-arborescence A.

Notice that every spanning tree T can be viewed as a strongly connected directed graph, which implies that it is a solution to MIN-POWER STRONG CONNECTIVITY, and thus

$$P(T) \geq opt_{mpsc}, \tag{6}$$

where opt_{mpsc} is the optimal value for MIN-POWER STRONG CONNECTIVITY. Now, suppose T_{mst} is a minimum-length spanning tree of the (undirected complete) graph G with edge weight $w(u, v) = d(u, v)^\alpha$. By inequality (3), (6), Lemmas 1 and 4, we have

$$P(T_{mst}) \leq 2 \sum_{(u,v)\in E(T_{mst})} w(u, v) \leq 2 \sum_{(u,v)\in E(A)} w(u, v) = 2w(A) \leq 2P(H).$$

$$\tag{7}$$

This means that the minimum-length spanning tree is a polynomial-time 2-approximation for MIN-POWER STRONG CONNECTIVITY.

Theorem 4 (Chen and Huang [16]) *The minimum-length spanning tree is a polynomial-time 2-approximation for MIN-POWER STRONG CONNECTIVITY.*

Calinescu [8] indicated that 2 is a tight upper bound in the above theorem. Actually, the example for MIN-POWER SYMMETRIC CONNECTIVITY (Figure 2) also works here. In fact, this example implies that there exists a spanning tree (namely the minimum power spanning tree) T with

$$\frac{P(T_{mst})}{P(T)} \to 2, \quad \text{as } \varepsilon \to 0 \text{ and } n \to \infty.$$

By $P(T) \geq opt_{mpsc}$ and $P(T_{mst}) \leq 2 \cdot opt_{mpsc}$, we have

$$2 \geq \frac{P(T_{mst})}{opt_{mpsc}} \geq \frac{P(T_{mst})}{P(T)} \to 2, \quad \text{as } \varepsilon \to 0, n \to \infty,$$

and thus

$$\frac{P(T_{mst})}{opt_{mpsc}} \to 2, \text{ as } \varepsilon \to 0, n \to \infty.$$

Although many efforts have been made to improve the performance of approximation algorithms [15, 24, 35] and the performance ratio less than 2 has been established in several special cases, it was a long-standing open problem whether there is a polynomial-time approximation algorithm with performance ratio less than 2 for MIN-POWER STRONG CONNECTIVITY in a general case. In 2010, this open problem was solved by Calinescu [8] who presented a greedy 1.992-approximation algorithm, the performance ratio of which was later improved to 1.85 in its journal version [9].

This algorithm is a greedy approximation with a monotone increasing submodular potential function. For an element set U, a function $f : 2^U \mapsto \mathbb{R}^+$ is *monotone increasing* if $A \subseteq B \subseteq U \Rightarrow f(A) \leq f(B)$. It is *submodular* if for any $A, B \subseteq U$,

$$f(A) + f(B) \geq f(A \cap B) + f(A \cup B).$$

There are a lot of equivalent conditions for f to be submodular and monotone increasing, one of which is the following (see Lemma 2.25 in [22]):

$$\forall A \subseteq B \subseteq U, \forall x \in U \Rightarrow \Delta_x f(A) \geq \Delta_x(B), \tag{8}$$

where $\Delta_x(A) = f(A \cup x) - f(A)$.

The submodular potential function used in [8, 9] is defined as follows: Let T be a minimum-length spanning tree of the given graph $G = (V, E)$. For any two nodes $u, v \in V$, denote by T_{uv} the unique path on T connecting u and v. For any vertex $u \in V$ and a real number $p \in \{w(u, v) \mid (u, v) \in E\}$ (recall that $w(u, v) = d(u, v)^\alpha$), denote by $S(u, p)$ the directed star with center u and all those arcs (u, v) with $w(u, v) \leq p$. Let $Q(u, p) = \bigcup_{x, y \in V(S(u,p))} E(T_{xy})$. For a collection $\mathscr{A}$ of directed stars, define $Q(\mathscr{A}) = \bigcup_{S(u,p) \in \mathscr{A}} Q(u, p)$ and $f(\mathscr{A}) = \sum_{e \in Q(\mathscr{A})} w(e)$.

Lemma 5 *The above function f is submodular and monotone increasing.*

Proof For any two collections $\mathscr{A}$ and $\mathscr{B}$ of directed stars with $\mathscr{A} \subseteq \mathscr{B}$ and any directed star S,

$$\Delta_S f(\mathscr{A}) = \sum_{e \in Q(\mathscr{A} \cup \{S\})} w(e) - \sum_{e \in Q(\mathscr{A})} w(e)$$

$$= \sum_{e \in Q(\{S\}) \setminus Q(\mathscr{A})} w(e)$$

$$\geq \sum_{e\in Q(\{S\})\setminus Q(\mathscr{B})} w(e)$$

$$= \Delta_S f(\mathscr{B}).$$

The lemma follows from (8).

Denote by $\overrightarrow{T}$ the directed graph obtained from T by replacing each edge with two opposite arcs. $\overrightarrow{T}$ is called the *bidirectional version of T* and T is called the *unidirectional version of $\overrightarrow{T}$*. For arc $(u, v) \in S(u, p)$, let $\overrightarrow{P}_{uv}$ be the directed path from u to v on $\overrightarrow{T}$. Set $\overrightarrow{Q}(u, p) = \bigcup_{v\in V(S(u,p))} E(\overrightarrow{P}_{uv})$. Clearly, the undirected version of $\overrightarrow{Q}(u, p)$ is $Q(u, p)$. Denote by $I_{\mathscr{A}}(S(u, p))$ the set of arcs in $\overrightarrow{Q}(u, p)$ whose unidirectional version is not in $Q(\mathscr{A})$. Calinescu's algorithm is presented in Algorithm 2.

Algorithm 2 Calinescu's Greedy Algorithm

Input: A connected graph $G = (V, E)$ with edge weight function w.
Output: An arc set $\widetilde{E}$ which induces a strongly connected subgraph of G.

1: Let T be a minimum spanning tree of G.
2: $\mathscr{A} \leftarrow \emptyset$.
3: $M \leftarrow \overrightarrow{T}$.
4: **while** $f(\mathscr{A}) < w(T)$ **do**
5:     $(u, p) \leftarrow \mathrm{argmax}_{(u', p')} \Delta_{S(u', p')} f(\mathscr{A})/p'$.
6:     $M \leftarrow M \setminus I_{\mathscr{A}}(S(u, p))$.
7:     $\mathscr{A} \leftarrow \mathscr{A} \cup \{S(u, p)\}$.
8: **end while**
9: Output $\widetilde{E} = \left(\bigcup_{S\in\mathscr{A}} E(S)\right) \cup M$.

Figure 4 illustrates the rough idea of Algorithm 2. Figure 4a is the bidirectional version of a minimum-length spanning tree. Adding a directed star $\overrightarrow{u_1u_4}$, arc (u_1, u_4) can play the role of the directed path $u_1u_2u_3u_4$ in connection (see Figure 4b), so arcs on this directed path can be removed. The situation is different if we further add a directed star $\overrightarrow{u_4u_2}$ (see Figure 4c), the removal of any arc on the directed path $u_4u_3u_2$ will break the strong connectivity. This is why the set of deleted arcs $I_{\mathscr{A}}(S(u, p))$ (see Line 6 of the algorithm) does not include those arcs whose unidirectional versions are in $Q(\mathscr{A})$.

The following lemma refines the above idea to show the correctness of Algorithm 2. A *digon* is a pair of anti-directional arcs between a same pair of nodes. Denote by $\overrightarrow{D}(M)$ the set of digons in M, and by $D(M)$ the set of edges which correspond to digons in $\overrightarrow{D}(M)$.

Lemma 6 *In each iteration of Algorithm 2, the directed graph H which is induced by the edge set $\left(\bigcup_{S\in\mathscr{A}} E(S)\right) \cup M$ is a strongly connected spanning subgraph.*

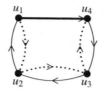

Fig. 4 An illustration for the idea of Algorithm 2. Blackened arcs indicate an added directed star. Dashed arcs can be deleted, while strong connectivity is still kept

Proof Removing $D(M)$ from T, T is broken into several subtrees. Call the subgraph of H induced by the node set of such a subtree as a *reduced component of H*. We shall show that every reduced component is a strongly connected subgraph of H. Then, adding back $\overrightarrow{D}(M)$, the lemma follows immediately.

Initially, $M = \overrightarrow{T}$. Thus, every edge of T corresponds to a digon in M. This means that each reduced component of $\overrightarrow{T}$ is a singleton, which is clearly strongly connected.

Now, suppose at some stage of the algorithm, every reduced component of H is strongly connected. Let $\widetilde{H}$ be the directed graph obtained from H after adding $S(u, p)$ to $\mathscr{A}$ and deleting arcs in $I_{\mathscr{A}}(S(u, p))$ from M. For each $v \in V(S(u, p)) - \{u\}$, suppose on the unique path of T from u to v, there are h edges $(x_1, y_1), (x_2, y_2), \ldots, (x_h, y_h)$ in $D(M)$. Since u and x_1 are in a same subtree of $T - D(M)$, by induction hypothesis, there is a strongly connected component H_1 of H containing both x_1 and u. Similarly, there are strongly connected components $H_2, \ldots, H_h, H_{h+1}$ of H such that H_2 contains both x_2 and $y_1, \ldots, H_h$ contains both x_h and y_{h-1}, and H_{h+1} contains both v and x_h. When arc (u, v) is added, arcs $(x_1, y_1), \ldots, (x_h, y_h)$ are deleted, and thus $V(H_1), \ldots, V(H_{h+1})$ are merged into the node set of a subtree of T with respect to the new M. Notice that the presence of arcs $\{(u, v), (y_h, x_h), \ldots, (y_1, x_2)\}$ connects $H_1, \ldots, H_{h+1}$ into a strongly connected subgraph of $\widetilde{H}$.

Recursively using the above argument for every arc in $S(u, p)$ would complete the proof. $\blacksquare$

The following properties hold for Algorithm 2.

Lemma 7 *In each iteration of Algorithm 2, the directed star $S(u, p)$ chosen in Line 5 satisfies*

$$\frac{\Delta_{S(u,p)} f(\mathscr{A})}{p} \geq 1. \tag{9}$$

Algorithm 2 terminates when $f(\mathscr{A}) = w(T)$ and M contains exactly one arc from each digon of $\overrightarrow{T}$.

Proof Suppose there is a digon $\{(x, y), (y, x)\}$ in M. For the directed star $S(x, p_x)$ with $p_x = w(x, y)$, arc $(x, y) \in Q(S(x, p_x)) \setminus Q(\mathscr{A})$, and thus $\Delta_{S(x,p_x)} f(\mathscr{A})/p_x \geq w(x, y)/p_x = 1$. Inequality (9) follows from the greedy choice of $S(u, p)$ in Line 5.

Since $Q(\mathscr{A}) \subseteq E(T)$, we have $f(\mathscr{A}) \leq w(T)$. As a consequence of the argument in the above paragraph, as long as there is a digon in M, the value of $f(\mathscr{A})$ can always be increased, and thus $f(\mathscr{A}) < w(T)$, the algorithm continues. On the other hand, by the deletion rule in Line 6 and the definition of $I_{\mathscr{A}}(S(u, p))$, if an arc in a digon is deleted, then its opposite arc remains to be in M throughout the algorithm. Hence, when the algorithm terminates, M contains exactly one arc from each digon of $\overrightarrow{T}$, and $f(\mathscr{A}) = w(T)$.

To analyze the performance of Algorithm 2, we first consider a closely related problem. For a monotone increasing, submodular function $f : 2^U \mapsto \mathbb{R}^+$, denote $\Omega(f) = \{A \subseteq U \mid f(A) = f(U)\}$. Let c be a non-negative cost function on U. The following problem is known as SUBMODULAR-COVER:

$$\min c(A) = \sum_{x \in A} c(x)$$

$$s.t. \ A \in \Omega(f).$$

Algorithm 3 employs a greedy strategy to compute an approximation for this problem, and the following is a general theorem for its performance (see Theorem 3.7 of [22]).

Algorithm 3 Greedy Algorithm for SUBMODULAR-COVER

1: $A \leftarrow \emptyset$.
2: **while** $f(A) < f(U)$ **do**
3:     Choose $x \in U$ to maximize $\Delta_x f(A)/c(x)$.
4:     $A \leftarrow A \cup \{x\}$.
5: **end while**
6: Output A.

Theorem 5 *Suppose in every iteration of Algorithm 3, the selected x always satisfies*

$$\frac{\Delta_x f(A)}{c(x)} \geq 1. \tag{10}$$

Then Algorithm 3 produces an approximation solution to SUBMODULAR-COVER *with performance ratio at most*

$$1 + \ln \frac{f(U) - f(\emptyset)}{opt},$$

where opt is the optimal value of SUBMODULAR-COVER.

For any directed star $S(u, p)$, denote $p(S) = \max\{w(u, v): (u, v) \in E(S(u, p))\}$. Consider the following problem called MIN-STARS:

$$\min \sum_{S \in \mathscr{A}} p(S)$$

$$s.t. \ f(\mathscr{A}) = w(T),$$

where T is a minimum spanning tree and $f(\mathscr{A})$ is defined as in Calinescu's algorithm. MIN-STARS is a special SUBMODULAR-COVER problem in which the ground set U consists of all possible directed stars and each directed star $S \in U$ has cost $p(S)$.

Corollary 8 *Suppose* $opt_{min\text{-}stars} \leq \alpha \cdot opt_{mpsc}$ *for some* $\alpha \leq 1$, *where* $opt_{min\text{-}stars}$ *and* opt_{mpsc} *are the optimal values for* MIN-STARS *and* MIN-POWER STRONG CONNECTIVITY, *respectively. Then Algorithm 2 approximates* MIN-POWER STRONG CONNECTIVITY *within performance ratio*

$$1 + \alpha(1 - \ln \alpha).$$

Proof Observe that ignoring M, Algorithm 2 is exactly Algorithm 3 for MIN-STARS. Furthermore, condition (10) in Theorem 5 is guaranteed by Lemma 7. Hence for the final set $\mathscr{A}$ of directed stars computed by Algorithm 2,

$$\sum_{S \in \mathscr{A}} p(S) \leq opt_{min\text{-}stars} \left(1 + \ln \frac{w(T)}{opt_{min\text{-}stars}}\right). \tag{11}$$

Let H be the directed graph induced by the edge set $\left(\bigcup_{S \in \mathscr{A}} E(S)\right) \cup M$ which is the output of Algorithm 2. Notice that each node $u \in V$ has power $p_H(u) \leq \sum_{(u,v) \in M} w(u, v) + \sum_{S(u,p) \in \mathscr{A}} p$. Summing over all nodes in V, we have

$$p(H) \leq w(M) + \sum_{S \in \mathscr{A}} p(S) = w(T) + \sum_{S \in \mathscr{A}} p(S),$$

where the equality follows from the observation that M contains exactly one arc from each digon of $\overrightarrow{T}$ (see Lemma 7). By Lemma 4, $w(T) \leq opt_{mpsc}$. Hence

$$p(H) \leq opt_{mpsc} + opt_{min\text{-}stars} \left(1 + \ln \frac{w(T)}{opt_{min\text{-}stars}}\right)$$

$$\leq opt_{mpsc} \left(1 + \frac{opt_{min\text{-}stars}}{opt_{mpsc}} \left(1 + \ln \frac{opt_{mpsc}}{opt_{min\text{-}stars}}\right)\right)$$

$$\leq opt_{mpsc} \left(1 + \alpha \left(1 + \ln \frac{1}{\alpha}\right)\right)$$

$$= opt_{mpsc} \left(1 + \alpha(1 - \ln \alpha)\right),$$

where the third inequality holds because function $x(1 + \ln(1/x))$ is monotone increasing for $0 < x \leq 1$. The corollary is proved.

Calinescu proved the following in [8].

Lemma 9 $opt_{min\text{-}stars} \leq \frac{7}{8} opt_{mpsc}$.

Performance ratio 1.992 for Algorithm 2 follows from Corollary 8 and Lemma 9.

Theorem 6 (Calinescu [8]) *There exists a polynomial-time approximation for* MIN-POWER STRONG CONNECTIVITY *with performance ratio at most* $1 + \frac{7}{8}(1 - \ln \frac{7}{8}) < 1.992$.

The proof of Lemma 9 employs T-joins in certain 2-edge-connected hypergraphs. It is interesting that by a simpler analysis, factor 1.992 can be improved to 1.85 [9]. For this purpose, we first adapt the proof of Theorem 5 to yield a general lemma, and then use it to obtain a theorem with a similar taste as Theorem 5. This lemma will also be used in the next section, when considering MIN-POWER BROADCAST.

Lemma 10 *For two positive numbers K and k, suppose $(a_0, a_1, \ldots, a_g)$ and $(c_1, \ldots, c_g)$ are two sequences of real numbers satisfying the following conditions:*

(*i*) $a_0 \geq a_1 \geq \ldots a_g$;

(*ii*) $a_0 \geq kK$ and $a_g < kK$;

(*iii*) *For each $i = 1, \ldots, g$,* $\dfrac{a_{i-1} - a_i}{c_i} \geq \dfrac{a_{i-1}}{K}$;

(*iv*) *For each $i = 1, \ldots, g$,* $\dfrac{a_{i-1} - a_i}{c_i} \geq k$.

Then the following inequality holds:

$$\sum_{i=1}^{g} c_i \leq K \left(\ln \left(\frac{a_0}{K}\right) - \ln k + 1\right) - \frac{a_g}{k}.$$

Proof By condition (*iii*),

$$a_i \leq a_{i-1} \left(1 - \frac{c_i}{K}\right) \text{ for } i = 1, \ldots, g. \tag{12}$$

By condition (*i*) and (*ii*), there exists an index r with $0 \leq r < g$ such that $a_{r+1} < kK \leq a_r$. Let $a' = kK - a_{r+1}$ and $a'' = a_r - kK$. Then $a' + a'' = a_r - a_{r+1}$. Let $c_{r+1} = c' + c''$ such that

$$\frac{a'}{c'} = \frac{a''}{c''} = \frac{a_r - a_{r+1}}{c_{r+1}}. \tag{13}$$

Combining this with condition (iii),

$$a'' \geq c'' \cdot \frac{a_r}{K},$$

and thus

$$kK = a_r - a'' \leq a_r \left(1 - \frac{c''}{K}\right).$$

Then by recursively using inequality (12),

$$kK \leq a_0 \left(1 - \frac{c''}{K}\right) \cdot \left(\prod_{i=1}^{r}\left(1 - \frac{c_i}{K}\right)\right) \leq a_0 \cdot e^{-(c'' + \sum_{i=1}^{r} c_i)/K}, \tag{14}$$

where the second inequality uses $1 + x \leq e^x$. It follows that

$$c'' + \sum_{i=1}^{r} c_i \leq K \cdot \ln\left(\frac{a_0}{kK}\right). \tag{15}$$

By condition (iv),

$$\sum_{i=r+2}^{g} c_i \leq \frac{1}{k} \sum_{i=r+2}^{g} (a_{i-1} - a_i) = \frac{1}{k}(a_{r+1} - a_g).$$

Notice that $a'/c' \geq a_r/K \geq k$ (by condition (iii), inequality (13), and the choice of a_r). Hence

$$c' + \sum_{i=r+2}^{g} c(x_i) \leq \frac{1}{k}(a' + a_{r+1} - a_g) = \frac{1}{k}(kK - a_g). \tag{16}$$

The lemma follows by summing up inequalities (15) and (16).

Theorem 7 *Suppose there exists some K with $0 < K \leq f(U) - f(\emptyset)$ such that for any $A \subseteq 2^U$, there exists a set of non-negative coefficients $\{z_y\}_{y \in U}$ with*

(i) $\sum_{y \in U} z_y \cdot \Delta_y f(A) \geq f(U) - f(A)$, *and*

(ii) $0 < \sum_{y \in U} z_y \cdot c(y) \leq K$.

Also suppose in every iteration of Algorithm 3, the selected x always satisfies inequality (10). Then Algorithm 3 produces a solution to SUBMODULAR-COVER *with cost at most*

$$K \left(1 + \ln \frac{f(U) - f(\emptyset)}{K} \right).$$

Proof Suppose the output of Algorithm 3 is $A = \{x_1, \ldots, x_g\}$. Elements are ordered according to the time they are selected into A. Denote by $A_i = \{x_1, \ldots, x_i\}$ for $i = 1, \ldots, g$ and $A_0 = \emptyset$. By the greedy criterion and the assumptions on $\{z_y\}_{y \in U}$,

$$\frac{\Delta_{x_i} f(A_{i-1})}{c(x_i)} \geq \max_{y \in U} \frac{\Delta_y f(A_{i-1})}{c(y)} \geq \frac{\sum_{y \in U} z_y \cdot \Delta_y f(A_{i-1})}{\sum_{y \in U} z_y \cdot c(y)} \geq \frac{f(U) - f(A_{i-1})}{K}.$$

Let $a_i = f(U) - f(A_i)$. By noticing that

$$\Delta_{x_i} f(A_{i-1}) = a_{i-1} - a_i, \tag{17}$$

we have

$$\frac{a_{i-1} - a_i}{c(x_i)} \geq \frac{a_{i-1}}{K}.$$

Hence condition (iii) of Lemma 10 is satisfied. By the monotonicity of f, sequence $(a_0, a_1, \ldots, a_g)$ is monotone decreasing, satisfying condition (i). Taking $k = 1$ in Lemma 10, condition (ii) is satisfied by observing

$$a_0 = f(U) - f(\emptyset) \geq K \text{ and } a_g = f(U) - f(A) = 0 < K. \tag{18}$$

Moreover, condition (iv) is guaranteed by inequality (10) and observation (17). Then, the theorem follows from Lemma 10 and (18).

Now, let us come back to Algorithm 2. In [9], Calinescu proved the following lemma.

Lemma 11 (Calinescu [9]) *Let U be the collection of all possible directed stars. For any sub-collection $\mathscr{A} \subseteq U$, there exists a set of non-negative coefficients $\{z_S\}_{S \in U}$ with*

(i) $\sum_{S \in U} z_S \cdot \Delta_S f(\mathscr{A}) \geq w(T) - f(\mathscr{A})$, *and*
(ii) $0 < \sum_{S \in U} z_S \cdot p(S) \leq opt_{mpsc}/2$.

Proof Let OPT be an optimal solution to MIN-POWER STRONG CONNECTIVITY. Let $z_S = 1/2$ if $S \in OPT$ and $z_S = 0$ otherwise. Hence $\sum_{S \in U} z_S \cdot p(S) = opt_{mpsc}/2$.

Since f is monotone increasing, we have $\sum_{S \in U} z_S \cdot \Delta_S f(\mathscr{A}) \geq 0$. Hence to prove (i), we may assume that $f(\mathscr{A}) < w(T)$. In this case, there is an edge $e \in E(T) \setminus Q(\mathscr{A})$. Denote by T_1 and T_2 the two subtrees of $T - e$. Since OPT is strongly connected, there is a star $S_1 \in OPT$ with center in $V(T_1)$ and another node in $V(T_2)$. By the definition of $Q(S)$, we have $e \in Q(S_1)$. Similarly, there is a star $S_2 \in OPT$ with center in $V(T_2)$ and another node in $V(T_1)$, and thus $e \in Q(S_2)$. Since S_1 and S_2 are two different stars in OPT, we have $z_{S_1} = z_{S_2} = 1/2$ and

$$\sum_{S \in U, e \in Q(S)} z_S \geq z_{S_1} + z_{S_2} = 1.$$

This inequality holds for any edge $e \in E(T) \setminus Q(\mathscr{A})$. Hence

$$\sum_{S \in U} z_S \cdot \Delta_S f(\mathscr{A}) = \sum_{S \in U} z_S \sum_{e \in Q(S) \setminus Q(\mathscr{A})} w(e) = \sum_{e \in E(T) \setminus Q(\mathscr{A})} w(e) \sum_{S \in U, e \in Q(S)} z_S$$

$$\geq \sum_{e \in E(T) \setminus Q(\mathscr{A})} w(e) = w(T) - f(\mathscr{A}).$$

The lemma is proved.

Combining this lemma with Theorem 7 yields the 1.85-approximation.

Theorem 8 *There exists a polynomial-time approximation for* MIN-POWER STRONG CONNECTIVITY *with performance ratio at most* $1 + \frac{1}{2}(1 - \ln \frac{1}{2}) < 1.85$.

Proof Take $K = opt_{mpsc}/2$ in Theorem 7. Since $\overrightarrow{T}$ is a feasible solution to MIN-POWER STRONG CONNECTIVITY, whose power-cost $p(\overrightarrow{T}) \leq 2w(T)$, we have $opt_{mpsc} \leq 2w(T)$. Hence $K \leq w(T)$. Combining this with Lemmas 7 and 11, all conditions in Theorem 7 are satisfied. Hence the set $\mathscr{A}$ of directed stars computed by Algorithm 2 satisfies

$$\sum_{S \in \mathscr{A}} p(S) \leq \frac{opt_{mpsc}}{2} \left(1 + \ln \frac{2w(T)}{opt_{mpsc}} \right).$$

Replacing inequality (11) by the above one in the proof of Corollary 8, the desired performance ratio follows.

4   Broadcast

A broadcasting routing is for transmitting data from a source node s to all other nodes. Mathematically, it is an out-arborescence T rooted at s, namely every node except the root has exactly one in-neighbor. The unique in-neighbor of u is called the *parent* of u, and the out-neighbors of u are called *children* of u. The power-cost of u in the routing equals $p_T(u) = \max\{w(u, v): v \text{ is a child of } u\}$. The power-cost of broadcasting routing T is the sum of power-costs over all nodes in T.

MIN-POWER BROADCAST: Given a network $G = (V, E)$, a source node s, and an edge weight function w, find a broadcasting routing T from s with the minimize power-cost.

In particular, if the network is induced by a set of points in d-dimensional space, in which the edge weight function $w(u, v) = \|uv\|^\alpha$ and $\|uv\|$ is the Euclidean distance between u and v, then the problem is called EUCLIDEAN MIN-POWER BROADCAST.

In general networks, it is unlikely to have a polynomial-time $(1 - \varepsilon) \ln n$-approximation ($\forall \varepsilon > 0$) for MIN-POWER BROADCAST [6, 18]. Approximation algorithms with performance ratio $O(\ln n)$ exist. The first one was given by Caragiannis et al. [12] by transforming the problem into MINIMUM NODE-WEIGHTED DOMINATING SET, achieving performance ratio $10.8 \ln n$. Calinescu et al. [11] presented an algorithm with performance ratio $2 + 2 \ln(n - 1)$, using an idea of "spider decomposition" which was used in the analysis of greedy algorithm for MINIMUM NODE-WEIGHTED STEINER TREE.

In geometric setting, the situation is different. For EUCLIDEAN MIN-POWER BROADCAST in d-dimensional space, the problem is polynomial-time solvable for $d = 1$ [12] or for $\alpha = 1$ [18]. When $d \geq 2$ and $\alpha > 1$, it is NP-hard [18].

It is interesting that the minimum Euclidean spanning tree T is also a constant-approximation for EUCLIDEAN MIN-POWER BROADCAST on the plane. This is because of the existence of the following lemma, which plays an important role for EUCLIDEAN MIN-POWER BROADCAST, similar to the role of Lemma 2 for MIN-POWER SYMMETRIC CONNECTIVITY and the role of Lemma 4 for MIN-POWER STRONG CONNECTIVITY.

Lemma 12 (Out-Arborescence Lemma) *For any out-arborescence T rooted at s and any minimum Euclidean spanning tree T_{mst},*

$$\sum_{e \in E(T_{mst})} \|e\|^\alpha \leq 6P(T).$$

The proof of Lemma 12 relies on the following fact.

Lemma 13 (Ambühl [2]) *Let C be a disk with center x and radius R, P be a set of points lying in C, and $\widetilde{T}$ be a minimum Euclidean spanning tree on P. Assume $x \in P$ and $\alpha \geq 2$. Then,*

$$\sum_{e \in E(\widetilde{T})} \|e\|^\alpha \leq 6R^\alpha.$$

With the aid of Lemma 13, the proof of Lemma 12 is as follows. For each node u of T, draw a smallest disk $D(u)$ centered at u which covers all children of u. Denote by $C(u)$ the set of children of u. Let $\widetilde{T}(u)$ be a minimum Euclidean spanning tree on $C(u) \cup \{u\}$. By Lemma 13,

$$\sum_{e \in E(\tilde{T}(u))} \|e\|^{\alpha} \leq 6R(u)^{\alpha},$$

where $R(u)$ is the radius of $D(u)$. Notice that $\bigcup_{u \in V} \tilde{T}(u)$ contains a spanning tree on the given node set V. Hence by Lemma 1,

$$\sum_{e \in E(T_{mst})} \|e\|^{\alpha} \leq \sum_{u \in V} \sum_{e \in E(\tilde{T}(u))} \|e\|^{\alpha} \leq \sum_{u \in V} 6R(u)^{\alpha} = 6P(T).$$

This completes the proof of Lemma 12.

Actually, this result was obtained through a sequence of efforts [2, 7, 25, 28, 30, 36], which was initiated by Wan et al. [36] and finally accomplished by Ambühl [2]. Wan et al. [36] also pointed out that the constant 6 is the best possible for MST heuristic. An example which asymptotically reaches this bound is shown in Figure 5. The minimum spanning tree is the path $sp_1p_2 \ldots p_6$, whose cost is 6. But the optimal solution is the directed spanning star centered at node s, whose power-cost is $(1 + \varepsilon)^{\alpha}$. Thus performance ratio in this example approaches 6 when $\varepsilon \to 0$.

The approach for establishing the upper bound for this performance ratio is quite interesting. The proof of Lemma 13 given by Ambühl [2] is quite complicated. Here, a proof of a weaker bound is included to show the main idea.

Lemma 14 ([30]) *Let C be a disk with center x and radius R, P be a set of points lying in C, and $\tilde{T}$ be a minimum Euclidean spanning tree on P. Assume $x \in P$ and $\alpha \geq 2$. Then,*

$$\sum_{e \in E(\tilde{T})} \|e\|^{\alpha} \leq 8R^{\alpha}.$$

Proof The assumption $x \in P$ implies that every edge in $\tilde{T}$ has length at most R. For $0 \leq r \leq R$, denote by $\tilde{T}(r)$ the forest with node set P and edge set $\{e \in E(\tilde{T}): \|e\| \leq r\}$. Let $n(\tilde{T}(r))$ be the number of connected components of $\tilde{T}(r)$. Define

Fig. 5 Ratio 6 is best possible for MST heuristic

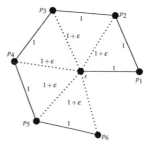

$$\chi_e^r = \begin{cases} 1, \, r < \|e\|, \\ 0, \, r \geq \|e\|. \end{cases}$$

Notice that $\sum_{e \in E(\widetilde{T})} \chi_e^r = |E(\widetilde{T})| - |E(\widetilde{T}(r))| = n(\widetilde{T}(r)) - 1$. Hence,

$$\sum_{e \in E(\widetilde{T})} \|e\|^\alpha = \sum_{e \in E(\widetilde{T})} \int_0^{\|e\|} dr^\alpha = \sum_{e \in E(\widetilde{T})} \int_0^R \chi_e^r dr^\alpha = \alpha \int_0^R (n(\widetilde{T}(r)) - 1) r^{\alpha-1} dr.$$
(19)

For each node $u \in P$, draw a disk $D_u^{r/2}$ with center u and radius $r/2$. For each connected component H of $\widetilde{T}(r)$, the union $\bigcup_{u \in V(H)} D_u^{r/2}$ forms a connected region, which will be denoted as $A(H)$. Furthermore, for different connected components H_1 and H_2 of $\widetilde{T}(r)$, regions $A(H_1)$ and $A(H_2)$ are disjoint. Since each connected region contains at least one disk with radius $r/2$, its area is at least $\pi(r/2)^2$. Let $a(P, r)$ be the total area of $\bigcup_{u \in P} D_u^{r/2}$. Then,

$$a(P, r) \geq n(\widetilde{T}(r))\pi(r/2)^2.$$

We claim that

$$a(P, R) \geq \frac{\pi}{2} \int_0^R n(\widetilde{T}(r)) r \, dr. \tag{20}$$

In fact,

$$a(P, R) = \int_0^R d(a(P, r)) \geq \int_0^R \left(\pi(r/2)^2 dn(\widetilde{T}(r)) + \frac{\pi r}{2} n(\widetilde{T}(r)) dr \right).$$

Suppose there are q distinct lengths between pairs of points of P, say $b_1 < b_2 < \ldots < b_q$. Let $b_0 = 0$. Notice that when $r \in (b_{i-1}, b_i)$, $n(\widetilde{T}(r))$ is a constant, and thus $dn(\widetilde{T}(r)) = 0$ in this interval. Since $\{b_0, b_1, \ldots, b_q\}$ is a discrete set of values whose measure is zero, we have

$$\int_0^R dn(\widetilde{T}(r)) = \sum_{i=1}^q \int_{b_{i-1}}^{b_i} dn(\widetilde{T}(r)) = 0.$$

Inequality (20) follows.

Combining (20) and (19) with $\alpha = 2$,

$$a(P, R) \geq \frac{\pi}{2} \int_0^R (n(\widetilde{T}(r)) - 1) r \, dr + \frac{\pi R^2}{4} = \frac{\pi}{4} \sum_{e \in E(\widetilde{T})} \|e\|^2 + \frac{\pi R^2}{4}.$$

Notice that $a(P, R)$ is contained in a disk with center x and radius $1.5R$. Thus,

$$\pi(1.5R)^2 \geq \frac{\pi}{4} \sum_{e \in E(\widetilde{T})} \|e\|^2 + \frac{\pi R^2}{4}.$$

Hence,

$$\sum_{e \in E(\widetilde{T})} \|e\|^2 \leq 8R^2.$$

Since for every $e \in E(\widetilde{T})$, $\|e\| \leq R$, one has

$$\sum_{e \in E(\widetilde{T})} (\|e\|/R)^\alpha \leq \sum_{e \in T} (\|e\|/R)^2 \leq 8.$$

The lemma is proved.

By Lemma 12, it is easy to see the following.

Theorem 9 *The minimum spanning tree induces a 6-approximation for* EUCLIDEAN MIN-POWER BROADCAST *on the plane.*

The above results are for networks on the plane. For networks with all nodes lying in a d-dimensional space with $d \geq 2$, the example in Figure 5 can be extended to show that a lower bound for MST heuristic is the d-dimensional kissing number n_d, which is the maximum number of non-overlapping unit balls touching a unit ball. It is known that $n_2 = 6$, $n_3 = 12$, and in general, $n_d = 2^{cd(1+o(1))}$, where $0.2075 \leq c \leq 0.401$ [20]. Determining the exact value of n_d for large d is quite hard [40]. Generalizing the method in the proof of Lemma 14 to higher dimensional space, Flammini et al. [30] proved that MST heuristic is a $(3^d - 1)$-approximation for EUCLIDEAN MIN-POWER BROADCAST with $\alpha \geq d$. In particular, this yields a 26-approximation for the 3-dimensional case, which was improved to 18.8 by Navarra [29]. For $d \geq 2$ and $\alpha < d$, Clementi et al. [18] indicated that the ratio of MST heuristic cannot be bounded by any function of α and d. Thus to solve the problem with $\alpha < d$ requires new techniques other than MST.

Problem 2 For $\alpha < d$, find an approximation algorithm for MIN-POWER BROADCAST.

Problem 3 Improve performance analysis of the MST heuristic for MIN-POWER BROADCAST in d-dimensional space with $d \geq 3$.

Is it possible to have an approximation with performance better than that of the minimum-length spanning tree for MIN-POWER BROADCAST? The answer is confirmative.

One possible candidate is the BIP (broadcasting incremental power) algorithm [31]. The algorithm starts from the source node. In each iteration, a new node is reached with the smallest increasing power. Such an increasing may have two ways. One way is to assign power to an unassigned node, and the other way is to increase the power of a node which has received a power assignment in previous iterations. Wan et al. [36] showed that for $d = 2$, the performance ratio of BIP is between $13/3 \approx 4.33$ and 12, while the MST heuristic has performance ratio between 6 and 12 (in both cases, upper bound 12 has been improved to 6 after Ambühl [2] proved Lemma 13). Because of the smaller lower bound of BIP, people believe that BIP is better than MST heuristic. Currently the best known lower bound for BIP is 4.598 [37]. However, it is still an open problem of whether one can obtain an upper bound better than 6 for BIP.

Problem 4 Improve analysis for the performance ratio of BIP.

A breakthrough was made by Caragiannis et al. [13, 14], who gave a greedy approximation which can be applied to any network (not just a network in an Euclidean space), saying that if the minimum-length spanning tree induces a ρ-approximation for MIN-POWER BROADCAST, then their greedy approximation can achieve performance ratio $2 \ln \rho - 2 \ln 2 + 2$. It is 4.2 for $d = 2$ (which is definitely better than both BIP and MST heuristic), 6.49 for $d = 3$ (improving previously best known ratio 18.8), and $2.2d + 0.61$ for $d \geq 4$ (reducing previous ratio $(3^d - 1)$ exponentially).

This greedy approximation is quite interesting, which is designed using the following property of spanning trees. For a spanning tree T of graph $G = (V, E)$, and an edge subset $F \subseteq E$, a *swap set* for (T, F) is an edge subset $A \subseteq E(T)$ such that $T + F - A$ is still a spanning tree of G. Suppose T is rooted at source node s. For any non-leaf node u in T, denote by S_u the set of *all* edges between u and its children in T.

Lemma 15 *Let T_1 and T_2 be two spanning trees over the same set of nodes V, and T_1 is rooted at s. There exists a one-to-one mapping $\sigma : E(T_1) \mapsto E(T_2)$ such that for any non-leaf node u of T_1, $\sigma(S_u)$ is a swap set for (T_2, S_u), where $\sigma(S_u) = \{\sigma(e): e \in S_u\}$.*

Proof The mapping is constructed by the following procedure. Order nodes of the rooted tree T_1 such that every node precedes its parent and the children of a same node are ordered consecutively. Such an ordering can be obtained by viewing T_1 as a breadth-first search tree in which nodes are divided into layers according to their distances from root s, and ordering nodes layer by layer in a "bottom-up" manner, with nodes in farther layers ordered first. Initially, all edges of T_2 are marked as *unused* and all nodes in V are marked as *alive*. Denote by $parent(u)$ the parent of node u in T_1. Establish the mapping σ for edges $\{(u, parent(u))\}_{u \in V, u \neq s}$ one by one in the above order. When it is the turn to consider node u, let $\sigma(u, parent(u))$ be the *last* unused edge on the unique path of T_2 from $parent(u)$ to u. Then, mark edge $\sigma(u, parent(u))$ as *used*, mark node u as *dead*, and iterate. An illustration is given in Figure 6.

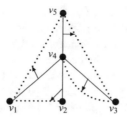

Fig. 6 An illustration for the construction of mapping σ. Solid edges are in T_1. Dashed edges are in T_2. Arrows indicate the mapping. The path on T_2 from $v_4 = parent(v_1)$ to v_1 is $v_4 v_3 v_5 v_1$, thus $\sigma(v_1, v_4) = (v_5, v_1)$. Similarly, $\sigma(v_2, v_4) = (v_1, v_2)$ and $\sigma(v_3 v_4) = (v_3, v_4)$ are established sequentially. Next, consider edge (v_4, v_5). The path on T_2 from $v_5 = parent(v_4)$ to v_4 is $v_5 v_3 v_4$. Since (v_3, v_4) is already used, we have $\sigma(v_4, v_5) = (v_3, v_5)$

To show that σ is as desired, we prove a stronger claim.

Claim For $i = 1, \ldots, n - 1$, in the i-th iteration:

(a) there always exists an unused edge on the path of T_2 from $parent(u)$ to u;
(b) every connected component of F^i contains exactly one alive node, where F^i is the *spanning* forest consisting of those used edges after the i-th iteration.

The claim can be proved by induction on i.

For $i = 1$, since no edge of T_2 is used, (a) is clearly true. By the rule of the mapping, v_1 is an end of $\sigma(v_1, parent(v_1))$. Suppose $\sigma(v_1, parent(v_1)) = (v_1, v_1') \in E(T_2)$. Then F^1 consists of $n - 2$ trivial connected components (each containing exactly one node and this node is alive) and one connected component with node set $\{v_1, v_1'\}$ in which v_1' is the only alive node.

Suppose the claim is true after the i-th iteration. In the $(i+1)$-th iteration, suppose it is the turn to consider node u. By the ordering method and the mapping rule, both u and $parent(u)$ are alive, and thus by induction hypothesis, they belong to two different connected components of F^i. Since T_2 is connected, the path of T_2 from $parent(u)$ to u contains some edges crossing components of F^i, which are unused by the definition of F^i (thus (a) follows). The last one of such edges serves as $\sigma(u, parent(u))$. Adding $\sigma(u, parent(u))$ merges two connected components of F^i, one of which contains u. Suppose these two connected components are C_1 and C_2 with $u \in C_1$. By induction hypothesis, after the i-th iteration, u is the unique alive node of C_1, and C_2 contains a unique alive node u'. After the $(i + 1)$-th iteration, node u is marked as dead, and thus the merged component contains exactly one alive node, namely u'. All other components remain the same. Thus (b) follows. This finishes the proof of the claim.

As a consequence of (a), σ is a one-to-one mapping from $E(T_1)$ to $E(T_2)$.

For a non-leaf node u of T_1, suppose u has ℓ children $u_1, \ldots, u_\ell$, and the mapping procedure has come to the moment just before the children of u are considered. Denote by F^u the spanning forest consisting of those used edges of T_2. Since at this moment, $u, u_1, \ldots, u_\ell$ are all alive, by (b), they belong to different connected components of F^u, say $C, C_1, \ldots, C_\ell$. For $i = 1, \ldots, \ell$, let P_i be the path of T_2 from node u to component C_i, and let e_i be the edge of P_i entering C_i. Notice that after nodes $u_1, \ldots, u_{i-1}$ are considered, edge e_i is still the last unused edge on the path of T_2 from u to u_i. Hence $\sigma(u, u_i) = e_i$ for $i = 1, \ldots, \ell$. Deleting $\sigma(S_u) = \{e_1, \ldots, e_\ell\}$ breaks T_2 into exactly $\ell + 1$ connected components, each component containing exactly one node of $\{u, u_1, \ldots, u_\ell\}$. Adding S_u connects these components, and thus $T_2 + S_u - \sigma(S_u)$ is a spanning tree. This finishes the proof that $\sigma(S_u)$ is a swap set for (T, S_u).

The algorithm is presented in Algorithm 4. Let $S(u, p) = \{(u, v) \in E(G) \mid w(u, v) \le p\}$ be the directed star centered at u which contains *all* those arcs whose existence is supported by power p at node u. Use $A_T(u, p)$ to denote a swap set for $(T, S(u, p))$. In each iteration, the algorithm greedily selects a directed star $S(u, p)$ which maximizes the ratio $w(A_T(u, p))/p$. As long as this ratio is greater than 2, construct a new tree by adding $S(u, p)$ and deleting $A_T(u, p)$. Reset the weight of every edge in $S(u, p)$ to be zero. Suppose the while loop is executed K times. Let $\{S(u_i, p_i)\}_{i=1}^{K}$ be the directed stars obtained in Line 4 and let $\{T_i\}_{i=1}^{K}$ be the trees obtained in Line 7 of Algorithm 4. Turn T_K into an out-arborescence $\widehat{T}$. The power assignment is induced by $\widehat{T}$.

Algorithm 4 CFM Algorithm

Input: A connected graph $G = (V, E)$ with edge weight function w, and a source node s.
Output: A power assignment p on V which induces a broadcasting routing from s.

 1: $T \leftarrow$ a minimum spanning tree of G.
 2: $flag \leftarrow 1, \mathscr{E} \leftarrow \emptyset$.
 3: **while** $flag = 1$ **do**
 4:    $S(u, p) \leftarrow \text{argmax}_{S(u', p')} w(A_T(u', p'))/p'$.
 5:    **if** $w(A_T(u, p))/p > 2$ **then**
 6:       $\mathscr{E} \leftarrow \mathscr{E} \cup \{S(u, p)\}$.
 7:       $T \leftarrow T + S(u, p) - A_T(u, p)$.
 8:       $w(e) \leftarrow 0$ for every $e \in E(S(u, p))$.
 9:    **else**
10:      $flag \leftarrow 0$.
11:    **end if**
12: **end while**
13: Orient edges of T away from s. Denote the resulting out-arborescence as $\widehat{T}$.
14: For each node $u \in V$, set $p(u) = \max_{(u,v) \in E(\widehat{T})} w(u, v)$.
15: Output $\{p(u)\}_{u \in V}$.

The algorithm can be executed in polynomial time. In fact, the most time-consuming step is in Line 4. Notice that there are $O(n)$ choices for node u and for each node u, it suffices to try every p in $\{w(u, v): v \ne u\}$. So there are $O(n^2)$ choices for $S(u, p)$. For a fixed $S(u, p)$, the maximum ratio of $w(A_T(u, p))/p$ is achieved by a maximum weight swap set of $S(u, p)$, which can be obtained by finding a minimum spanning tree of $T/S(u, p)$, and the maximum weight swap set is exactly set of edges not in this minimum spanning tree. Since the number of edges

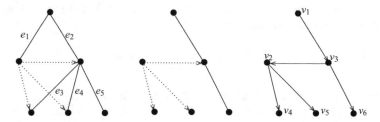

Fig. 7 An illustration for Algorithm 4

in the spanning tree with zero weights is strictly increasing (see Line 8), the number of iterations is at most $n - 1$.

An illustration for this algorithm is given in Figure 7. In figure (a), solid lines are in the minimum spanning tree T, and the directed star $S(u, p)$ is indicated by the dashed arrows. Suppose the swap set for $(T, S(u, p))$ is $A_T(u, p) = \{e_1, e_3, e_4\}$. Graph $T + S(u, p) - A_T(u, p)$ is depicted in (b). The out-arborescence induced by this spanning tree is depicted in (c).

Lemma 16 *The power-cost determined by Algorithm 4 is upper bounded by*

$$2 \sum_{S(u,p)\in\mathscr{E}} p + w(T_K), \tag{21}$$

where K is the number of iterations of the while loop and T_K is the tree T obtained when the algorithm jumps out of the while loop.

To obtain an intuition for the upper bound, consider Figure 7c. Node v_2 has power at most p, where p is the power of directed star $S(v_2, p) \in \mathscr{E}$. For node v_1, since it is not incident with any directed star, its power is determined by the weight on the edges of T_K which connect v_1 to its children of $\widehat{T}$. Node v_3 has power $\max\{w(v_3, v_6), w(v_3, v_2)\}$. Since arc $(v_2, v_3) \in S(u_2, p)$, by the definition of $S(u, p)$ and the symmetric assumption on weights, we have $w(v_3, v_2) \leq p$. Hence $p(v_3) \leq p + w(v_3, v_6)$. The first term p corresponds to an increase of power in the first term of inequality (21), and the second term $w(v_3, v_6)$ is included in $w(T_k)$.

For a directed star $S(u, p)$, call those nodes of degree one in $S(u, p)$ as *feet of* $S(u, p)$. From the above example, it can be observed that an increase of power occurs at a node v such that v is a foot of some directed star $S(u, p)$ and v is the parent of u in $\widehat{T}$. Call such an arc $(u, v) \in S(u, p)$ as a *back arc*. In the above example, (v_2, v_3) is a back arc. A key observation is that

$$\text{every directed star has at most one back arc,} \tag{22}$$

since otherwise T_K will contain a cycle. The proof of the lemma is given below.

Proof (Proof of Lemma 16) For each node u, denote by $CT_K(u)$ the set of edges in T_K which connect u to its children in $\widehat{T}$. Define

$$p_{T_K}(u) = \max_{(u,v) \in CT_K(u)} w(u,v).$$

Suppose u is incident with directed stars $S(u_{i_1}, p_{i_1}), \ldots, S(u_{i_q}, p_{i_q}), \ldots,$ $S(u_{i_\ell}, p_{i_\ell})$, where $u_{i_j} = u$ for $j = 1, \ldots, q$ (that is, u is the center for the first q directed stars), and u is a foot of $S(u_{i_j}, p_{i_j})$ for $j = q + 1, \ldots, \ell$. Define

$$p_{center}(u) = \max_{j=1,\ldots,q} p_{i_j},$$

$$p_{foot}(u) = \max_{j=q+1,\ldots,\ell} p_{i_j}.$$

It can be seen that the power assigned to node u is upper bounded by

$$\max\{p_{T_K}(u), p_{center}(u), p_{foot}(u)\} \le p_{T_K}(u) + p_{center}(u) + p_{foot}(u).$$

So,

$$p(\widehat{T}) \le \sum_{u \in V} p_{T_K}(u) + \sum_{u \in V} p_{center}(u) + \sum_{u \in V} p_{foot}(u). \tag{23}$$

Notice that $CT_K(u) \cap CT_K(v) = \emptyset$ for $u \ne v$, since they belong to different child-edge-set. Hence the first term of (23) is upper bounded by $w(T_K)$. The second term of (23) is clearly upper bounded by $\sum_{S(u,p) \in \mathcal{E}} p$. By observation (22), the power of each directed star is summed at most once in the third term of (23); hence, the third term is also upper bounded by $\sum_{S(u,p) \in \mathcal{E}} p$. Then the lemma follows.

The performance of CFM Algorithm is given in the following theorem.

Theorem 10 *Suppose $\rho \ge 2$ is the performance ratio of the minimum spanning tree as an approximation for* MIN-POWER BROADCAST. *Then the performance ratio of the approximation solution obtained from CFM Algorithm is at most* $2 \ln \rho - 2 \ln 2 + 2$.

Proof As before, suppose the while loop is iterated K times, and in the i-th iteration, $S(u_i, p_i)$ is the directed star chosen in Line 4, $A_{T_{i-1}}(u_i, p_i)$ is the corresponding swap set for $(T_{i-1}, S(u_i, p_i))$, and $T_i = T_{i-1} + S(u_i, p_i) - A_{T_{i-1}}(u_i, p_i)$. Let T_0 denote the minimum spanning tree at the initial step. Since edges in $E(S(u, p))$ have zero weights, so for $i = 1, \ldots, K$,

$$w(A_{T_{i-1}}(u_i, p_i)) = w(T_{i-1}) - w(T_i). \tag{24}$$

Denote by *opt* the optimal value of MIN-POWER BROADCAST. If $w(T_0) \leq 2opt$, then T_0 already has the desired performance ratio. Hence in the following, we assume $w(T_0) > 2opt$.

We shall prove that the problem satisfies all four conditions in Lemma 10 with $a_i = w(T_i)$, $c_i = p_i$, $K = opt$, and $k = 2$.

Suppose T^* is an optimal broadcast tree. For each node u, denote by S_u the directed star consisting of all arcs of T^* from u to its children. By Lemma 15, there exists a one-to-one onto mapping $\sigma : E(T^*) \to E(T_{i-1})$ such that $\sigma(S_u)$ is a swap set for (T_{i-1}, S_u). As a consequence, $\{\sigma(S_u)\}$ form a partition of $E(T_{i-1})$, and thus

$$w(T_{i-1}) = \sum_{u \in V} w(\sigma(S_u)).$$

By

$$\frac{\sum_{u \in V} w(\sigma(S_u))}{\sum_{u \in V} p_{T^*}(u)} = \frac{w(T_{i-1})}{opt}$$

and the pigeonhole principle, there exists a node u such that

$$\frac{w(\sigma(S_u))}{p_{T^*}(u)} \geq \frac{w(T_{i-1})}{opt}.$$

Since $S_u \subseteq S(u, p_{T^*}(u))$ and thus $\sigma(S_u) \subseteq A_{T_{i-1}}(u, p_{T^*}(u))$, it follows from the greedy choice of $S(u_i, p_i)$ that

$$\frac{w(A_{T_{i-1}}(u_i, p_i))}{p_i} \geq \frac{w(A_{T_{i-1}}(u, p_{T^*}(u)))}{p_{T^*}(u)} \geq \frac{w(\sigma(S_u))}{p_{T^*}(u)} \geq \frac{w(T_{i-1})}{opt}. \tag{25}$$

Substituting (24) into (25), we have

$$\frac{w(T_{i-1}) - w(T_i)}{p_i} \geq \frac{w(T_{i-1})}{opt},$$

condition (iii) of Lemma 10 is proved.

By (25), we must have $w(T_K) \leq 2opt$, since otherwise there exists a directed star $S(u, p)$ with $w(A_{T_K}(u, p))/p \geq w(T_K)/opt > 2$, and thus the algorithm will not jump out of the while loop at the K-th iteration. Combining this with our assumption that $w(T_0) > 2opt$, condition (ii) follows. Condition (i) is obvious since T_i is obtained from T_{i-1} by adding some edges of zero weights and removing some edges. By Line 5 of the algorithm, $(w(T_{i-1}) - w(T_i))/p_i = w(A_{T_{i-1}}(u_i, p_i))/p_i > 2$ for $i = 1, \ldots, K$, which is condition (iv).

Then by Lemma 10, the output of Algorithm 4 satisfies

$$\sum_{i=1}^{K} p_i \le opt \left(\ln \frac{w(T_0)}{opt} - \ln 2 + 1 \right) - \frac{w(T_K)}{2}.$$

By Lemma 16, the power of $\widehat{T}$ is upper bounded by

$$2 \sum_{i=1}^{K} p_i + w(T_k) \le 2opt \left(\ln \frac{w(T_0)}{opt} - \ln 2 + 1 \right) \le 2opt \left(\ln \rho - \ln 2 + 1 \right),$$

since minimum spanning tree is a ρ-approximation. The performance ratio follows.

5   Asymmetric Power Requirement

Recall that with an asymmetric power requirement, the power to support an arc (u, v) might be different from the power to support arc (v, u), namely one may have $w(u, v) \ne w(v, u)$.

Althaus et al. [1] studied the following problem:

MIN-POWER SYMMETRIC CONNECTIVITY WITH ASYMMETRIC POWER REQUIREMENT: Given a complete graph $G = (V, E)$ and asymmetric power requirement $w : V \times V \to \mathbb{R}^+$, find a minimum power-cost spanning tree T in which an undirected edge (u, v) exists if and if $p_T(u) \ge w(u, v)$ and $p_T(v) \ge w(v, u)$.

Althaus et al. [1] showed that the minimum spanning tree with respect to edge weight $c(u, v) = w(u, v) + w(v, u)$ is a 2-approximation. However, their proof has a flaw. In fact, their proof uses a claim that for any tree T, $c(T) \le 2p(T)$. But this is not true. Consider a complete graph on node set $\{u, u_1, \ldots, u_{n-1}\}$, in which $w(u, u_i) = M$ and $w(u_i, u) = 1$ for $i = 1, \ldots, n-1$, and $w(u_i, u_j) = \infty$ for $i \ne j$. The spanning star S centered at u has cost $c(S) = (n-1)(1+M)$, while the power-cost of S is $p(S) = (n-1) + M$. For sufficiently large M, ratio $c(S)/p(S)$ is $\Theta(n)$. The reason why a cost-over-power ratio cannot be bounded lies in the difficulty to bound the power of an out-arborescence from below in terms of its cost. This can be observed by considering a directed star with unit weight on every edge, for which the cost is $n-1$ and the power-cost is 1. This is a sharp contrast to an in-arborescence whose power-cost is always lower bounded by its cost.

Problem 5 Does MIN-POWER SYMMETRIC CONNECTIVITY WITH ASYMMETRIC POWER REQUIREMENT admits a polynomial-time constant approximation?

Calinescu et al. [11] gave a positive answer to this problem when the asymmetric power requirement is generated by the following special way. Let $r : V \mapsto \mathbb{R}^+$ be a node function which reflects *transmission efficiency* at nodes. The power requirement to support arc (u, v) is

$$w(x, y) = d(x, y)^\alpha / r(x), \tag{26}$$

where $d(x, y)$ is the distance between x and y in a metric space.

Theorem 11 *The minimum spanning tree with respect to edge weight $c(u, v) = w(u, v) + w(v, u)$ induces an approximation solution within a factor of*

$$\min_{q > 1} \left\{ 2^\alpha + (q + 1)^\alpha + \frac{q^\alpha}{q^\alpha - 1} \right\}$$

for MIN-POWER SYMMETRIC CONNECTIVITY WITH ASYMMETRIC POWER REQUIREMENT.

This theorem follows from the following lemma.

Lemma 17 (Rooted-Spanning Tree Lemma) *For any rooted-spanning tree T and any number $q > 1$, there exists a rooted-spanning tree T_q such that*

$$c(T_q) \leq \left(2^\alpha + (q + 1)^\alpha + \frac{q^\alpha}{q^\alpha - 1} \right) P(T),$$

where $c(T_q) = \sum_{(u,v) \in E(T_q)} (w(u, v) + w(v, u))$.

Proof For each node u, let C_u be the set of children of u, and let T^u be the subtree of T induced by $\{u\} \cup C_u$. For each T^u, construct a tree T_q^u by modifying T^u in the following way:

- Sort all children in C_u as $x_1, x_2, \ldots, x_k$ such that $d(u, x_1) \geq d(u, x_2) \geq \cdots \geq d(u, x_k)$.
- For each $i = 1, \ldots, k-1$, if $d(u, x_i) \leq q \cdot d(u, x_{i+1})$, then replace edge (u, x_{i+1}) by edge (x_i, x_{i+1}).

Notice that for each edge (x_i, x_{i+1}) in T_q^u,

$$d(x_i, x_{i+1}) \leq d(u, x_i) + d(u, x_{i+1}) \leq (1 + q)d(u, x_{i+1})$$

and

$$d(x_i, x_{i+1}) \leq d(u, x_i) + d(u, x_{i+1}) \leq 2d(u, x_i).$$

By the definition of w in (26),

$$w(x_{i+1}, x_i) \leq (1 + q)^\alpha w(x_{i+1}, u),$$

and

$$w(x_i, x_{i+1}) \leq 2^\alpha w(x_i, u).$$

Therefore,

$$c(x_i, x_{i+1}) \leq (1+q)^\alpha w(x_{i+1}, u) + 2^\alpha w(x_i, u). \tag{27}$$

Let (u, x_{i_1}), (u, x_{i_2}), ..., (u, x_{i_h}) be those edges of T^u which remain in T_q^u, where $i_1 < i_2 < \ldots i_h$. Since (u, x_{i_j}) is not replaced by (x_{i_j}, x_{i_j+1}), we have $d(u, x_{i_j}) > q \cdot d(u, x_{i_j+1}) \geq q \cdot d(u, x_{i_{j+1}})$. Hence

$$d(u, x_{i_1}) \geq q d(u, x_{i_2}) \geq q^2 d(u, x_{i_3}) \geq \cdots \geq q^{h-1} d(u, x_{i_h}),$$

and thus

$$w(u, x_{i_1}) \geq q^\alpha w(u, x_{i_2}) \geq q^{2\alpha} w(u, x_{i_3}) \geq \cdots \geq q^{(h-1)\alpha} w(u, x_{i_h}).$$

Therefore,

$$\sum_{(u,x)\in E(T_q^u)} w(u, x) = \sum_{j=1}^{h} w(u, x_{i_j}) \leq w(u, x_{i_1}) \sum_{j=0}^{h-1} q^{-j\alpha} \leq w(u, x_1) \frac{q^\alpha}{q^\alpha - 1}. \tag{28}$$

Combining (27) and (28), we obtain

$$c(T_q^u) = \sum_{(u,x_i)\in E(T_q^u)} c(u, x_i) + \sum_{(x_i,x_{i+1})\in E(T_q^u)} c(x_i, x_{i+1}) \tag{29}$$

$$\leq w(u, x_1) \frac{q^\alpha}{q^\alpha - 1} + \sum_{(u,x_i)\in E(T_q^u)} w(x_i, u)$$

$$+ \sum_{(x_i,x_{i+1})\in E(T_q^u)} \left((1+q)^\alpha w(x_{i+1}, u) + 2^\alpha w(x_i, u)\right).$$

For each $(x_i, x_{i+1}) \in E(T_q^u)$, if $(u, x_i) \in E(T_q^u)$, then $w(x_i, u)$ is added $1+2^\alpha$ times in (29); if $(u, x_i) \notin E(T_q^u)$, then (x_{i-1}, x_i) is also in $E(T_q^u)$, in which case $w(x_i, u)$ is added $(1 + q)^\alpha + 2^\alpha$ times in (29). The weight $w(x_{i+1}, u)$ is added $(1 + q)^\alpha$ times if x_{i+1} is a leaf of T_q^u. Otherwise, (x_{i+1}, x_{i+2}) is also in $E(T_q^u)$, and thus $w(x_{i+1}, u)$ is added $2^\alpha + (1+q)^\alpha$ times in (29). In any case, the repetition of weight $w(x, u)$ for any $x \in C_u$ is at most $(1 + q)^\alpha + 2^\alpha$. Furthermore, $w(u, x_1) \leq p_T(u)$. Hence

$$c(T_q^u) \leq p_T(u) \frac{q^\alpha}{q^\alpha - 1} + (2^\alpha + (1+q)^\alpha) \sum_{x\in C_u} w(x, u).$$

Now, let $T_q = \bigcup_{u \in V} T_q^u$. Then

$$c(T_q) \le \frac{q^\alpha}{q^\alpha - 1} \sum_{u \in V} p_T(u) + (2^\alpha + (1 + q)^\alpha) \sum_{u \in V} \sum_{x \in C_u} w(x, u). \tag{30}$$

Orient edges in T towards its root to obtain an in-arborescence T_{in}. Notice that $C_u \cap C_v = \emptyset$ for $u \ne v$. Furthermore, $w(x, u) \le p_T(x)$ since the existence of (x, u) is supported by $p_T(x)$. Hence,

$$\sum_{u \in V} \sum_{x \in C_u} w(x, u) = \sum_{(x,u) \in E(T_{in})} w(x, u) \le \sum_{x \in V} p_T(x).$$

Substituting this into (30), the lemma is proved.

Proof (Proof of Theorem 11) Let T_{mst} be a minimum spanning tree with respect to weight $c(u, v) = w(u, v) + w(v, u)$. To avoid ambiguity, we use (u, v) to denote an edge and $\overrightarrow{(u, v)}$ to denote an arc. Let $\overrightarrow{T_{mst}}$ be the directed graph obtained from T_{mst} by replacing every edge by two opposite arcs. Then

$$P(T_{mst}) \le \sum_{u \in V} \sum_{v: \overrightarrow{(u,v)} \in \overrightarrow{T_{mst}}} w(u, v) = \sum_{(u,v) \in E(T_{mst})} \left(w(u, v) + w(v, u) \right) = c(T_{mst}).$$

Then the theorem follows from Lemma 17 and the arbitrariness of $q > 1$.

References

1. Althaus, E., Calinescu, G., M.andoiu, I.I., Prasad, S., Tchervenski, N., Zelikovsky, A.: Power efficient range assignment in ad-hoc wireless networks. In: Proceedings of the IEEE Wireless Communications and Networking Conference (WCNC), pp. 1889–1894. IEEE, New Orleans (2003)
2. Ambühl, C.: An optimal bound for the MST algorithm to compute energy efficient broadcast trees in wireless networks. Lect. Notes Comput. Sci. **3580**, 1139–1150 (2005)
3. Blough, D.M., Leoncini, M., Resta, G., Santi, P.: On the symmetric range assignment problem in wireless ad hoc networks. In: Proc. of the IFIP 17th World Computer Congress TC1 Stream/2nd IFIP International Conference on Theoretical Computer Science (TCS), pp. 71–82 (2002)
4. Borchers, A., Du, D.-Z.: The k-Steiner ratio in graphs. SIAM J. Comput. **26**, 857–869 (1997)
5. Byrka, J., Grandoni, F., Rothvoss, T., Sanitá, L.: An improved LP-based approximation for Steiner tree. In: STOC 2010, pp. 583–592 (2010)
6. Cagalj, M., Hubaux, J., Enz, C.: Minimum-energy broadcast in all-wireless networks: NP-completeness and distribution issues. In: Proceedings of the 8th Annual International Conference on Mobile Computing and Networking (MobiCom'02), pp. 172–182. ACM Press, New York (2002)
7. Cai, H., Zhao, Y.: On approximation ratios of minimum-energy multicast routing in wireless networks. J. Comb. Optim. **3**, 243–262 (2005)

8. Calinescu, G.: Minimum-power strong connectivity. Lect. Notes Comput. Sci. **6302**, 67–80 (2010)
9. Calinescu, G.: Approximate min-power strong connectivity. SIAM J. Discret. Math. **27**(3), 1527–1543 (2013)
10. Calinescu, G., Mandoiu, I., Zelikovsky, A.: Symmetric connectivity with minimum power consumption in radio networks. In: 2nd IFIP International Conference on Theoretical Computer Science (TCS 2002), pp. 119–130. Kluwer Academic Publishers, Dordrecht (2002)
11. Calinescu, G., Kapoor, S., Olshevsky, A., Zelikovsky, A.: Network lifetime and power assignment in ad-hoc wireless networks. Lect. Notes Comput. Sci. **2832**, 114–126 (2003)
12. Caragiannis, I., Kaklamanis, C., Kanellopulos, P.: New results for energy-efficient broadcasting in wireless networks. Lect. Notes Comput. Sci. **2518**, 332–343 (2002)
13. Caragiannis, I., Flammini, M., Moscardelli, L.: An exponential improvement on the mst heuristic for minimum energy broadcasting in ad hoc wireless networks. In: ICALP 2007, pp. 447–458 (2007)
14. Caragiannis, I., Flammini, M., Moscardelli, L.: An exponential improvement to the MST heuristic for minimum energy broadcasting in ad hoc wireless networks. IEEE/ACM Trans. Netw. **21**(4), 1322–1331 (2013)
15. Carmi, P., Katz, M.J.: Power assignment in radio networks with two power levels. Algorithmica **47**, 183–201 (2007)
16. Chen, W.T., Huang, N.F.: The strongly connection problem on multihop packet radio networks. IEEE Trans. Commun. **37**(3), 293–295 (1989)
17. Clementi, A.E.F., Penna, P., Silvestri, R.: Hardness results for the power range assignment problem in packet radio networks. Lect. Notes Comput. Sci. **1671**, 197–208 (1999)
18. Clementi, A.E.F., Crescenzi, P., Penna, P., Rossi, G., Vocca, P.: On the complexity of computing minimum energy consumption broadcast subgraphs. Lect. Notes Comput. Sci. **2010**, 121–131 (2001)
19. Clementi, A.E.F., Penna, P., Silvestri, R.: On the power assignment problem in radio networks. Mobile Netw. Appl. **9**(2), 125–140 (2004)
20. Conway, J.H., Sloane, N.J.A.: "The kissing number probe" and "bounds on kissing numbers". In: Sphere Packings, Lattices, and Groups, 3rd edn. Springer, New York (1998)
21. Du, D.-Z., Zhang, Y.-J.: On better heuristic for Steiner minimum trees. Math. Program. Ser. B **57**, 193–202 (1992)
22. Du, D.-Z., Ko, K.-I., Hu, X.: Design and Analysis of Approximation Algorithms. Springer, New York (2011)
23. Hajiadhayi, M.T., Kortsarz, G., Mirrokni, V.S., Nutov, Z.: Power optimization for connectivity problems. Math. Program **110**, 195–208 (2007)
24. Kirousis, L.M., Kranakis, E., Krizanc, D., Pelc, A.: Power consumption in packer radio networks. Theor. Comput. Sci. **243**, 289–305 (2000)
25. Klasing, R., Navarra, A., Papadopoulos, A., Perennes, S.: Adaptive broadcast consumption (ABC), a new heuristic and new bounds for the minimum energy broadcast routing problem. Lect. Notes Comput. Sci. **3042**, 866–877 (2004)
26. Kortsarz, G., Mirrokni, V.S., Nutov, Z., Tsanko, E.: Approximating minimum-power degree and connectivity problems. Lect. Notes Comput. Sci. **4957**, 423–435 (2008)
27. Li, M., Huang, S.L., Sun, X., Huang, X.: Performance evaluation for energy efficient topologic control in ad hoc wireless networks. Theor. Comput. Sci. **326**(1-3), 399–408 (2004)
28. Navarra, A.: Tighter bounds for the minimum energy broadcasting problem. In: Proceedings of the 3rd International Symposium on Modeling and Optimization in Mobile, Ad Hoc, and Wireless Networks (WIOPT '05), pp. 313–322 (2005)
29. Navarra, A.: 3-D minimum energy broadcasting. Lect. Notes Comput. Sci. **4056**, 240–252 (2006)
30. Navarra, A., Flammini, M., Klasing, R., Perennes, S.: Improved approximation results for the minimum energy broadcasting problem. Algorithmica **49**, 318–336 (2007)
31. Nguyen, G.D., Wieselthier, J.E., Ephremides, A.: On the construction of energy-efficient broadcast and multicast trees in wireless networks. In: IEEE INFOCOM 2000 (2000)

32. Prömel, H.J., Steger, A.: A new approximation algorithm for the Steiner tree problem with performance ratio 5/3. J. Algorithms **36**, 89–101 (2000)
33. Rappaport, T.: Wireless Communications: Principles and Practice. Prentice-Hall, Englewood Cliff
34. Robin, G., Zelikovsky, A.: Improved Steiner tree approximation in graphs. In: SODA 2000, pp. 770–779 (2000)
35. Sun, J., Cheng, X., Wang, L., Xu, Y., Cheng, M.X., Cardei, M., Du, D.-Z.: Topology control of ad hoc wireless networks for energy efficiency. IEEE Trans. Comput. **53**(12), 1629–1635 (2004)
36. Wan, P.-J., Calinescu, G., Li, X.-Y., Frieder, O.: Minimum energy broadcast routing in static ad-hoc wireless networks. In: IEEE INFOCOM 2001, pp. 1162–1171 (2001)
37. Yuan, D., Bauer, J., Haugland, D.: New results on the time complexity and approximation ratio of the broadcast incremental power algorithm. Inf. Process. Lett. **109**, 615–619 (2009)
38. Zelikovsky, A.: Better approximation bounds for the network and Euclidean Steiner tree problems. Department of Computer Science, University of Virginia (1996)
39. Zelikovsky, A.: An 11/6-approximation algorithm for the network Steiner tree problem. Algorithmica **9**, 463–470 (1993)
40. Zhong, C.: Sphere Packing. Springer, New York (1999)

Discrete Newton Method

Zhao Zhang and Xiaohui Huang

Abstract Newton method is a classic and powerful method in continuous nonlinear optimization. However in this chapter, we introduce its counterpart in combinatorial optimization: discrete Newton method, and show that there exists a strong polynomial time algorithm for finding the root of a piecewise linear decreasing function, where the number of pieces is exponential. Then we show how to apply it in solving linear fractional combinatorial optimization problem, inverse combinatorial problem, and bottleneck expansion problem.

1 Discrete Newton Method

Let U be a finite set, $\mathscr{F}$ be a family of subsets of U (called *feasible domain*), and k be a constant. Suppose f, b are two functions on $\mathscr{F}$ which are *linear* in the following sense: there exist two real vectors α, β such that $f(F) = \alpha^T x_F$ and $b(F) = \beta^T \hat{x}_F$, where x_F and $\hat{x}_F$ are 0–1 vectors of length n_f and length n_b, respectively, which are associated with F. Also assume that

$$f(F) > k \text{ for some } F \in \mathscr{F} \text{ and } b(F) > 0 \text{ for all } F \in \mathscr{F}. \tag{1}$$

Notice that for any fixed $F \in \mathscr{F}$, $f(F) - \lambda b(F) - k$ is a linear decreasing function on λ. Hence,

$$h(\lambda) = \max_{F \in \mathscr{F}} \{f(F) - \lambda b(F) - k\} \tag{2}$$

is a piecewise linear decreasing function with a unique root λ^* (see Figure 1). The problem is how to find this root.

Z. Zhang (✉) · X. Huang
Zhejiang Normal University, Jinhua, Zhejiang, China
e-mail: hxhzz@sina.com

© Springer Nature Switzerland AG 2019
D.-Z. Du et al. (eds.), *Nonlinear Combinatorial Optimization*,
Springer Optimization and Its Applications 147,
https://doi.org/10.1007/978-3-030-16194-1_2

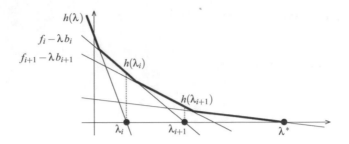

Fig. 1 An illustration for discrete Newton method. The thick lines indicate function $h(\lambda)$

ROOT OF PIECEWISE LINEAR DECREASING FUNCTION (RPLDF): Find the unique root of the above piecewise linear decreasing function $h(\lambda)$.

One difficulty in solving this problem is that the number of pieces might be exponential. If all input numbers are integral, then the binary search method solves the problem with $O(nK)$ iterations, where $n = \max\{n_f, n_b\}$ and K is the maximum value of input numbers. So, if for any λ, $h(\lambda)$ can be determined in polynomial time, then binary search gives a pseudo-polynomial time algorithm.

In 1993, Radzik [7] proposed a discrete Newton method for the linear fractional combinatorial optimization problem, the core of which is to solve a RPLDF in such a way that the number of iterations is polynomial in the input size but independent of the input numbers, and thus might possibly provide a strong polynomial time algorithm.

Discrete Newton method requires a solver for the parameterized combinatorial optimization problem (2). To be more concrete, there is an algorithm $\mathscr{A}$ such that for any fixed parameter λ, algorithm $\mathscr{A}$ returns a pair (h_λ, F_λ), where $F_\lambda \in \mathscr{F}$ reaches the maximum value of (2) and h_λ is the maximum value. With the aid of algorithm $\mathscr{A}$, discrete Newton method solves RPLDF as in Algorithm 1. Notice that λ_{i+1} is the root of the linear function $f(F_i) - \lambda b(F_i) - k$. This operation has a similar taste as the classic Newton method in finding the root of a derivable function g, which iteratively uses the root of a tangent line to approximate the root of g. For an illustration of this algorithm, see Figure 1.

To estimate the time efficiency of Algorithm 1, the following lemma is needed.

Lemma 1 (Michel Goemans [7]) *Let* $\delta = (\delta_1, \ldots, \delta_n)^T$ *be a vector in* $\mathbb{R}^n$ *with nonnegative coordinates, and let* $y_1, \ldots, y_q$ *be a sequence of vectors in* $\{-1, 0, 1\}^n$. *If*

$$0 < y_{i+1}^T \delta \le \frac{1}{2} y_i^T \delta$$

holds for any $i = 1, \ldots, q - 1$, *then* $q = O(n \log n)$.

Proof The condition of this lemma implies that the following linear program is feasible:

$$\begin{cases} (y_i^T - 2y_{i+1}^T)z \geq 0, \\ y_q^T z = 1, \\ z \geq 0. \end{cases} \tag{3}$$

Algorithm 1 Discrete Newton method

Input: $(U, \mathscr{F}, f, b, k)$.
Output: The unique root of $h(\lambda)$.
 1: $i = 1, \lambda_i = 0, flag \leftarrow 1$.
 2: **while** $flag = 1$ **do**
 3:     $(h_i, F_i) = \mathscr{A}(\lambda_i)$.
 4:     **if** $h_i = 0$ **then**
 5:         $flag \leftarrow 0$.
 6:     **else**
 7:         $\lambda_{i+1} = (f(F_i) - k)/b(F_i)$.
 8:         $i \leftarrow i + 1$.
 9:     **end if**
10: **end while**
11: Output λ_i.

In fact, (3) has at least one feasible solution $\delta/y_q^T \delta$. Let z^* be an extreme vertex of the above polytope. By the theory of linear programming, there is an invertible matrix A and a vertex b such that $Az^* = b$. By Crammer's rule, every component of z^* has the form $z_j^* = \det(\Lambda_j)/\det(A)$, where A_j is the matrix obtained from A by replacing the j-th column with vector b. By the assumption on y-vectors, every component in A and b is an integer from $[-3, 3]$. Thus, it follows from the definition of determinant that $\det(A_j) \leq n!3^n$. Since A is an integral matrix which is invertible, we have $\det(A) \geq 1$. So $|z_j^*| \leq n!3^n$ for $j = 1, \ldots, n$, and thus

$$1 = y_q^T z^* \leq \frac{1}{2^{q-1}} y_1^T z^* \leq \frac{n \cdot n!3^n}{2^{q-1}}.$$

By Stirling's formula which says that $n! \sim \sqrt{2\pi n} \left(\frac{n}{e}\right)^n$, we have $q = O(n \log n)$.

The following lemma lists some properties for the algorithm. It should be noted that for the last iteration, strict inequalities of some properties might degenerate into equalities. However, such degenerations are not essential to the analysis and can be dealt with by paying a little more attention. To avoid deviating from the main idea, we choose to state the properties in a less stringent manner, ignoring the details for the last iteration.

Lemma 2 *For the i-th iteration, denote by $f_i = f(F_i) - k$ and $b_i = b(F_i)$.*

(i) $h_i = f_i - \lambda_i b_i = \max\limits_{F \in \mathcal{F}} \{f(F) - \lambda_i b(F) - k\}$.

(ii) $\lambda_{i+1} = \dfrac{f_i}{b_i}$.

(iii) $\lambda_{i+1} - \lambda_i = \dfrac{h_i}{b_i}$.

(iv) $\dfrac{h_{i+1}}{h_i} + \dfrac{b_{i+1}}{b_i} \leq 1$.

(v) $h_i > 0$ *and* $b_i > 0$ *during the whole iterations except for the last iteration which has $h_i = 0$.*

(vi) *Both $\{h_i\}$ and $\{b_i\}$ are strictly decreasing sequences before the last iteration.*

(vii) *Sequence $\{\lambda_i\}$ is strictly increasing.*

Proof The first two equalities are by line 3 and line 7 of Algorithm 1, respectively.
Equality (iii) follows easily from (i) and (ii).
By (i) and (iii),

$$h_i \geq f(F_{i+1}) - \lambda_i b(F_{i+1}) - k = f_{i+1} - \lambda_{i+1} b_{i+1} + (\lambda_{i+1} - \lambda_i) b_{i+1} = h_{i+1} + \frac{h_i}{b_i} \cdot b_{i+1}.$$

Dividing both sides by h_i, inequality (iv) follows.

By assumption (1), $b_i > 0$ for all i and $h_0 > 0$. Suppose i is the first iteration with $h_i \leq 0$. Then, $h_i = \max_{F \in \mathcal{F}} \{f(F) - \lambda_i b(F) - k\} \leq 0$. Since $f(F_{i-1}) - \lambda_i b(F_{i-1}) - k = f_{i-1} - \lambda_i b_{i-1} = 0$ (by (ii)), we have $h_i = 0$. At this point, the algorithm terminates. Property (v) is proved.

Combining (iv) and (v), before the last iteration,

$$\frac{h_{i+1}}{h_i} < 1 \text{ and } \frac{b_{i+1}}{b_i} < 1.$$

So, $\{h_i\}$ and $\{b_i\}$ are both strictly decreasing sequences.
Combining (iii) and (v), $\lambda_{i+1} - \lambda_i > 0$, (vi) follows.

Theorem 1 *The number of iterations in Algorithm 1 is $O((n_f n_b) \log(n_f n_b))$.*

Proof By the linearity assumption of f and b, we may assume that

$$f_i = \alpha^T x_{F_i} - k \text{ and } b_i = \beta^T \hat{x}_{F_i}, \tag{4}$$

where $\alpha = (\alpha_1, \ldots, \alpha_{n_f})^T$, $\beta = (\beta_1, \ldots, \beta_{n_b})^T$ are real vectors, and $x_{F_i} = (x_{F_i,1}, \ldots, x_{F_i,n_f})^T$, $\hat{x}_{F_i} = (\hat{x}_{F_i,1}, \ldots, \hat{x}_{F_i,n_b})^T$ are 0–1 vectors.

By property (iv) in Lemma 2, for each iteration, either $h_{i+1}/h_i \leq 1/2$ or $b_{i+1}/b_i \leq 1/2$.

Claim 1 There are $O(n_b \log n_b)$ iterations satisfying $b_{i+1}/b_i \leq 1/2$.

Suppose the iterations satisfying $b_{i+1}/b_i \leq 1/2$ are indexed by $i_1 < \ldots < i_q$. Since $\{b_i\}$ is a positive decreasing sequence, we have

$$0 < b_{i_{j+1}} \le b_{i_j+1} \le \frac{h_{i_j}}{2}.$$

Notice that $b_i = \beta^T \hat{x}_{F_i} = \sum_{j=1}^{n_b} \beta_j \hat{x}_{F_i,j}$ can be written as $b_i = y_i^T \delta$, where $\delta = (|\beta_1|, \ldots, |\beta_{n_b}|)^T$, and $y_i = (sign(\beta_1)\hat{x}_{F_i,1}, \ldots, sign(\beta_{n_b})\hat{x}_{F_i,n_b})^T$. Since x_{F_i} is a 0–1 vector, each component of y_i takes a value from $\{-1, 0, 1\}$. So, δ and $y_{i_1}, \ldots, y_{i_q}$ satisfy the conditions in Lemma 1, and thus $q = O(n_b \log n_b)$.

Claim 2 There are $O((n_f n_b) \log(n_f n_b))$ iterations satisfying $h_{i+1}/h_i \le 1/2$.

Suppose the iterations satisfying $h_{i+1}/h_i \le 1/2$ are indexed by $i_1 < \ldots < i_q$. Similarly to the above, since $\{h_i\}$ is a positive decreasing sequence, we have

$$0 < h_{i_{j+1}} \le h_{i_j+1} \le \frac{h_{i_j}}{2}.$$

Notice that

$$h_i = f_i - \lambda_i b_i = f_i - \frac{f_{i-1}}{b_{i-1}} \cdot b_i = \frac{f_i b_{i-1} - f_{i-1} b_i}{b_{i-1}}. \tag{5}$$

Denote $g_i = f_i b_{i-1} - f_{i-1} b_i$. Then

$$\frac{g_{i_{j+1}}}{b_{i_{j+1}-1}} \le \frac{g_{i_j}}{2b_{i_j-1}}.$$

By the monotonicity of $\{b_i\}$, we have $0 < b_{i_{j+1}-1} < b_{i_j-1}$. Hence

$$g_{i_{j+1}} \le \frac{g_{i_j}}{2}.$$

Furthermore, by (5) and property (v) of Lemma 2, $g_i = h_i b_{i-1} > 0$. Substituting (4) into the definition of g_i, we have

$$g_i = (\alpha^T x_{F_i} - k)\beta^T \hat{x}_{F_{i-1}} - (\alpha^T x_{F_{i-1}} - k)\beta^T \hat{x}_{F_i}$$

$$= \left(\sum_{j=1}^{n_f} \alpha_j x_{F_i,j}\right)\left(\sum_{l=1}^{n_b} \beta_l \hat{x}_{F_{i-1},l}\right) - \left(\sum_{s=1}^{n_f} \alpha_s x_{F_{i-1},s}\right)\left(\sum_{t=1}^{n_b} \beta_t \hat{x}_{F_i,t}\right)$$

$$+ k\beta^T (\hat{x}_{F_i} - \hat{x}_{F_{i-1}})$$

$$= \sum_{j=1}^{n_f}\sum_{l=1}^{n_b} \alpha_j \beta_l x_{F_i,j} \hat{x}_{F_{i-1},l} - \sum_{s=1}^{n_f}\sum_{t=1}^{n_b} \alpha_s \beta_t x_{F_{i-1},s} \hat{x}_{F_i,t}$$

$$+ k\sum_{m=1}^{n_b} \beta_m (\hat{x}_{F_i,m} - \hat{x}_{F_{i-1},m}).$$

The above g_i can be written as an inner product

$$g_i = y_i^T \delta,$$

where y_i and δ are vectors of length $2n_f n_b + n_b$ defined as follows: $\delta = (\delta_{1,1}^{(1)}, \ldots, \delta_{1,n_b}^{(1)}, \delta_{2,1}^{(1)}, \ldots, \delta_{2,n_b}^{(1)}, \ldots, \quad \delta_{n_f,1}^{(1)}, \ldots, \delta_{n_f,n_b}^{(1)}, \delta_{1,1}^{(2)}, \ldots, \delta_{1,n_b}^{(2)}, \delta_{2,1}^{(2)}, \ldots,$ $\delta_{2,n_b}^{(2)}, \ldots, \delta_{n_f,1}^{(2)}, \ldots, \delta_{n_f,n_b}^{(2)}, \delta_1, \ldots, \delta_{n_b})^T$ with

$$\begin{cases} \delta_{j,l}^{(1)} = |\alpha_j \beta_l| \text{ for } j = 1, \ldots, n_f; \ l = 1, \ldots, n_b; \\ \delta_{s,t}^{(2)} = |\alpha_s \beta_t| \text{ for } s = 1, \ldots, n_f; \ t = 1, \ldots, n_b; \\ \delta_m = k|\beta_m| \text{ for } m = 1, \ldots, n_b. \end{cases}$$

$$y_i = (y_{i,1,1}^{(1)}, \ldots, y_{i,1,n_b}^{(1)}, y_{i,2,1}^{(1)}, \ldots, y_{i,2,n_b}^{(1)}, \ldots, y_{i,n_f,1}^{(1)}, \ldots, y_{i,n_f,n_b}^{(1)}, y_{i,1,1}^{(2)}, \ldots,$$
$$y_{i,1,n_b}^{(2)}, y_{i,2,1}^{(2)}, \ldots, y_{i,2,n_b}^{(2)}, \ldots, y_{i,n_f,1}^{(2)}, \ldots, y_{i,n_f,n_b}^{(2)}, y_{i,1}, \ldots, y_{i,n_b})^T \text{ with}$$

$$\begin{cases} y_{i,j,l}^{(1)} = sign(\alpha_j \beta_l) x_{F_i,j} \hat{x}_{F_{i-1},l} \text{ for } j = 1, \ldots, n_f; \ l = 1, \ldots, n_b; \\ y_{i,s,t}^{(2)} = -sign(\alpha_s \beta_t) x_{F_{i-1},s} \hat{x}_{F_i,t} \text{ for } s = 1, \ldots, n_f; \ t = 1, \ldots, n_b; \\ y_{i,m} = sign(\beta_m)(\hat{x}_{F_i,m} - \hat{x}_{F_{i-1},m}) \text{ for } m = 1, \ldots, n_b. \end{cases}$$

Because $\{x_{F_i}\}$ and $\{\hat{x}_{F_i}\}$ are 0-1 vectors, it can be seen that every $y_i \in \{-1, 0, 1\}^{2n_f n_b + n_b}$. Then, δ and $y_{i_1}, \ldots, y_{i_q}$ satisfy the conditions of Lemma 1, and thus $q = O((2n_f n_b + n_b) \log(2n_f n_b + n_b)) = O((n_f n_b) \log(n_f n_b))$.

Combining Claim 1 and Claim 2, the theorem follows.

Suppose algorithm $\mathcal{A}$ has time complexity T. Then the time complexity of Algorithm 1 is $O((n_f n_b) \log(n_f n_b) \cdot T)$. Hence we have the following corollary.

Corollary 1 *If for any fixed parameter λ, there exists a (strong) polynomial time algorithm to solve the optimization problem (2), then* ROOT OF PIECEWISE LINEAR DECREASING FUNCTION *can be solved in (strong) polynomial time.*

If $\beta = (1, \ldots, 1)^T$, the problem is said to be *uniform*, which has a better time complexity.

Theorem 2 *For uniform RPLDF, the number of iterations of Algorithm 1 is at most n_b.*

Proof Notice that for uniform RPLDF, b_i is a positive integer and $b_i \leq n_b$. Then the theorem follows from the property that $\{b_i\}$ is strictly decreasing.

Whether (2) can be solved in (strong) polynomial time depends on the structure of X. In the following section, we consider some problems which can be transformed to ROOT OF PIECEWISE LINEAR DECREASING FUNCTION.

2 Applications

2.1 Linear Fractional Combinatorial Optimization

In a LINEAR FRACTIONAL COMBINATORIAL OPTIMIZATION problem (LFCO), one is to

$$\max_{x \in X} \left\{ \frac{\alpha^T x - k}{\beta^T x} \right\},$$

where α, β are vectors in $\mathbb{R}^n$, x is a vector in $X \subseteq \{0, 1\}^n$, and

$$\alpha^T x > k \text{ for some } x \in X \text{ and } \beta^T x > 0 \text{ for all } x \in X.$$

This problem is equivalent to

$$\min_{\lambda \geq 0} \lambda \tag{6}$$
$$s.t. \ \alpha^T x - \lambda \beta^T x - k \leq 0, \ \forall x \in X.$$

Define

$$h(\lambda) = \max_{x \in X} \{\alpha^T x - \lambda \beta^T x - k\}. \tag{7}$$

Then $h(\lambda)$ is a piecewise linear decreasing function and the problem (6) is equivalent to finding the unique root of $h(\lambda)$. Hence, LFCO is a special RPLDF with $f(x) = \alpha^T x$, $b(x) = \beta^T x$. Then by Section 1, we have the following result for LFCO.

Theorem 3 *Discrete Newton method solves LFCO in time $O(n^2 \log n \cdot T)$, where T is the time complexity to solve (7) for any fixed parameter λ.*

A lot of combinatorial optimization problems can be modeled as an LFCO problem. We introduce the MAXIMUM MEAN-WEIGHT-SURPLUS CUT problem (MMWSC) considered in [7] as an example.

Let $G = (V, E)$ be a directed graph, $c : E \mapsto \mathbb{R}^+$ be a capacity function on E, and $d : V \mapsto \mathbb{R}^+$ be a demand function on V such that $\sum_{v \in V} d(v) = 0$. A vertex v with $d(v) > 0$ is called a *sink* and a vertex v with $d(v) < 0$ is called a *source*. A sink demands some supply, and a source provides some supply. For a vertex set $S \subseteq V$, the demand of S is $d(S) = \sum_{v \in S} d(v)$. For a nonempty proper vertex subset S of V, the set of arcs with tails in $\overline{S}$ and heads in S is called a *cut* of G, denoted by $(\overline{S}, S)$, where $\overline{S}$ is the complement of S in V. For a cut $(\overline{S}, S)$, its *capacity* is $c(\overline{S}, S) = \sum_{e \in (\overline{S}, S)} c(e)$, and its *surplus* is $s(\overline{S}, S) = d(S) - c(\overline{S}, S)$. Furthermore, let $w : E \mapsto \mathbb{R}^+$ be a weight function on E. The *weight* of cut $(\overline{S}, S)$ is $w(\overline{S}, S) = \sum_{e \in (\overline{S}, S)} w(e)$.

MAXIMUM MEAN-WEIGHT-SURPLUS CUT: Given (G, c, d, w) such that there exists a positive surplus cut in G, find a cut $(\overline{S}, S)$ of G with a maximum mean-weight-surplus

$$mws(\overline{S}, S) = \frac{s(\overline{S}, S)}{w(\overline{S}, S)}.$$

This problem is a special LFCO. In fact, suppose $V = \{v_1, \ldots, v_n\}$ and $E = \{e_1, \ldots, e_m\}$. Let $x_S^T = (y_S^T, z_S^T)$, where y_S is the characteristic vector of S and z_S is the characteristic vector of $(\overline{S}, S)$ (that is, y_S is a vector of length n such that $y_{S,j} = 1$ if $v_j \in S$ and $y_{S,j} = 0$ otherwise, z_S is a vector of length m such that $z_{S,l} = 1$ if $e_l \in (\overline{S}, S)$ and $z_{S,l} = 0$ otherwise). Let $\alpha = (d(v_1), \ldots, d(v_n), -c(e_1), \ldots, -c(e_m))^T$ and $\beta = (0, \ldots, 0, w(e_1), \ldots, w(e_m))^T$. Then $s(\overline{S}, S) = \alpha^T x_S$ and $w(\overline{S}, S) = \beta^T x_S$. Hence discrete Newton method solves MMWSC in time $O((m+n)^2 \log(m+n) \cdot T)$, where T is the time to solve

$$\max_{(\overline{S}, S)} \{s(\overline{S}, S) - \lambda w(\overline{S}, S)\} \tag{8}$$

for any fixed λ.

Next, we consider how to solve (8) for any fixed λ. Denote by G_c the directed graph with capacity c, and let $s_c(\overline{S}, S)$ be the surplus of cut $(\overline{S}, S)$ in G_c. Notice that

$$s_c(\overline{S}, S) - \lambda w(\overline{S}, S) = d(S) - (c + \lambda w)(\overline{S}, S) = s_{c+\lambda w}(\overline{S}, S). \tag{9}$$

Hence solving (8) is equivalent to finding a cut with the maximum surplus in $G_{c+\lambda w}$. In the following, we focus on how to find a maximum surplus cut.

For directed graph $G = (V, E)$ with capacity c, denote by SR_G and SN_G the set of sources and the set of sinks, respectively. Construct an auxiliary directed graph $G' = (V', E')$ with $V' = V \cup \{s, t\}$ and $E' = E \cup \{(s, v)\}_{v \in SR_G} \cup \{(v, t)\}_{v \in SN_G}$. The capacity c' on E' is defined to be

$$c'(e) = \begin{cases} c(e), & e \in E, \\ d(v), & e = (v, t), v \in SN_G \\ -d(v), & e = (s, v), v \in SR_G. \end{cases}$$

Let f is a maximum s-t flow in G'. For each vertex $v \in V$, let $e^f(v) = \sum_{(u,v) \in E} f(u, v) - \sum_{(v,u) \in E} f(v, u)$ be the net flow into v, called the *excess* at vertex v. Notice that arcs of the form (s, v) and (v, t) are not counted in the definition of excess, they are merely virtual arcs which paly an auxiliary role in finding flow f. It is not difficult to see that for any vertex subset S,

$$e^f(S) = \sum_{v \in S} e^f(v) = f(\overline{S}, S) - f(S, \overline{S}). \tag{10}$$

Define a *residual* directed graph G^f on vertex set V and arc set E, with residual capacity function $c^f = c - f$ and residual demand function $d^f = d - e^f$.

For any cut $(\overline{S}, S)$, by (10),

$$e^f(S) = f(\overline{S}, S) - f(S, \overline{S}) \le f(\overline{S}, S) \le c(\overline{S}, S).$$

Hence if the demand of every vertex is satisfied, that is, if $e^f(v) = d(v)$ for any $v \in V$, then $d(S) = e^f(S) \le c(\overline{S}, S)$, and thus there is no positive surplus cut in G. So, under the assumption that G has some positive surplus cut, not all demands are satisfied. It follows that $SR_{G^f} \ne \emptyset$ and $SN_{G^f} \ne \emptyset$. Let S^0 be the vertex subset of V such that $\overline{S^0} = \{v\colon v$ is reachable from some source vertex of G^f by an f-augmenting path$\}$ (a path P is f-augmenting if every forward arc e on P has $f(e) < c(e)$ and every backward arc e on P has $f(e) > 0$). Call $(\overline{S^0}, S^0)$ a *blocking cut* of G.

Lemma 3 *The blocking cut $(\overline{S^0}, S^0)$ has the following properties:*

  (*i*) $SR_{G^f} \subseteq \overline{S^0}$ *and* $SN_{G^f} \subseteq S^0$.
 (*ii*) $c^f(\overline{S^0}, S^0) = 0$.
(*iii*) $f(S^0, \overline{S^0}) = 0$.
 (*iv*) $s^f(\overline{S^0}, S^0) = d^f(S^0)$.
       Denote the surplus of cut $(\overline{S}, S)$ in G and G^f by $s(\overline{S}, S)$ and $s^f(\overline{S}, S)$, respectively.
  (*v*) *For any cut $(\overline{S}, S)$, $s^f(\overline{S}, S) \ge s(\overline{S}, S)$.*
 (*vi*) $s^f(\overline{S^0}, S^0) = s(\overline{S^0}, S^0)$.
(*vii*) $(\overline{S^0}, S^0)$ *is a maximum surplus cut, both in G and G^f.*

Proof It is clear by the definition of S^0 that $SR_{G^f} \subseteq \overline{S^0}$. If $SN_{G^f} \cap \overline{S^0} \ne \emptyset$, then there is an f-augmenting path from some source vertex of G^f to some sink vertex of G^f, and thus the flow can be augmented through this path, contradicting that f is a maximum flow. So, property (*i*) is true.

Also by the definition of S^0, every arc $e \in (\overline{S^0}, S^0)$ has $f(e) = c(e)$ and every arc $e \in (S^0, \overline{S^0})$ has $f(e) = 0$. Properties (*ii*) and (*iii*) follow.

Property (*iv*) is a consequence of property (*ii*) and the definition of surplus. By (10), we have

$$s^f(\overline{S}, S) = d^f(S) - c^f(\overline{S}, S) = \big(d(S) - c(\overline{S}, S)\big) + \big(f(\overline{S}, S) - e^f(S)\big)$$

$$= s(\overline{S}, S) + f(S, \overline{S}) \ge s(\overline{S}, S).$$

Property (*v*) is proved.

It can be seen from the above inequality that $s^f(\overline{S}, S) = s(\overline{S}, S)$ if and only if $f(S, \overline{S}) = 0$. Hence by property (*iii*), property (*vi*) is true.

Since $SN_{G^f} \subseteq S^0$, we have $d^f(S^0) = \max_{S \subseteq V} d^f(S)$. Then by property (*iv*), for any cut $(\overline{S}, S)$,

$$s^f(\overline{S^0}, S^0) = d^f(S^0) \geq d^f(S) \geq d^f(S) - c^f(\overline{S}, S) = s^f(\overline{S}, S).$$

Hence $(\overline{S^0}, S^0)$ is a maximum surplus cut in G^f. Combining this with properties (v) and (vi), it is also a maximum surplus cut in G. Property (vii) is proved.

By property (vii) of Lemma 3, for any fixed parameter λ, a maximum surplus cut in $G_{c+\lambda w}$ can be found by implementing a maximum-flow algorithm once. Hence the parameterized combinatorial problem (8), and thus MMWSC, can be solved in strong polynomial time.

In fact, by considering special structures of MMWSC, Radzik [7] showed that the number of iterations can be further reduced to $O(m)$, where m is the number of edges. To illustrate the idea, we include in this section a simpler analysis, showing that the number of iterations for MMWSC can be reduced to $O(m \log m)$. For the more complicated analysis of $O(m)$ iterations, the interested readers may refer to [7].

Theorem 4 *Discrete Newton method solves MMWSC in time $O(m \log m \cdot T)$, where T is the time complexity for the maximum-flow problem.*

Proof We first interpret the variables in the analysis of discrete Newton method in terms of the MMWSC problem. In the i-th iteration, a maximum surplus cut $(\overline{S}_i, S_i)$ of $G_{c+\lambda_i w}$ is found by a maximum-flow algorithm. Clearly,

$$b_i = w(\overline{S}_i, S_i) \tag{11}$$

is the weight of this cut. By (9) and property (vi), property (iv) of Lemma 3

$$h_i = s_{c+\lambda_i w}(\overline{S}_i, S_i) = s^{f_i}_{c+\lambda_i w}(\overline{S}_i, S_i) = d^{f_i}(S_i) \tag{12}$$

is the maximum surplus of $G_{c+\lambda_i w}$, as well as the demand of S_i in the residual graph $G^{f_i}_{c+\lambda_i w}$, where f_i is a maximum flow in $G_{c+\lambda_i w}$. By (ii) of Lemma 2, $\lambda_{i+1} = s(\overline{S}_i, S_i)/w(\overline{S}_i, S_i)$. So,

$$(c + \lambda_{i+1} w)(\overline{S}_i, S_i) = c(\overline{S}_i, S_i) + s(\overline{S}_i, S_i) = d(S_i).$$

By (10) and property (iii) of Lemma 3,

$$d^{f_i}(S_i) = d(S_i) - e^{f_i}(S_i) = d(S_i) - f_i(\overline{S}_i, S_i) + f_i(S_i, \overline{S}_i) = d(S_i) - f_i(\overline{S}_i, S_i).$$

Then by (12), the residual capacity of $(\overline{S}_i, S_i)$ in $G_{c+\lambda_{i+1} w}$ satisfies

$$(c + \lambda_{i+1} w)^{f_i}(\overline{S}_i, S_i) = d(S_i) - f_i(\overline{S}_i, S_i) = d^{f_i}(S_i) = h_i. \tag{13}$$

An arc e is called *unessential* at iteration i if

$$\text{either } w(e) > b_i \text{ or } (c + \lambda_{i+1} w)^{f_i}(e) > h_i.$$

The idea behind this definition is as follows. By (11) and (13), an unessential arc at iteration i cannot belong to cut $(\overline{S}_i, S_i)$. So, if it can be proved that after every $O(\log m)$ iterations, an essential arc turns into an unessential arc and remains to be unessential in the latter iterations, then after $K = O(m \log m)$ iterations, all arcs are unessential, and thus $G_{c+\lambda_K w}$ no longer has a positive surplus cut, at which point, the iteration terminates.

To realize the above idea, we first claim that

$$(c + \lambda_{i+l} w)^{f_{i+l-1}}(\overline{S}_i, S_i) \geq h_{i+1} \text{ for any } l \geq 2. \tag{14}$$

In fact, since $\{\lambda_i\}$ is monotone increasing, we have $\lambda_{i+l} \geq \lambda_{i+2}$. Notice that $f_{i+l-1}(\overline{S}_i, S_i) - f_i(\overline{S}_i, S_i)$ is the increase of the flow across cut $(\overline{S}_i, S_i)$, which is upper bounded by the residual demand $d^{f_i}(S_i)$. Hence by (13), (12), the monotonicity of $\{b_i\}$, and (iii) of Lemma 2,

$$(c + \lambda_{i+1} w)^{f_{i+l-1}}(\overline{S}_i, S_i)$$

$$= (c + \lambda_{i+1} w)^{f_i}(\overline{S}_i, S_i) + (\lambda_{i+l} - \lambda_{i+1}) w(\overline{S}_i, S_i) - (f_{i+l-1}(\overline{S}_i, S_i) - f_i(\overline{S}_i, S_i))$$

$$\geq h_i + (\lambda_{i+l} - \lambda_{i+1}) b_i - d^{f_i}(S_i)$$

$$= (\lambda_{i+l} - \lambda_{i+1}) b_i$$

$$> (\lambda_{i+2} - \lambda_{i+1}) b_{i+1}$$

$$= h_{i+1}.$$

By Lemma 2 (iv),

$$\frac{h_{i+1}}{h_i} \cdot \frac{b_{i+1}}{b_i} \leq \left(\frac{\frac{h_{i+1}}{h_i} + \frac{b_{i+1}}{b_i}}{2} \right)^2 \leq \frac{1}{4}.$$

For any $t \geq \lceil \log m \rceil + 1$, we have

$$h_{i+t} b_{i+t} \leq \left(\frac{1}{4} \right)^{t-1} h_{i+1} b_{i+1} \leq \frac{h_{i+1} b_{i+1}}{m^2}. \tag{15}$$

In the case that $b_{i+t} \leq b_{i+1}/m$, since $\{b_i\}$ is strictly decreasing, we have

$$m \cdot b_{i+t} \leq b_{i+1} < b_i = w(\overline{S}_i, S_i) \leq m \max_{e \in (\overline{S}_i, S_i))} w(e).$$

In the case that $b_{i+t} > b_{i+1}/m$, we have $h_{i+t} < h_{i+1}/m$ by (15). Taking $l = t + 1$ in (14),

$$m h_{i+t} < h_{i+1} \leq m \max_{e \in (\overline{S}_i, S_i)} (c + \lambda_{i+t+1} w)^{f_{i+t}}(e).$$

In any case, the arc in $(\overline{S}_i, S_i)$ with the maximum weight is unessential at iteration $i + t$. Notice that any arc in $(\overline{S}_i, S_i)$ is essential, hence after every $\lceil \log m \rceil + 1$ iterations, a *new* unessential arc emerges. As we have explained before, this implies that the number of iterations is $O(m \log m)$. The time complexity follows.

2.2 Inverse Combinatorial Problems

Suppose $U = \{e_1, \ldots, e_n\}$, $\mathscr{F}$ is a family of subsets of U, and $c \colon U \mapsto \mathbb{R}^+$ is a cost function on U. For any $F \subseteq U$, its cost is $c(F) = \sum_{e \in F} c(e)$. Consider a combinatorial optimization problem

$$(P) \quad \min_{F \in \mathscr{F}} c(F). \tag{16}$$

Its inverse problem can be formulated as follows.

INVERSE COMBINATORIAL PROBLEM: Given a feasible solution $F^0 \in \mathscr{F}$ which is not optimal, modify c to a new cost function $\tilde{c}$ such that F^0 is an optimal solution with respect to $\tilde{c}$ and the modification from c to $\tilde{c}$ is as small as possible, i.e.,

$$(IP) \quad \min_{\tilde{c}} \|\tilde{c} - c\|$$
$$s.t. \ \tilde{c}(F^0) \leq \tilde{c}(F) \text{ for all } F \in \mathscr{F}, \tag{17}$$

where $\|\cdot\|$ is a norm measuring the difference between c and $\tilde{c}$.

Inverse combinatorial problem was first proposed by Burton and Toint and is extensively studied in the literature. A summary of the works before 2004 can be found in [5]. For some new trends in this field as well as some studies on NP-hard inverse problems, the readers may refer to [2].

We show in this section that if $\|\cdot\|$ is the weighted l_∞ norm, then (17) can be transformed to a ROOT OF PIECEWISE LINEAR DECREASING FUNCTION problem. Under l_∞ norm, (17) can be written as

$$\min_{\tilde{c}} \max_{e \in U} p_e |\tilde{c}(e) - c(e)|$$
$$s.t. \ \sum_{e \in F^0} \tilde{c}(e) \leq \sum_{e \in F} \tilde{c}(e) \text{ for all } F \in \mathscr{F}, \tag{18}$$

where $p_e > 0$ is the price of changing one unit of cost on edge e.

Since the goal is to make $\tilde{c}(F^0)$ being minimum, a natural idea is to decrease $c(e)$ for $e \in F^0$ and increase $c(e)$ for $e \in U \setminus F^0$. This intuition can be certified by the following lemma.

Lemma 4 *There exists an optimal solution $\tilde{c}$ to (18) such that $\tilde{c}(e) \leq c(e)$ for any $e \in F_0$ and $\tilde{c}(e) \geq c(e)$ for any $e \in U \setminus F_0$.*

Proof Suppose $\tilde{c}$ is an optimal solution to (18) such that $|\{e : \tilde{c}(e) = c(e)\}|$ is as large as possible. If there exists an element $e_0 \in F_0$ such that $\tilde{c}(e_0) > c(e_0)$, or an element $e_0 \in U \setminus F_0$ such that $\tilde{c}(e_0) < c(e_0)$, let

$$\bar{c}(e) = \begin{cases} \tilde{c}(e), \ e \neq e_0, \\ c(e), \ e = e_0. \end{cases}$$

It can be verified that $\bar{c}$ also satisfies the constraint in (18). Furthermore,

$$\max_{e \in U} p_e |\bar{c}(e) - c(e)| \leq \max_{e \in U} p_e |\tilde{c}(e) - c(e)|.$$

Hence $\bar{c}$ is also an optimal solution to (18). However, $|\{e : \bar{c}(e) = c(e)\}| > |\{e : \tilde{c}(e) = c(e)\}|$, contradicting the choice of $\tilde{c}$. The lemma is proved.

Motivated by the above lemma, we may suppose

$$\tilde{c}(e) = \begin{cases} c(e) - \theta_e, \ \text{for } e \in F^0, \\ c(e) + \theta_e, \ \text{for } e \in U \setminus F^0, \end{cases}$$

where $\{\theta_e\}$ are nonnegative real numbers. Then, (18) can be written as

$$\min_{\{\theta_e\}_{e \in U}} \max_{e \in U} p_e \theta_e$$

$$s.t. \ \sum_{e \in F^0 \setminus F} c(e) - \sum_{e \in F \setminus F^0} c(e) - \sum_{e \in F \Delta F^0} \theta_e \leq 0 \text{ for all } F \in \mathscr{F}, \tag{19}$$

$$\theta_e \geq 0, \ \text{for all } e \in U,$$

where $F \Delta F^0 = (F \setminus F^0) \cup (F^0 \setminus F)$ is the symmetric difference of F and F^0. The objective function of (19) can be linearized as follows:

$$\min_{\lambda \geq 0} \lambda$$

$$s.t. \ \sum_{e \in F^0 \setminus F} c(e) - \sum_{e \in F \setminus F^0} c(e) - \lambda \sum_{e \in F \Delta F^0} \frac{1}{p_e} \leq 0 \text{ for all } F \in \mathscr{F}. \tag{20}$$

In fact, for any feasible solution $\{\theta_e\}_{e \in U}$ of (19), $\lambda = \max_{e \in U} p_e \theta_e$ is a feasible solution of (20). Conversely, any feasible solution λ of (20) determines a feasible solution of (19) by setting $\theta_e = \lambda / p_e$ $(e \in U)$. Furthermore, the objective value is the same.

For $F \in \mathscr{F}$, define $f(F) = \sum_{e \in F^0 \setminus F} c(e) - \sum_{e \in F \setminus F^0} c(e)$ and $b(F) = \sum_{e \in F \Delta F^0} \frac{1}{p_e}$. Then (20) is equivalent to finding the unique root of the piecewise linear decreasing function

$$h(\lambda) = \max_{F \in \mathscr{F}} \{f(F) - \lambda b(F)\}. \tag{21}$$

Notice that $f(F)$ and $b(F)$ can be written as $f(F) = \alpha^T x_F$ and $b(F) = \beta^T x_F$ where $\beta_j = 1/p_{e_j}$ for $j = 1, \ldots, n$,

$$\alpha_j = \begin{cases} c(e_j), & \text{if } e_j \in F^0, \\ -c(e_j), & \text{if } e_j \in U \setminus F^0, \end{cases} \tag{22}$$

and

$$x_{F,j} = \begin{cases} 1, & \text{if } e_j \in F \Delta F^0, \\ 0, & \text{otherwise.} \end{cases} \tag{23}$$

Since F^0 is not an optimal solution with respect to c, there exists another feasible solution $F^1 \in \mathscr{F}$ such that $c(F^0) > c(F^1)$, which implies that $f(F^1) = c(F^0 \setminus F^1) - c(F^1 \setminus F^0) > 0$. Hence the first part of assumption (1) is satisfied with $k = 0$. Since $p_e > 0$ for any $e \in U$, $b(F) > 0$ for any $F \in \mathscr{F}$ with $F \neq F^0$. Although the second part of assumption (1) is not completely satisfied because of $b(F_0) = 0$, discrete Newton method can still be applied here. In fact, what is needed in the argument of Section 1 is $b_i > 0$ *before the termination of the algorithm*. When F^0 becomes an optimal solution, the algorithm terminates, and thus $b(F^0)$ is never used. By Theorem 1, discrete Newton method solves the inverse problem (18) in time $O(n^2 \log n \cdot T)$, where T is the time complexity of solving (21) for any fixed λ.

Notice that for any fixed λ,

$$h(\lambda) = \max_{F \in \mathscr{F}} \{c_\lambda(F^0) - c_\lambda(F)\} = c_\lambda(F^0) - \min_{F \in \mathscr{F}} c_\lambda(F), \tag{24}$$

where

$$c_\lambda(e) = \begin{cases} c(e) - \frac{\lambda}{p_e}, & e \in F^0, \\ c(e) + \frac{\lambda}{p_e}, & e \in U \setminus F^0. \end{cases}$$

Hence solving (21) for fixed λ is equivalent to finding an $F \in \mathscr{F}$ with a minimum value of $c_\lambda(F)$. For example, if one is considering the inverse minimum spanning tree problem, then $\mathscr{F}$ is the set of all spanning trees, and (24) is to find a minimum spanning tree with respect to weight function c_λ. It is well known that a simple greedy algorithm finds a minimum spanning tree in time $O(|E| \log |E|)$, where $|E|$ is the number of edges; hence, the inverse minimum spanning tree problem can be solved in time $O(|E|^3 (\log |E|)^2)$.

It should be noted that similar deduction applies to the inverse combinatorial optimization problems in which the objective is maximization. In this case, the costs of elements in F^0 are increased and the costs of the other elements are decreased. To unify the statement, we use opt to stand for either min or max.

Theorem 5 *If the optimization problem $\mathrm{opt}_{F \subset \mathscr{F}} c(F)$ can be solved in (strong) polynomial time for any cost function c, then so is its corresponding inverse problem under l_∞ norm.*

As a consequence of Theorem 5, a lot of inverse combinatorial optimization problems under l_∞ norm can be solved in strong polynomial time, including inverse minimum basis of a matroid (which includes the inverse minimum spanning tree problem), inverse maximum intersection of two matroids (which includes the inverse maximum matching problem), inverse minimum cost flow (which includes the inverse shortest path problem), etc.

Some variations of (18) can also be solved in strong polynomial time. For example, Guan et al. [4] considered the inverse max+sum spanning tree problem.

INVERSE MAX+SUM SPANNING TREE: In an Max+Sum Spanning Tree problem, every edge e is assigned a weight $w(e)$ and a cost $c(e)$, the goal is to find a spanning tree T with the minimum $p_{w,c}(T) = \max_{e \in T} w(e) + \sum_{e \in T} c(e)$. For its corresponding inverse problem in which *only costs* are modified, given a spanning tree T^0, the goal is to modify c to $\tilde{c}$ as small as possible such that $p_{w,\tilde{c}}(T^0)$ is minimum among all spanning trees.

The deduction is almost the same, except that (21) becomes

$$h(\lambda) = \max_{F \in \mathscr{F}} \{ f(F) - \lambda b(F) - k \}$$

with $k = -\max_{e \in F^0} w(e)$ and

$$f(F) = \sum_{e \in F^0 \setminus F} c(e) - \sum_{e \in F \setminus F^0} c(e) - \max_{e \in F} w(e).$$

In this case, $f(F)$ can be written as $f(F) = \tilde{\alpha}^T \tilde{x}_F$, where $\tilde{\alpha}$ and $\tilde{x}_F$ are vectors of length $2|E|$ defined as follows: $\tilde{\alpha}^T = (\alpha^T, \bar{\alpha}^T)$, $\alpha \subset \mathbb{R}^{|E|}$ is defined in (22) and $\bar{\alpha} \in \mathbb{R}^{|E|}$ is defined by $\bar{\alpha}_j = -w(e_j)$ for $j = 1, \ldots, |E|$; $\tilde{x}_F = (x_F, \bar{x}_F)$, $x_F \in \{0, 1\}^{|E|}$ is defined in (23) and $\bar{x}_F \in \{0, 1\}^{|E|}$ has a unique nonzero element 1 at position j with $e_j = \arg \max_{e \in F} \{ w(e) \}$.

Notice that the model in Section 1 involves a parameter k and the functions $f(F), b(F)$ are assumed to be inner products of *possibly different lengths*. Hence, using the theory in Section 1 and the fact that max-sum spanning tree problem can be solved in time $O(|E| \log |V|)$ [6], INVERSE MAX-SUM SPANNING TREE can be solved in time $O(|E|^3 \log |E| \log |V|)$.

Our model with parameter k also enables us to deal with the requirement that $\tilde{c}(F^0) \le \tilde{c}(F) + k$ for any $F \in \mathscr{F}$, which occurs if one only needs F^0 to be a *near optimal* solution.

For the general inverse problem with cost function $c(F) = c^T x_F$, it was indicated by Ahuja and Orlin [1] that under some regularity conditions, if (16) can be solved in polynomial time, then its inverse problem (17) under L_1 norm and L_∞ norm can also be solved in polynomial time. This result uses the ellipsoid algorithm. Roughly speaking, given a polyhedron $X \subseteq \mathbb{R}^n$ which is defined by rational inequalities with rationals of size at most ϕ, and a rational vector $c \in \mathbb{Q}^n$, the optimization problem $\max\{c^T x : x \in X\}$ has a corresponding feasibility problem: given a vector x, does x belong to X? A separating oracle answers the feasibility problem by either claiming that $x \in X$ or finding out a violated constraint. The ellipsoid algorithm solves the optimization problem by iteratively solving the feasibility problem, and the time complexity is polynomially bounded by n, ϕ, c, and the running time of the separating oracle [3]. To apply the ellipsoid algorithm on inverse problems, consider the inverse problem (17) under l_∞ norm, which can be written as

$$\min \lambda$$

$$s.t. \ \tilde{c}^T x_F^0 \le \tilde{c}^T x_F, \ \text{for all } F \in \mathcal{F}, \tag{25}$$

$$\tilde{c}_i - c_i \le \lambda, \ \text{for all } i = 1, \ldots, n, \tag{26}$$

$$c_i - \tilde{c}_i \le \lambda, \ \text{for all } i = 1, \ldots, n, \tag{27}$$

whose variables are $\tilde{c}$ and λ. For a proposed solution $(\tilde{c}, \lambda)$, it is easy to check whether (26) and (27) are satisfied. To check constraint (25), it suffices to solve $\min_{F \in \mathcal{F}} \tilde{c}^T x_F$ to obtain an optimal solution F^* and then compare $\tilde{c}^T x_{F^0}$ and $\tilde{c}^T x_{F^*}$. If $\tilde{c}^T x_{F^0} \le \tilde{c}^T x_{F^*}$, then $(\tilde{c}, \lambda)$ is an optimal solution to the inverse problem. Otherwise, we have found a violated constraint $\tilde{c}^T x_{F^0} > \tilde{c}^T x_{F^*}$. So, an algorithm to (16) provides a separating oracle, and thus as long as (16) can be solved in polynomial time, so can its inverse problem. However, the time complexity of using ellipsoid algorithm depends on input numbers, which is not strongly polynomial, while discrete Newton method drastically reduces the time complexity.

2.3   Bottleneck Expansion

Suppose U is a finite set and $\mathcal{F}$ is a family of subsets of U. For a positive cost function c defined on U, the *capacity* of a subset $F \subseteq U$ is defined to be $\text{cap}_c(F) = \min_{e \in U} c(e)$. Element $e \in F$ with $c(e) = \text{cap}_c(F)$ is called a *bottleneck* of F. The capacity of $\mathcal{F}$ is $\text{cap}_c(\mathcal{F}) = \max_{F \in \mathcal{F}} \text{cap}_c(F)$. Let w be a positive weight function on U, indicating the price of altering one unit of cost on an element. A bottleneck expansion problem is to raise capacities of elements under a given budget B such that the capacity of the family is as large as possible.

BOTTLENECK EXPANSION: Given a budget B, find a new cost function $\tilde{c}$ satisfying

$$\max_{\tilde{c}} \text{cap}_{\tilde{c}}(\mathcal{F})$$

$$s.t. \ \sum_{e \in U} w(e)(\tilde{c}(e) - c(e)) \le B, \tag{28}$$

$$\tilde{c}(e) \ge c(e), \ \text{for any } e \in U.$$

First, we derive a characterization for the optimal value of (28). The following special cost function plays an important role: for a real number λ and a subset $F \subseteq U$, define

$$c_{\lambda,F}(e) = \begin{cases} \max\{\lambda, c(e)\}, & e \in F, \\ c(e), & e \in U \setminus F. \end{cases} \tag{29}$$

Clearly, $c_{\lambda,F}(e) \geq c(e)$ for any $e \in U$ and $c_{\lambda,F}(e) \geq \lambda$ for any $e \in F$. As a consequence, $c_{\lambda,F}$ satisfies the second constraint of (28) and $\mathrm{cap}_{c_{\lambda,F}}(\mathscr{F}) \geq \mathrm{cap}_{c_{\lambda,F}}(F) = \min_{e \in F} c_{\lambda,F}(e) \geq \lambda$.

Lemma 5 *If $\tilde{\lambda}$ is the optimal value, then $g(\tilde{\lambda}) = B$, where*

$$g(\lambda) = \min_{F \in \mathscr{F}} \sum_{e \in F} w(e) \cdot \max\{\lambda - c(e), 0\}. \tag{30}$$

Proof Let $\tilde{c}$ be an optimal solution to (28) and let $\tilde{F}$ be a subset of U with $\mathrm{cap}_{\tilde{c}}(\tilde{F}) = \tilde{\lambda}$. We first show that $c_{\tilde{\lambda},\tilde{F}}$ is also an optimal solution to (28).

Since $\tilde{\lambda} = \min_{e \in \tilde{F}} \tilde{c}(e) \leq \tilde{c}(e)$ for any $e \in \tilde{F}$ and $\tilde{c}(e) \geq c(e)$ for any $e \in U$, we have

$$\sum_{e \in U} w(e)(c_{\tilde{\lambda},\tilde{F}}(e) - c(e))$$

$$= \sum_{e \in \tilde{F}} w(e)(\max\{\tilde{\lambda}, c(e)\} - c(e)) \leq \sum_{e \in \tilde{F}} w(e)(\tilde{c}(e) - c(e)) \leq B.$$

Hence $c_{\tilde{\lambda},\tilde{F}}$ is a feasible solution of (28). By

$$\tilde{\lambda} = \max_{\hat{c}} \mathrm{cap}_{\hat{c}}(\mathscr{F}) \geq \mathrm{cap}_{c_{\tilde{\lambda},\tilde{F}}}(\mathscr{F}) \geq \mathrm{cap}_{c_{\tilde{\lambda},\tilde{F}}}(\tilde{F}) \geq \tilde{\lambda},$$

we have $\mathrm{cap}_{c_{\tilde{\lambda},\tilde{F}}}(\mathscr{F}) = \tilde{\lambda}$, and thus $c_{\tilde{\lambda},\tilde{F}}$ is also optimal.

Next, we show that $g(\tilde{\lambda}) = B$. By the definition of $g(\lambda)$ and the feasibility of $c_{\tilde{\lambda},\tilde{F}}$,

$$g(\tilde{\lambda}) \leq \sum_{e \in \tilde{F}} w(e) \cdot \max\{\tilde{\lambda} - c(e), 0\}$$

$$= \sum_{e \in \tilde{F}} w(e)(\max\{\tilde{\lambda}, c(e)\} - c(e)) = \sum_{e \in \tilde{F}} w(e)(c_{\tilde{\lambda},\tilde{F}}(e) - c(e)) \leq B.$$

If $g(\tilde{\lambda}) < B$, then for a sufficiently small real number $\varepsilon > 0$, $g(\tilde{\lambda} + \varepsilon) \leq B$. Hence cost function $c_{\tilde{\lambda}+\varepsilon,\tilde{F}}$ is also a feasible solution to (28). But $\mathrm{cap}_{c_{\tilde{\lambda}+\varepsilon,\tilde{F}}}(\mathscr{F}) \geq \tilde{\lambda} + \varepsilon$, contradicting that $\tilde{\lambda}$ is the optimal value. Hence $g(\tilde{\lambda}) = B$.

Lemma 6 *If $\tilde{\lambda}$ satisfies $g(\tilde{\lambda}) = B$, then $\tilde{\lambda}$ is the optimal value of (28). Furthermore, for the subset $\tilde{F} \subseteq U$ reaching the minimum value of $g(\tilde{\lambda})$ in the definition of (30), the cost function $c_{\tilde{\lambda}, \tilde{F}}$ is an optimal solution to (28).*

Proof Suppose $\tilde{\lambda}$ is not optimal. Let λ^* be the optimal value of (28). Then $\lambda^* > \tilde{\lambda}$. By Lemma 5, $g(\lambda^*) = B$. Suppose $g(\lambda^*)$ is reached by subset $F^* \subseteq U$. We have

$$
\begin{aligned}
B = g(\lambda^*) &= \sum_{e \in F^*} w(e) \cdot \max\{\lambda^* - c(e), 0\} \\
&> \sum_{e \in F^*} w(e) \cdot \max\{\tilde{\lambda} - c(e), 0\} \\
&\geq \min_{F \in \mathscr{F}} \sum_{e \in F} w(e) \cdot \max\{\tilde{\lambda} - c(e), 0\} \\
&= g(\tilde{\lambda}) = B,
\end{aligned}
$$

a contradiction. Hence $\tilde{\lambda}$ is the optimal value. By noticing that

$$
\sum_{e \in \tilde{F}} w(e)(c_{\tilde{\lambda}, \tilde{F}}(e) - c(e)) = g(\tilde{\lambda}) = B,
$$

$c_{\tilde{\lambda}, \tilde{F}}$ is a feasible solution to (28). By $\mathrm{cap}_{c_{\tilde{\lambda}, \tilde{F}}}(\mathscr{F}) \geq \tilde{\lambda}$ and the optimality of $\tilde{\lambda}$, we see that $c_{\tilde{\lambda}, \tilde{F}}$ is an optimal solution to (28).

By Lemma 6, to solve BOTTLENECK EXPANSION, it suffices to find a root of $g(\lambda) = B$. Then an optimal solution can be obtained by (29). In the following, we focus on how to find the root of $g(\lambda) = B$. Notice that $g(\lambda)$ is monotone increasing, but it is not a piecewise linear function. So, discrete Newton method cannot be applied to it directly.

By a binary search method, Zhang et al. [9] proved that the root can be found in $O(\log K)$ iterations, where K is the maximum value of the input numbers and the input size. Hence a pseudo-polynomial time algorithm exists if (30) can be solved in polynomial time. Later, Zhang and Liu [8] improved the time complexity by making use of discrete Newton method. The algorithm is divided into two phases. In the first phase, binary search is used to narrow the location of $\tilde{\lambda}$ to an interval. This interval is such that $g(\lambda)$ on this interval can be written as a piecewise linear function and thus discrete Newton method can be applied to it in the second phase.

To be more concrete, it is not difficult to see that the optimal value $\tilde{\lambda}$ for (28) is bounded by

$$
\min_{e \in U} c(e) \leq \tilde{\lambda} \leq \max_{e \in U} c(e) + \frac{B}{\min_{F \in \mathscr{F}} |F| \cdot \min_{e \in U} w(e)}.
$$

Order the elements of U as $e_1, \ldots, e_n$ such that $c(e_1) \le c(e_2) \le \ldots \le c(e_n)$. Denote $c_i = c(e_i)$ for $i = 1, \ldots, n$, and denote the upper bound for $\tilde{\lambda}$ in the above inequality as c_{n+1}. Using binary search on the subscripts of $c_1, \ldots, c_{n+1}$ (see Algorithm 2), one can either find an index m such that $g(c_m) = B$, implying that the optimal value $\tilde{\lambda} = c_m$, or find an index l with $g(c_l) < B$ and $g(c_{l+1}) > B$, implying that $\tilde{\lambda} \in (c_l, c_{l+1})$. In the first case, $\tilde{\lambda}$ is found in $O(\log n)$ iterations. In the second case, $g(\lambda)$ on interval (c_l, c_{l+1}) can be rewritten as

$$g(\lambda) = \min_{F \in \mathscr{F}} \sum_{e \in F, c(e) < \lambda} w(e)(\lambda - c(e)) = \min_{F \in \mathscr{F}} \{-\alpha^T x_F + \lambda \cdot \beta^T x_F\},$$

where x_F is the characteristic vector of F and

$$\alpha_j = \begin{cases} w(e_j)c(e_j), & 1 \le j \le l, \\ 0, & l+1 \le j \le n. \end{cases}$$

$$\beta_j = \begin{cases} w(e_j), & 1 \le j \le l, \\ 0, & l+1 \le j \le n. \end{cases}$$

Applying discrete Newton method on the piecewise linear decreasing function

$$h(\lambda) = -g(\lambda) + B = \max_{F \in \mathscr{F}} \{\alpha^T x_F - \lambda \cdot \beta^T x_F + B\},$$

another $O(n^2 \log n)$ iterations are sufficient to find the optimal value $\tilde{\lambda}$.

Algorithm 2 Narrow the location of $\tilde{\lambda}$

1: $l = 1, u = n+1$.
2: **while** $u - l > 1$ **do**
3:    $m = \lfloor (l+u)/2 \rfloor$.
4:    **if** $g(c_m) = B$ **then**
5:        output $\tilde{\lambda} = c_m$.
6:    **else**
7:        **if** $g(c_m) < B$ **then**
8:            $l = m$.
9:        **else**
10:           $u = m$.
11:        **end if**
12:    **end if**
13: **end while**
14: Output $\tilde{\lambda} \in (c_l, c_{l+1})$.

Theorem 6 *If the optimization problem* (30) *can be solved in* (strong) *polynomial time for any fixed* λ, *then* BOTTLENECK EXPANSION *can also be solved in* (strong) *polynomial time.*

References

1. Ahuja, R.K., Orlin, J.B.: Inverse optimization. Oper. Res. **49**, 771–783 (2001)
2. Demange, M., Monnot, J.: An introduction to inverse combinatorial problems. In: Paschos, V.T. (eds.) Paradigms of Combinatorial Optimization: Problems and New Approaches, pp. 547–586. Wiley, Hoboken (2013)
3. Grotschel, M., Lovasz, L., Schrijver, A.: Geometric Algorithms and Combinatorial Optimization. Springer, Heidelberg (1986)
4. Guan, X., Pardalos, P.M., Zuo, X.: Inverse max+sum spanning tree problem by modifying the sum-cost vector under weighted l_∞ norm. J. Glob. Optim. **61**(1), 165–182 (2015)
5. Heuberger, C.: Inverse combinatorial optimization: a survey on problems, methods, and results. J. Comb. Optim. **8**, 329–361 (2004)
6. Punnen, A.P., Nair, K.P.K.: An $o(m \log n)$ algorithm for the max+sum spanning tree problem. Eur. J. Oper. Res. **89**, 423–426 (1996)
7. Radzik, T.: Parametric flows, weighted means of cuts, and fractional combinatorial optimization. In: Complexity in Numerical Optimization. World Scientific, Singapore (1993)
8. Zhang, J., Liu, Z.: An oracle strongly polynomial algorithm for bottleneck expansion problems. Optim. Methods Softw. **17**(1), 61–75 (2002)
9. Zhang, J., Yang, C., Lin, Y.: A class of bottleneck expansion problems. Comput. Oper. Res. **28**(6), 505–519 (2001)

An Overview of Submodular Optimization: Single- and Multi-Objectives

Donglei Du, Qiaoming Han, and Chenchen Wu

Abstract We offer an overview on submodular optimization for both single- and multiple-objectives, with the moderate goal to highlight the different angles in interpreting submodularity and associated concepts.

1 Submodular Function

Submodularity is a phenomenon that is ubiquitous in almost all disciplines of the natural and social science, engineering, business and economics, computer sciences, etc. The importance of submodularity comes from its wide applications and its rich theory. Submodularity captures the idea of economies of scales (a.k.a., decreasing marginal return) and is a simple yet powerful condition that holds on numerous occasions.

We look at submodular function from two point of views: the set function view (Section 1.1) and the pseudo-Boolean function view (Section 1.2), offering two angles to comprehend the concept of submodularity. It turns out that one of these two different point of views (set vs pseudo-Boolean) can be more convenient than the other depending on particular applications. For example, the pseudo-Boolean definition can be used to define more general submodular functions over lattices, which has wide application in economics, in particular in comparative statics in

D. Du (✉)
Faculty of Business Administration, University of New Brunswick, Fredericton, NB, Canada
e-mail: ddu@unb.ca

Q. Han
School of Mathematics & Statistics, Zhejiang University of Finance & Economics, Hangzhou, People's Republic of China
e-mail: qmhan@zufe.edu.cn

C. Wu
College of Science, Tianjin University of Technology, Tianjin, China
e-mail: wu_chenchen_tjut@163.com

© Springer Nature Switzerland AG 2019   57
D.-Z. Du et al. (eds.), *Nonlinear Combinatorial Optimization*,
Springer Optimization and Its Applications 147,
https://doi.org/10.1007/978-3-030-16194-1_3

both optimization and game settings (e.g., [12]). On the other hand, the set function interpretation has been widely adopted in combinatorial optimization (e.g., [7, 9]).

1.1 Submodular Set Function

Given a ground set $E = \{1, \ldots, n\}$, a set function $f : 2^E \mapsto \mathbb{R}$ is *submodular* if one of the following equivalent conditions holds:

1. The zero-order definition:

$$f(S) + f(T) \geq f(S \cap T) + f(S \cup T), \ \forall S, T \subseteq E. \tag{1}$$

2. The first-order definition:

$$f(S \cup \{i\}) - f(S) \geq f(T \cup \{i\}) - f(T), \ \forall S \subset T, i \notin T. \tag{2}$$

3. The second-order definition:

$$f(S \cup \{i, j\}) + f(S) \leq f(S \cup \{i\}) + f(S \cup \{j\}), \ \forall S \subset E, i, j \notin S. \tag{3}$$

A set function $f : 2^E \mapsto \mathbb{R}$ is *supermodular* if $-f$ is submodular; and it is *modular* if it is both submodular and supermodular.

1.2 Submodular Pseudo-Boolean Function

1.2.1 Pseudo-Boolean function

A pseudo-Boolean function f is a real-valued function over the n-dimensional Boolean lattice, denoted $\mathbb{B}^n = \{0, 1\}^n$

$$f : \mathbb{B}^n \to \mathbb{R}.$$

The elements of $\mathbb{B}^n$ are in a one-to-one correspondence with the subsets of the ground set $E = \{1, 2, \ldots, n\}$, every pseudo-Boolean function can also be viewed as a real-valued set function defined on 2^N, the power set of N. By considering functions defined on $\mathbb{B}^n$ rather than on 2^N, the pseudo-Boolean approach provides an algebraic viewpoint, which sometimes carries clear advantages over the set-theoretic description. Any set $A \in 2^E$ can be identified by its characteristic/indicator vector $1_A \in \mathbb{B}^n = \{0, 1\}^n$:

$$1_A(s) = \begin{cases} 1, & \text{if } s \in A \\ 0, & \text{if } s \notin A. \end{cases}$$

Geometrically, the characteristic vectors are the 2^n vertices of the hypercube $\{0, 1\}^n$.

Example 1 For $n = 2$ assume the ground set is given by $E = \{1, 2\}$.
Then we have the following:

$$1_\emptyset = \begin{pmatrix} 0 \\ 0 \end{pmatrix}; \ 1_{\{1\}} = \begin{pmatrix} 1 \\ 0 \end{pmatrix}; \ 1_{\{2\}} = \begin{pmatrix} 0 \\ 1 \end{pmatrix}; \ 1_{\{1,2\}} = \begin{pmatrix} 1 \\ 1 \end{pmatrix}.$$

There is a correspondence between set operations and Boolean operations. Define the following Boolean operations for any $x, y \in \mathbb{B}^n$:

$$\text{meet/conjunction}: \ x \wedge y = (\min\{x_1, y_1\}, \ldots, \min\{x_n, y_n\})$$

$$\text{join/disjunction}: \ x \vee y = (\max\{x_1, y_1\}, \ldots, \max\{x_n, y_n\})$$

$$\text{complementation}: \ \bar{x} = (\bar{x}_1, \ldots, \bar{x}_n) = (1 - x_1, \ldots, 1 - x_n).$$

Then it is easy to see that we have the following correspondence and the set operations carry over to the Boolean operations:

$$\wedge \Longleftrightarrow \cap; \quad \vee \Longleftrightarrow \cup$$

$$A \cap B \Longleftrightarrow 1_A \wedge 1_B; A \cup B \Longleftrightarrow 1_A \vee 1_B$$

$$\bar{A} \Longleftrightarrow 1 - 1_A; A \subseteq B \Longleftrightarrow 1_A \leq 1_B; A - B \Longleftrightarrow 1_A \wedge (1 - 1_B).$$

Therefore any set function $f : 2^E \mapsto \mathbb{R}$ can be equivalently viewed as a pseudo-Boolean function $f : \mathbb{B}^n \mapsto \mathbb{R}$ such that $f(A) = f(1_A)$.

1.2.2 Partial Derivative for Pseudo-Boolean Function

For any pseudo-Boolean function $f : \mathbb{B}^n \mapsto \mathbb{R}$, we define its ith partial derivative at $x \in \mathbb{B}^n$ along the element $i \in E$, for some ground set $E = \{1, \ldots, n\}$:

$$\Delta_i(x) = f_i(x) = \frac{\partial f(x)}{\partial x_i} = f(x \vee e_i) - f(x \wedge \bar{e}_i).$$

We further define its second-order partial derivative at $x \in \mathbb{B}^n$ along the elements $i, j \in E$,

$$\Delta_{ij}(x) = f_i(x) = \frac{\partial^2 f(x)}{\partial x_i \partial x_i} = f(x \vee e_i \vee e_j) - f(x \vee e_i \wedge \bar{e}_j) - f(x \wedge \bar{e}_i \vee e_j) + f(x \wedge \bar{e}_i \wedge \bar{e}_j).$$

These first and second-order difference operators are analogous to the differentiations in the standard calculus (the real space $\mathbb{R}^n$).

1.2.3 Submodular Pseudo-Boolean Function

With the pseudo-Boolean function language, (1)–(3) are equivalent to

1. The zero-order definition:

$$f(x) + f(y) \geq f(x \wedge y) + f(x \vee y), \ \forall x, y \in \mathbb{B}^n. \tag{4}$$

2. The first-order definition: Nonincreasing partial derivative

$$f_i(x) \geq f_i(y), \forall x \leq y, \forall i \in E. \tag{5}$$

3. The second-order definition: Non-positive second-order partial derivative

$$f_{ij}(x) = \frac{\partial^2 f}{\partial x_i \partial x_j} \leq 0, \ \forall x \in \mathbb{B}^n, \ \forall i, j \in E. \tag{6}$$

(5)–(6) remind us concavity in the continuous domain.

2 Extensions

We discuss one of the most important mechanism in submodular optimization, namely how to convert discrete problems to continuous ones via the extension or relaxation technique. Historically, extensions are so-called in the set function world and the same concepts are more naturally called as the relaxation technique in the field of approximation algorithm.

In the following discussion of various extensions, we will take an unusual (not more natural) route via the pseudo-Boolean language instead of the oft-opted set language. One immediate benefit from this pseudo-Boolean point of view is the intuitive geometric interpretation of the extension/relaxation technique: pseudo-Boolean functions are only defined on the corner of the hypercube $\mathbb{B}^n = \{0, 1\}^n$, while an extension/relaxation will be defined on the entire hypercube $[0, 1]^n$.

2.1 Convex Envelope/Closure

We consider the convex underestimators of set function $f : \{0, 1\}^n \to \mathbb{R}$. One motivation is that, whenever an optimization problem involves minimizing $f(x)$ or contains an inequality $f(x) \leq r$, replacing $f(x)$ by a convex underestimator yields a convex relaxation of the problem. The convex envelope is the largest convex underestimator.

The convex closure $\underline{f}(x) : [0, 1]^n \rightarrow \mathbb{R}$ of any pseudo-Boolean function $f : \mathbb{B}^n \rightarrow \mathbb{R}$ is defined as: $\forall x \in [0, 1]^n$,

$$\underline{f}(x) = \max_{\begin{pmatrix} y_0 \\ y \end{pmatrix} \in \mathbb{R}^{n+1}} \left\{ y_0 + x^T y : y_0 + a^T y \leq f(a), \forall a \in \mathbb{B}^n \right\} \tag{7}$$

$$= \min_{z \in \mathbb{R}_+^{2^n}} \left\{ \sum_{a \in \mathbb{B}^n} f(a) z_a : \sum_{a \in \mathbb{B}^n : e_i^T a = 1} z_a = x_i, i \in \{1, \ldots, n\}, \sum_{a \in \mathbb{B}^n} z_a = 1 \right\}. \tag{8}$$

Example 2 (Convex Closure for $n = 2$) Assume the pseudo-Boolean function $f : \mathbb{B}^2 \rightarrow \mathbb{R}$ is defined as follows:

$$f(0, 0) = 0, \ f(1, 0) = f(0, 1) = f(1, 1) = 1.$$

The primal program (7) becomes

$$\underline{f}(x_1, x_2) = \max \ y_0 + x_1 y_1 + x_2 y_2$$
$$\text{s.t.} \quad y_0 \leq f(0, 0) = 0$$
$$y_0 + y_1 \leq f(1, 0) = 1$$
$$y_0 + y_2 \leq f(0, 1) = 1$$
$$y_0 + y_1 + y_2 \leq f(1, 1) = 1,$$

along with its dual program (8)

$$\underline{f}(x_1, x_2) = \min \ f(0, 0) z_{(0,0)} + f(1, 0) z_{(1,0)} + f(0, 1) z_{(0,1)} + f(1, 1) z_{(1,1)}$$
$$\text{s.t.} \quad z_{(1,0)} + z_{(1,1)} = x_1$$
$$z_{(0,1)} + z_{(1,1)} = x_2$$
$$z_{(0,0)} + z_{(1,0)} + z_{(0,1)} + z_{(1,1)} = 1$$
$$z \geq 0.$$

The optimal solutions of the primal and dual are given as follows:

(i) For $1 \geq x_1 \geq x_2 \geq 0$: Primal solution $y_1 = 1$, $y_0 = y_2 = 0$ with objective value x_1. Dual solution $z_{(0,0)} = 1 - x_1$, $z_{(1,0)} = x_1 - x_2$, $z_{(0,1)} = 0$ and $z_{(1,1)} = x_2$ with objective value x_1. Therefore both are optimal.

(ii) Symmetrically, when $1 \geq x_2 \geq x_1 \geq 0$, the optimal value is x_2.

The convex closure for any $x = (x_1, x_2) \in [0, 1]^2$ is therefore

$$\underline{f}(x_1, x_2) = \max\{x_1, x_2\}.$$

Table 1 Dual variable z as
joint distribution and x as
marginals

(X_1, X_2)	0	1	
0	$z_{(0,0)}$	$z_{(0,1)}$	$1 - x_1$
1	$z_{(1,0)}$	$z_{(1,1)}$	x_1
	$1 - x_2$	x_2	

2.1.1 Interpretation of Convex Closure from Primal Point of View

From (7), $\underline{f}$ is the (point-wise) largest convex function below f.

2.1.2 Interpretation of Convex Closure from Dual Point of View

Let $z_{(x_1,\ldots,x_n)} \in \{0, 1\}^E$ be a feasible solution of (8). For each index $x_i \in \{0, 1\}$ ($i = 1, \ldots, n$), define a Bernoulli random variable X_i such that $P(X_i = 1) = x_i$ (Table 1). So $z = z_{X_1, X_2}$ is a joint probability distribution with X_i as marginal distributions.

The convex closure is to find such a joint distribution z such that the expectation of f is minimized:

$$\underline{f}(x) = \min_z \mathbb{E}_z[f(X_1, X_2)].$$

Extending the dual interpretation, we have $\forall x \in [0, 1]^n$, the convex closure can be formulated as an optimization problem which preserves marginals:

$$\underline{f}(x) = \min_{z_{(X_1,\ldots,X_n)}} \mathbb{E}[f(X_1, \ldots, X_n) : P(X_i = 1) = x_i, i = 1, \ldots, n].$$

2.2 Lovász Extension

Lovász extension is well-defined for any set function, not just submodular function and it has many facets, which are equivalent for submodular functions, such as convex closure/envelope (value of parametric optimization problem), Choquet expectation, Möbius transform, and interpolation via Kuhn's triangulation [1]

$$f : \{0, 1\}^n \to \mathbb{R} \Longrightarrow f^L : [0, 1]^n \to \mathbb{R}.$$

We define Lovász extension for any set function in Section 2.2.1. We show its geometric view in terms of Kuhn's triangulation in Section 2.2.2. We interpret Lovász extension as a Choquet expectation in Section 2.2.3. We present an alternative definition of Lovász extension via the multilinear representation of pseudo-Boolean function in Section 2.2.5. We show that Lovász extension is exactly the convex closure for any submodular function in Section 2.2.7. We look at the properties of Lovász extension in Section 2.3.

Throughout this chapter, we emphasize different representations of Lovász extension, offering perspectives on this important concept from various angles. Some representations are more convenient and handy to use than others in particular applications.

2.2.1 Definition

The Lovász extension [4, 9] $f^L(x) : [0, 1]^n \to \mathbb{R}$ of any pseudo-Boolean function $f : \{0, 1\}^n \to \mathbb{R}$ is defined as follows. Let $x \in [0, 1]^n$ be sorted such that $1 = x_0 \geq x_1 \geq \ldots \geq x_n \geq x_{n+1} = 0$:

$$f^L(x) := \sum_{i=0}^{n} f(\vee_{k=1}^i e_k)(x_i - x_{i+1}) \tag{9}$$

$$= \sum_{i=0}^{n} x_i \left[f(\vee_{k=1}^i e_k) - f(\vee_{k=1}^{i-1} e_k) \right], \tag{10}$$

where $e_0 \in \mathbb{R}^n$ is the all-zero vector and $f(e_{-1}) = 0$. The last equality follows by integration by parts, or Abel summation formula.

Example 3 (Lovász Extension for $n = 2$) Assume the pseudo-Boolean function $f : \mathbb{B}^2 \to \mathbb{R}$ is defined as follows:

$$f(0, 0) = 0, f(1, 0) = f(0, 1) = f(1, 1) = 1.$$

For $x_1 \geq x_2$, from (9) to (10), we have

$$\begin{aligned}
f^L(x_1, x_2) &= f(e_0)(1 - x_1) + f(e_1)(x_1 - x_2) + f(e_1 \vee e_2)x_2 \\
&= f(e_0) + (f(e_1) - f(e_0))x_1 + (f(e_1 \vee e_2) - f(e_1))x_2 \\
&= f(0, 0) + (f(1, 0) - f(0, 0))x_1 + (f(1, 1) - f(1, 0))x_2 \\
&= x_1.
\end{aligned}$$

Due to symmetry, the Lovász extension is

$$f^L(x_1, x_2) = \max\{x_1, x_2\}.$$

For this example, the convex closure and Lovász extension are one of the same! This equivalence is not coincidental as f in this example is submodular (please verify f is indeed submodular!).

Example 4 (Lovász Extension for n = 2 and the Cut Function of an Edge) Assume the pseudo-Boolean function $f : \mathbb{B}^2 \to \mathbb{R}$ is defined as follows:

$$f(0,0) = f(1,1) = 0, \, f(1,0) = f(0,1) = 1.$$

For $x_1 \geq x_2$, from (9) to (10), we have

$$\begin{aligned}
f^L(x_1, x_2) &= f(e_0)(1 - x_1) + f(e_1)(x_1 - x_2) + f(e_1 \vee e_2)x_2 \\
&= f(e_0) + (f(e_1) - f(e_0))x_1 + (f(e_1 \vee e_2) - f(e_1))x_2 \\
&= x_1 - x_2.
\end{aligned}$$

Due to symmetry, the Lovász extension is

$$f^L(x_1, x_2) = |x_1 - x_2|.$$

For this example, the convex closure and Lovász extension coincide again because cut functions are submodular.

2.2.2 Kuhn's Triangulation: Geometric View of (9)–(10)

A triangulation $K = \{\Delta, \ldots, \Delta_{n!}\}$ is said to be the Kuhn's triangulation of the hypercube, $[0, 1]^n$, if the simplices of K are in a one-to-one correspondence with the permutations of $\{1, \ldots, n\}$ as follows: given a permutation, π of $\{1, \ldots, n\}$, the $n + 1$ vertices of the corresponding simplex Δ_π are

$$\left\{ \begin{pmatrix} 0 \\ \vdots \\ 0 \end{pmatrix} + \sum_{j=1}^{k} e_{\pi_j} \right\}_{k=0,1,\ldots,n}. \tag{11}$$

Example 5 $n = 2$ (see Figure 1 for the case of $n = 2$)

$$\Delta_1 = \{(x : x_1 \geq x_2\}$$
$$\Delta_2 = \{(x : x_2 \geq x_1\}.$$

For $\Delta_1 = \{(x : x_1 \geq x_2\}$, then (11) becomes

$$\begin{pmatrix} 0 \\ 0 \end{pmatrix}, \begin{pmatrix} 1 \\ 0 \end{pmatrix}, \begin{pmatrix} 1 \\ 1 \end{pmatrix}.$$

Fig. 1 Illustration of Kuhn's triangulation when $n = 2$

$x = (1 - x_1)\begin{pmatrix} 0 \\ 0 \end{pmatrix} + (x_1 - x_2)\begin{pmatrix} 1 \\ 0 \end{pmatrix} + (x_2 - 0)\begin{pmatrix} 1 \\ 1 \end{pmatrix} \in \Delta_1$

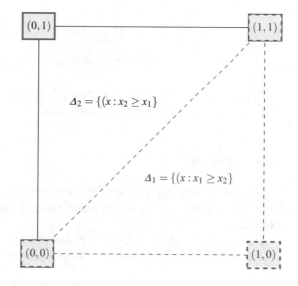

Therefore any $x \in \Delta_1$ can be written as the interpolation of the function on the simplices in the Kuhn triangulation:

$$x = (1 - x_1)\begin{pmatrix} 0 \\ 0 \end{pmatrix} + (x_1 - x_2)\begin{pmatrix} 1 \\ 0 \end{pmatrix} + (x_2 - 0)\begin{pmatrix} 1 \\ 1 \end{pmatrix}.$$

Therefore, Lovász extension is an affine interpolation of the function on the simplices in the Kuhn triangulation.

2.2.3 Lovász Extension as Choquet Expectation

$$f^L(x) = \mathbb{E}_{\lambda \sim U[0,1]}[f(i : x_i \geq \lambda)] \tag{12}$$

Proof Assume that $1 = x_0 \geq x_1 \geq \ldots \geq x_n \geq x_{n+1} = 0$. Then

$$f^L(x) = \sum_{k=0}^{n} f\left(\vee_{k=1}^{i} e_k\right)(x_k - x_{k+1})$$

$$= \sum_{k=0}^{n} \int_{x_{k+1}}^{x_k} f(\vee_{k=1}^{i} e_k) d\lambda$$

$$= \sum_{k=0}^{n} \int_{x_{k+1}}^{x_k} f(i : x_i \geq \lambda) d\lambda$$

$$= \int_0^1 f(i : x_i \geq \lambda) d\lambda$$

$$= \mathbb{E}_{\lambda \sim U[0,1]}[f(i : x_i \geq \lambda)].$$

This expression has a nice probabilistic interpretation: correlated rounding of elements with respect to uniform choice of $\lambda \sim U[0, 1]$.

2.2.4 Application

We now show one application of this Choquet expectation expression by designing and analyzing a 2-approximation algorithm for the submodular vertex cover problem.

For any nonnegative submodular function f, the submodular vertex cover problem can be formulated as the following submodular program:

$$\min \left[f(x) : x_i + x_j \geq 1, \forall (i, j) \in E, x \in \{0, 1\} \right]. \tag{13}$$

We apply Lovász extension to obtain a convex relaxation of (13). After solving this convex relaxation, we round the resulting fractional solution to an integer solution. Finally we use the expectation representation of the Lovász extension in the performance analysis.

Algorithm

Step 1.   Solve the following convex relaxation of (13) to obtain an optimal fractional solution x^*:

$$\min \left[f^L(x) : x_i + x_j \geq 1, \forall (i, j) \in E, x \geq 0 \right].$$

Step 2.   Choose $\lambda \in [0, 0.5]$ uniformly random and let

$$S_\lambda = \left\{ i : x_i^* \geq \lambda \right\}.$$

Analysis: (1) feasibility: Evidently S_λ is a feasible vertex cover because

$$x_1^* + x_2^* \geq 1 \Rightarrow \max\{x_1^*, x_2^*\} \geq 0.5,$$

and hence for each edge (i, j), either i or j is selected to be included into S.
(2) Approximation ratio: the expected value of this random vertex cover is

$$\mathbb{E}_\lambda[f(S_\lambda)] = 2 \int_0^{0.5} f(S_\lambda) d\lambda$$

$$\leq 2 \int_0^{0.5} f(S_\lambda) d\lambda + 2 \int_{0.5}^1 f(S_\lambda) d\lambda$$

$$= 2 \int_0^1 f(i : x_i^* \geq \lambda) d\lambda = 2 f^L(x^*).$$

2.2.5 Lovász Extension of Pseudo-Boolean Function Represented in Multilinear Form

Any pseudo-Boolean function $f : \{0, 1\}^n \mapsto \mathbb{R}$ can be represented as the following multilinear form (e.g., [5]):

$$f(x_1, \ldots, x_n) = \sum_{A \subseteq E} \hat{f}(A) \prod_{i \in A} x_i, \tag{14}$$

where the Möbius transform is given by

$$\hat{f}(A) = \sum_{S \subseteq A} (-1)^{|A|-|S|} f(1_S). \tag{15}$$

Then the Lovász extension is given by

$$f^L(x_1, \ldots, x_n) = \sum_{A \subseteq E} \hat{f}(A) \min_{i \in A} x_i. \tag{16}$$

Example 6 (Pseudo-Boolean Function for n = 2) Assume the pseudo-Boolean/set function $f : \mathbb{B}^2 \to \mathbb{R}$ is defined as follows:

$$f(0, 0) = f(\emptyset) = 0, f(1, 0) = f(\{1\}) = 1, f(0, 1) = f(\{2\}) = 1, f(1, 1) = f(\{1, 2\}) = 1.$$

Then (15) implies that

$$\hat{f}(\emptyset) = (-1)^{0-0} f(\emptyset) = 0$$
$$\hat{f}(\{1\}) = (-1)^{1-0} f(\emptyset) + (-1)^{1-} f(\{1\}) = 1$$
$$\hat{f}(\{2\}) = (-1)^{1-0} f(\emptyset) + (-1)^{1-} f(\{2\}) = 1$$
$$\hat{f}(\{1, 2\}) = (-1)^{2-0} f(\emptyset) + (-1)^{2-1} f(\{1\}) + (-1)^{2-1} f(\{2\}) + (-1)^0 f(\{1, 2\}) = -1.$$

So $\forall x_1, x_2 \in \mathbb{B}^2$, (16) implies that

$$f^L(x_1, x_2) = \sum_{A \subseteq \{1,2\}} \hat{f}(A) \min_{i \in A} x_i$$
$$= \hat{f}(\emptyset) + \hat{f}(\{1\}) x_1 + \hat{f}(\{2\}) x_2 + \hat{f}(\{1, 2\}) \min\{x_1, x_2\}$$
$$= x_1 + x_2 - \min\{x_1, x_2\}.$$

We obtain the same result as before via (9)–(10).

2.2.6 Various Representations of Lovász Extension: $n = 2$

We can summarize the various equivalent representations of Lovász extension for the two-dimensional case $n = 2$, derived from different angles.

- Definitions (9)–(10):

$$f^L(x_1, x_2) = \begin{cases} f(\emptyset)(1 - x_1) + f(1)(x_1 - x_2) + f(1, 2)x_2, & \text{If } x_1 \geq x_2 \\ f(\emptyset)(1 - x_2) + f(1)(x_2 - x_1) + f(1, 2)x_1, & \text{If } x_2 \geq x_1 \end{cases}$$

Or

$$f^L(x_1, x_2) = \begin{cases} f(\emptyset) + (f(1) - f(\emptyset))x_1 + (f(1, 2) - f(1))x_2, & \text{If } x_1 \geq x_2 \\ f(\emptyset) + (f(2) - f(\emptyset))x_2 + (f(1, 2) - f(1))x_1, & \text{If } x_2 \geq x_1 \end{cases}$$

- Pseudo-Boolean (16):

$$f^L(x_1, x_2) = \hat{f}(\emptyset) + \hat{f}(\{1\})x_1 + \hat{f}(\{2\})x_2 + \hat{f}(\{1, 2\}) \min\{x_1, x_2\}$$
$$= x_1 + x_2 - \min\{x_1, x_2\}$$

2.2.7 Lovász Extension of Submodular Function

Theorem 1 *If f is submodular, then Lovász extension is the same as the convex closure; namely*

$$f^L(x) = \underline{f}(x), \; x \in [0, 1]^E. \tag{17}$$

Proof (Greedy Algorithm + LP Duality) Recall the convex closure definition (7)–(8):

$$\underline{f}(x) = \min \left\{ \mathbb{E}_{y_A}[f(A)] : y \in \mathbb{R}_+^{2^n}, \mathbb{E}_{y_A}[\mathbf{1}_A] = x, \mathbb{E}_{y_A}[1] = 1 \right\}$$

$$= \min \left\{ \sum_{A \subseteq N} f(A)y_A : y \in \mathbb{R}_+^{2^n}, \sum_{A \subseteq N : i \in A} y_A = x_i, i = 1, \ldots, n, \sum_{A \subseteq N} y_A = 1 \right\} \text{ (Dual)}$$

$$= \max\{y_0 + x^T y : y_0 + y(A) \leq f(A), \forall A \subseteq N\} \text{ (Primal)}.$$

The optimal primal and dual solutions are given as follows by the greedy algorithm: sort elements in N in decreasing order such that $1 = x_0 \geq x_1 \geq \ldots \geq x_n \geq \mathbf{x_{n+1}} = 0$. Define $[i] = \{1, \ldots, i\}, i = 0, \ldots, n$, where $[-1] = [0] = \emptyset$

$$\begin{cases} y_i = f([i]) - f([i - 1]), & i = 0, \ldots, n; \\ z_{[i]} = x_i - x_{i+1}, & i = 0, \ldots, n. \end{cases}$$

Before we show (y, z) is an optimal primal–dual pair, note the greedy nature. In the primal LP: order the elements such that $1 = x_0 \geq x_1 \geq \ldots \geq x_n \geq \mathbf{x_{n+1}} = 0$.

1. Make y_1 as large as possible:

$$y_1^* = \max\{y_1 : y_0 + y(A) \leq f(A), \forall A \subseteq N\}$$
$$= f([1]) - f([0]).$$

2. Make y_2 as large as possible:

$$y_2^* = \max\{y_2 : y_1 = y_1^*, y_0 + y(A) \leq f(A), \forall A \subseteq N\}$$
$$= f([2]) - f([1]).$$

3. $\cdots$
4. Make y_n as large as possible:

$$y_j^* = \max\{y_j : y_i = y_i^*, i = 1, \ldots, j - 1, y_0 + y(A) \leq f(A), \forall A \subseteq N\}$$
$$= f([j]) - f([j - 1]).$$

5. $\cdots$

We first show that (y, z) are feasible solutions to the primal and dual and then we show they have the same objective values.

First, y is primal-feasible iff $y(A) \leq f(A)$ for each $A \subseteq N$. By induction on $|A|$, $y([0]) = 0 \leq f([0])$ due to the non-negativeness of f, implying the base case. Consider $A \neq [0]$, and let $i \in A$ with largest index according to the sorted order $\{1, \ldots, n\}$; that is, i is the index with smallest weight w_i in A. Then by induction, we have

$$y(A - i) = y(A) - y_i \leq f(A - i).$$

Moving y_i to the RHS:

$$y(A) \leq f(A - i) + y_i = f(A - i) + f([i]) - f([i - 1]) \overset{\text{submodularity}}{\leq} f(A),$$

where the second inequality follows from submodularity (since $A \subseteq [i]$).
Second, z is dual-feasible since $z \geq 0$, $\sum_{i=0}^{n} z_i = 1$ and for each i:

$$\sum_{A \ni i} z_A = \sum_{j \geq i} z_{[j]} = x_i.$$

Finally, the value

$$y_0 + x^T y = \sum_{i=0}^{n} x_i y_i = \sum_{i=0}^{n} x_i \left(f([i]) - f([i-1]) \right)$$

$$= \sum_{i=0}^{n} f([i])(x_i - x_{i+1}) = \sum_{i=0}^{n} f([i]) z_{[i]}$$

$$= \sum_{A \subseteq N} f(A) z_A. \tag{18}$$

The third equality follows from the Abel transformation (or summation by parts, integral by parts for continuous function): $\int f \, dg = fg - \int g \, df$.

We can characterize the submodular function via its extension.

Theorem 2 ([9]) *A set function f is submodular iff its Lovász extension f^L convex.*

2.3 Properties of Lovász Extension

- The Lovász extension f^L attains its maximum at a vertex of the hypercube $[0, 1]^n$:

$$\max_{x \in \mathbb{B}^n} f(x) = \max_{x \in [0,1]} f^L(x).$$

Proof First, $LHS \leq RHS$ is evident. For the other direction, note that

$$f^L(x) = \mathbb{E}_{\lambda \sim U[0,1]}[f(i : x_i \geq \lambda)],$$

implying that

$$f^L(x) \leq \max_{x \in \mathbb{B}^n} f(x).$$

- If f is submodular, then $f^L(x)$ is piece-wise linear convex.
- $(af + bg)^L(x) = af^L(x) + bg^L(x), \forall a, b \geq 0$.
- $f^L(x)$ is positively homogeneous:

$$f^L(\lambda x) = \lambda f^L(x), \quad \forall \lambda \in \mathbb{R}^+.$$

- $f^L(x)$ coincides with f at all 0–1 points:

$$f(S) = f^L(1_S), \forall S \subseteq E.$$

- $f^{I'}(x)$ can be evaluated in P-time via the greedy algorithm (even strongly P-time $O(E^4)$ [8, 10]).
- $f^L(x)$ has its minimum over the unit cube $[0, 1]^E$ attained at a vertex and hence can be minimized over the unit cube in P-time (by the ellipsoid method [8] weakly or [10] strongly).
- f is submodular $\Longrightarrow f^L(x)$ is submodular on lattice $w \in [0, 1]^n$ (proof: Choquet expectation representation (12)).

2.4 Concave Closure

We consider the concave overestimators of set function $f : \{0, 1\}^n \rightarrow \mathbb{R}$. One motivation is that, whenever an optimization problem involves maximizing $f(x)$ or contains an inequality $f(x) \geq r$, replacing $f(x)$ by a concave overestimator yields a convex relaxation of the problem. The concave envelope is the lowest concave overestimator.

The concave closure $\overline{f}(x) : [0, 1]^E \rightarrow \mathbb{R}$ of any pseudo-Boolean function $f : \{0, 1\}^n \rightarrow \mathbb{R}$ is defined as follows: $\forall x \in [0, 1]$:

$$\overline{f}(x) = \min_{\binom{y_0}{y} \in \mathbb{R}^{n+1}} \left\{ y_0 + x^T y : y_0 + a^T y \geq f(a), \forall a \in \mathbb{B}^n \right\} \tag{19}$$

$$= \max_{z \in \mathbb{R}_+^{2^n}} \left\{ \sum_{a \in \mathbb{B}^n} f(a) z_a : \sum_{a \in \mathbb{B}^n : e_i^T a = 1} z_a = x_i, i \in \{1, \ldots, n\}, \sum_{a \in \mathbb{B}^n} z_a = 1 \right\}. \tag{20}$$

Similar to the convex closure, the concave closure can be formulated as an optimization problem which preserves marginals:

$$\overline{f}(x) = \max_{Z_{(X_1, \ldots, X_n)}} \mathbb{E}\left[f(X_1, \ldots, X_n) : P(X_i = 1) = x_i, i = 1, \ldots, n \right].$$

However, this optimization problem has no compact representation, even for submodular functions.

Example 7 (Concave Closure for $n = 2$) Assume the pseudo-Boolean function $f : \mathbb{B}^2 \rightarrow \mathbb{R}$ is defined as follows:

$$f(0, 0) = 0, f(1, 0) = f(0, 1) = f(1, 1) = 1.$$

The program (19) becomes

$$\overline{f}(x_1, x_2) = \min y_0 + x_1 y_1 + x_2 y_2$$

$$\text{s.t.} \quad y_0 \geq f(0, 0) = 0$$

$$y_0 + y_1 \geq f(1, 0) = 1$$

$$y_0 + y_2 \geq f(0, 1) = 1$$

$$y_0 + y_1 + y_2 \geq f(1, 1) = 1$$

along with its dual program (20)

$$\overline{f}(x_1, x_2) = \max f(0, 0)z_{(0,0)} + f(1, 0)z_{(1,0)} + f(0, 1)z_{(0,1)} + f(1, 1)z_{(1,1)}$$

$$\text{s.t.} \quad z_{(1,0)} + z_{(1,1)} = x_1$$

$$z_{(0,1)} + z_{(1,1)} = x_2$$

$$z_{(0,0)} + z_{(1,0)} + z_{(0,1)} + z_{(1,1)} = 1.$$

$$z \geq 0$$

The optimal solutions of the primal and dual are given as follows:

(i) $x_1 + x_2 \geq 1, 0 \leq x_1 \leq 1, 0 \leq x_2 \leq 1$: Primal solution $y_0 = 1$, $y_1 = y_2 = 0$ with objective value 1. Dual solution $z_{00} = 0$, $z_{(1,0)} = 1 - x_2$, $z_{(0,1)} = 1 - x_1$, and $z_{(1,1)} = x_1 + x_2 - 1$ with objective value 1. Therefore both are optimal.

(ii) For $x_1 + x_2 \leq 1, 0 \leq x_1 \leq 1, 0 \leq x_2 \leq 1$: Primal solution $y_0 = 0$, $y_1 = y_2 = 1$ with objective value $x_1 + x_2$. Dual solution $z_{(0,0)} = 1 - x_1 - x_2$, $z_{(1,0)} = x_1$, $z_{(0,1)} = x_2$, and $z_{(1,1)} = 0$ with objective value $x_1 + x_2$. Therefore both are optimal.

The concave closure for any $x = (x_1, x_2) \in [0, 1]^2$ is therefore

$$\overline{f}(x_1, x_2) = \min\{1, x_1 + x_2\}.$$

2.5 Multilinear Extension

Multilinear extension of submodular functions has many applications, including the Shapley value (e.g, [11]), and approximation algorithms (e.g., [2, 3, 13]). Similar to the treatment of Lovász extension, we emphasize different representations of multilinear extension.

2.5.1 Definition

The multilinear extension $f^M : [0, 1]^n \to \mathbb{R}$ of any pseudo-Boolean function $f : \{0, 1\}^n \to \mathbb{R}$ over a ground set $E = \{1, \ldots, n\}$ is defined as:

$$f^M(x) = \sum_{S \subseteq E} f(S) \prod_{i \in S} x_i \prod_{i \notin S} (1 - x_i) = \mathbb{E}[f(R(x))], \tag{21}$$

where $R(x)$ is a random set independently containing each element i with probability x_i. Therefore, it has a simple probabilistic interpretation: independent rounding of elements to $\{0, 1\}$. In general, f^M is neither convex nor concave.

Example 8 (Multilinear Extension for n = 2) Assume the pseudo-Boolean function $f : \mathbb{B}^2 \to \mathbb{R}$ is defined as follows:

$$f(0, 0) = 0, f(1, 0) = f(0, 1) = f(1, 1) = 1.$$

The multilinear extension for any $x = (x_1, x_2) \in [0, 1]^2$ is therefore

$$f^M(x_1, x_2) = f(0, 0)(1 - x_1)(1 - x_2) + f(1, 0)x_1(1 - x_2) + f(0, 1)(1 - x_1)x_2 + f(1, 1)x_1 x_2$$

$$= x_1(1 - x_2) + (1 - x_1)x_2 + x_1 x_2 = x_1 + x_2 - x_1 x_2.$$

2.5.2 Properties of the Multilinear Extension

1. The multilinear extension f^M attains its maximum at a vertex of the hypercube $[0, 1]^n$:

$$\max_{x \in \mathbb{B}^n} f(x) = \max_{x \in [0,1]^n} f^M(x).$$

Proof First, $LHS \leq RHS$ is evident. For the other direction, note that

$$f^M(x) = \mathbb{E}[f(R(x))],$$

implying that

$$f^M(x) = \mathbb{E}[f(R(x))] \leq \max_{x \in \mathbb{B}^n} f(x).$$

2. Linear combination

$$(\alpha f + \beta g)^M(x) = \alpha f^M(x) + \beta g^M(x).$$

3. Composition: Given two set functions $f : 2^L \mapsto \mathbb{R}$ and $g : 2^N \mapsto \mathbb{R}$ where $L \cap N = \emptyset$, define the composition $(f \oplus g) : 2^{L \cup N} \mapsto \mathbb{R}$ of f and g as

$$(f \oplus g)^M(S) = f^M(L \cap S) + g^M(N \cap S), \forall S \subseteq L \cup N.$$

Then

$$(f \oplus g)^M(x, y) = f^M(x) + g^M(y), \forall x \in [0, 1]^m; \forall y \in [0, 1]^n.$$

4. Derivative: The ith partial derivative of f^M:

$$f_i^M(x) = \mathbb{E}[f(R \cup \{i\}) - f(R)]$$

$$= f^M(x_1, \ldots, x_{i-1}, 1, x_{i+1}, \ldots, x_n) - f^M(x_1, \ldots, x_{i-1}, 0, x_{i+1}, \ldots, x_n)$$

$$= \sum_{S: i \notin S \subseteq E} \prod_{j \in S} x_j \prod_{j: i \neq j \notin S} (1 - x_j)(f(S \cup \{i\}) - f(S)),$$

where $R \subseteq E - i$ is a random set whose elements are chosen independently with probability

$$\mathbb{P}(j \in R) = x_j, \forall j \in E - i.$$

5. Shapley value: Given a set function $f : 2^N \mapsto \mathbb{R}$, its Shapley value is given as

$$\varphi_i[f] = \int_0^1 f_i^M(t, \ldots, t), i = 1, \ldots, n.$$

Proof Consider the ith partial derivative of f^M:

$$f_i^M(x) = \sum_{S: i \notin S \subseteq E} \prod_{j \in S} x_j \prod_{j: i \neq j \notin S} (1 - x_j)(f(S \cup \{i\}) - f(S)).$$

Let $x = (t, \ldots, t)$, then

$$f_i^M(t, \ldots, t) = \sum_{S: i \notin S \subseteq E} t^{|S|}(1 - t)^{n - |S| - 1}(f(S \cup \{i\}) - f(S)).$$

Integrating on both side

$$\int_0^1 f_i^M(t, \ldots, t)dt = \sum_{S:i\notin S\subseteq E} \left(\int_0^1 t^{|S|}(1-t)^{n-|S|-1}dt \right) (f(S\cup\{i\}) - f(S))$$

$$= \sum_{S:i\notin S\subseteq E} \frac{|S|(n-|S|-1)}{n!} (f(S\cup\{i\}) - f(S)) = \varphi_i[f].$$

Example 9 (Multilinear Extension for n = 2) Assume the pseudo-Boolean function $f : \mathbb{B}^2 \to \mathbb{R}$ is defined as follows:

$$f(0, 0) = 0, \ f(1, 0) = f(0, 1) = f(1, 1) = 1.$$

The multilinear extension for any $x = (x_1, x_2) \in [0, 1]^2$ is therefore

$$f^M(x_1, x_2) = f(0, 0)(1 - x_1)(1 - x_2) + f(1, 0)x_1(1 - x_2) + f(0, 1)(1 - x_1)x_2 + f(1, 1)x_1x_2$$

$$= x_1(1 - x_2) + (1 - x_1)x_2 + x_1x_2 = x_1 + x_2 - x_1x_2.$$

So

$$f_1^M(x_1, x_2) = (1 - x_2) - x_2 + x_2 = 1 - x_2$$
$$f_2^M(x_1, x_2) = -x_1 + (1 - x_1) + x_1 = 1 - x_1$$

and the Shapley value

$$\varphi_1[f] = \int_0^1 F_1(t, t) = \int_0^1 (1 - t)dt = \frac{1}{2}$$

$$\varphi_2[f] = \int_0^1 F_2(t, t) = \int_0^1 (1 - t)dt = \frac{1}{2}.$$

6. Monotonicity preserving

$$f \text{ monotone} \iff f^M \text{ monotone}$$

Proof

$$\frac{\partial f^M(x)}{\partial x_i} = \mathbb{E}[f(R \cup \{i\}) - f(R\setminus\{i\})] \geq 0$$

7. Submodularity preserving

$$f \text{ submodular} \iff f^M \text{ submodular}$$

Proof

$$\frac{\partial^2 f^M}{\partial x_i \partial x_j} = \mathbb{E}[f(R \cup \{i, j\}) - f(R \cup \{i\} \setminus \{j\})$$

$$- f(R \setminus \{i\} \cup \{j\}) + f(R \setminus \{i, j\})] \leq 0$$

8. Up-monotonic: f increasing $\Longrightarrow \varphi(t) := f^M(x + tu)$ increasing of t if $u \geq 0$

$$\varphi'(t)\Big|_{x+tu} = \sum_{i \in E} u_i \frac{\partial f^M(x)}{\partial x_i} \geq 0.$$

9. Up-concave: f submodular $\Longrightarrow \varphi(t) := f^M(x + tu)$ concave of t if $u \geq 0$

$$\varphi''(t)\Big|_{x+tu} = \sum_{i \in E} u_i u_j \frac{\partial^2 f^M(x)}{\partial x_i \partial x_j} \leq 0.$$

10. Cross-convex: f submodular $\Longrightarrow \psi(t) : f^M(x + t(e_i - e_j))$ convex of t for any $i \neq j$

$$\psi''(t)\Big|_{x+t(e_i-e_j)} = \frac{\partial^2 f^M(x)}{\partial x_i^2} + \frac{\partial^2 f^M(x)}{\partial x_j^2} - 2\frac{\partial^2 f^M(x)}{\partial x_i \partial x_j}$$

$$= -2\frac{\partial^2 f^M(x)}{\partial x_i \partial x_j} \geq 0.$$

2.6 Relationship Among Extensions

Any extension can be described as $\mathbb{E}[f(R)]$ where R is a n-dimensional joint random vector with $x_1, \ldots, x_n$ as marginals.

• Concave closure maximizes expectation but is hard to compute.
• Convex closure minimizes expectation and leads to Lovász extension.
• Multilinear extension is somewhere in the "middle."

One benefit of this expectation representation is that limit theorems of probabilistic theory can be applied, such as the central limit theorem and simulation.

Let us recall the example where the ground set is $E = \{1, 2\}$ and a submodular function $f : 2^E \mapsto \mathbb{R}$ with

$$f(\emptyset) = 0, \ f(\{1\}) = f(\{2\}) = f(\{1, 2\}) = 1.$$

Earlier calculations showed that

$$\overline{f}(x_1, x_2) = \min\{1, x_1 + x_2\}$$
$$f^M(x_1, x_2) = x_1 + x_2 - x_1 x_2$$
$$\underline{f}(x_1, x_2) = \max\{x_1, x_2\}$$
$$f^L(x_1, x_2) = \max\{x_1, x_2\}.$$

Note that

$$\min\{1, x_1 + x_2\} \geq x_1 + x_2 - x_1 x_2 \geq \max\{x_1, x_2\}$$

.

Therefore we have

$$\overline{f}(x) \geq f^M \geq \underline{f}(x) = f^L(x).$$

This order on the concave, multilinear, convex, and Lovász extensions holds for any submodular function.

Theorem 3 *If $f : 2^E \to \mathbb{R}$ is a submodular function, then*

$$\underbrace{\overline{f}(x)}_{\text{concave closure}} \geq \underbrace{f^M}_{\text{multilinear ext.}} \geq \underbrace{\underline{f}(x)}_{\text{convex closure}} = \underbrace{f^L(x)}_{\text{Lovasz ext.}}$$

Please see Lemma A.4 in [14, p. 32] for a proof of the second inequality above.

3 Simultaneous Approximation of Multi-Criteria Submodular Function Maximization [6]

Let $f_j : 2^X \to \mathbb{R}^+$ $(j = 1, \ldots, k)$ be k nonnegative symmetric submodular functions. The main focus of this section is on maximizing multiple nonnegative symmetric submodular functions, namely solving the following k-criteria submodular function maximization problem:

$$(P): \quad \max_{S \subseteq X}\{f_1(S), \ldots, f_k(S)\}.$$

Sometimes we also need to refer to the jth ($j \in \{1, \ldots, k\}$) mono-criterion problem:

$$(P_j): \quad \max_{S \subseteq X} f_j(S).$$

Let S_j^* ($j = 1, \ldots, k$) be the optimal solution for the jth mono-criterion problem (P_j).

Definition 1 For any $\alpha \in [0, 1]$,

(i) A subset $S \subseteq X$ is an α-deterministic solution for the problem (P) if

$$f_j(S) \geq \alpha f_j(S_j^*), \ \forall j = 1, \ldots, k.$$

(ii) A random subset $T \subseteq X$ is an α-randomized solution for the problem (P) if

$$\mathbb{E}[f_j(T)] \geq \alpha f_j(S_j^*), \ \forall j = 1, \ldots, k.$$

Du et al. [6] obtain the following tight result for symmetric submodular functions.

Theorem 4 *Assume the submodular functions are nonnegative, symmetric, and submodular in problem (P). Then*

(i) *there exists an α-randomized solution such that*

$$\alpha = \frac{2^{k-1}}{2^k - 1},$$

and this quantity approaches to 0.5 when $k \to \infty$;

(ii) *For the k-criteria* MAX-CUT *problem, there exists an instance such that no β-randomized solution exists such that $\beta > \alpha$.*

Acknowledgements The first author's research is supported by the Natural Sciences and Engineering Research Council of Canada (NSERC) (No. 06446), and the Natural Science Foundation of China (NSFC) (No. 11771386 and No. 11728104). The second author's research is supported by the NSFC (No. 11771386 and No. 11728104). The third author's research was supported by the NSFC (No. 11501412).

References

1. Bach, F.: Learning with submodular functions: a convex optimization perspective. Found. Trends Mach. Learn. **6**(2-3), 145–373 (2013)
2. Buchbinder, N., Feldman, M., Naor, J.S., Schwartz, R.: A tight linear time (1/2)-approximation for unconstrained submodular maximization. In: Foundations of Computer Science (FOCS), 2012 IEEE 53rd Annual Symposium on, pp. 649–658. IEEE, Piscataway (2012)
3. Calinescu, G., Chekuri, C., Pál, M., Vondrák, J.: Maximizing a submodular set function subject to a matroid constraint. In: Integer Programming and Combinatorial Optimization, pp. 182–196. Springer, Berlin (2007)
4. Choquet, G.: Theory of capacities. In: Annales de l'institut Fourier, vol. 5, pp. 131–295. Institut Fourier, Chartres (1954)
5. Crama, Y., Hammer, P.L.: Boolean Functions: Theory, Algorithms, and Applications. Cambridge University, Cambridge (2011)

6. Du, D.L., Li, Y., Xiu, N.H., Xu, D.C.: Simultaneous approximation of multi-criteria submodular function maximization. J. Oper. Res. Soc. China **2**(3), 271–290 (2014)
7. Fujishige, S.: Submodular Functions and Optimization, vol. 58. Elsevier, Boston (2005)
8. Grötschel, M., Lovász, L., Schrijver, A.: The ellipsoid method and its consequences in combinatorial optimization. Combinatorica **1**(2), 169–197 (1981)
9. Lovász, L.: Submodular functions and convexity. In: Mathematical Programming The State of the Art, pp. 235–257. Springer, New York (1983)
10. McCormick, S.T.: Submodular function minimization. Handbooks Oper. Res. Management Sci. **12**, 321–391 (2005)
11. Owen, G.: Game Theory. Academic, San Diego (1995)
12. Topkis, D.M.: Supermodularity and Complementarity. Princeton University, Princeton (1998)
13. Vondrak, J.: Optimal approximation for the submodular welfare problem in the value oracle model. In: Proceedings of the 40th Annual ACM Symposium on Theory of Computing, pp. 67–74. ACM, New York (2008)
14. Vondrák, J.: Symmetry and approximability of submodular maximization problems. SIAM J. Comput. **42**(1), 265–304 (2013)

Discrete Convex Optimization and Applications in Supply Chain Management

Shengyu Cao and Simai He

Abstract In supply chain management and other operations management applications, various discrete convexities are important tools in modeling complementary or supplementary behaviors. Furthermore, the discrete nature of many decision scenarios also requires optimization tools from discrete convex theory.

In this chapter, we aim at introducing the classical discrete convex theory from the perspective of supply chain applications. We illustrate some direct applications and connections in supply chain applications. Certain proofs are modified/shortened, to fit into the scope of this chapter.

1 Introduction

Many practical problems are of discrete nature. For example, in inventory management retailers need to place orders in discrete quantity or even large batches. In scheduling, transportation planning, and production planning one needs to use discrete assignment variables $x_{ij} = 1$ or 0 to model whether a job or a truck should be assigned to a machine or a route. In combinatorial optimization theory, there are many brilliant problem-based algorithms developed. However, it remains an important question that whether there exists a framework for a general class of problems, like convex optimization theory in continuous optimization. For this purpose, one needs to extend the **Separation Lemma**, which implies strong duality and global optimality. Luckily, Separation Lemma holds for **submodular set functions**, and the so-called $L^{\natural}$ **functions** [23].

In economic theory and operations management applications, an important question arises from practice: whether two decisions have conflict against each other. This question belongs to the area of **comparative statics**, and is often related

S. Cao · S. He (✉)
Department of Management Science, Shanghai University of Finance and Economics, Shanghai, China
e-mail: shengyu.cao@163.sufe.edu.cn; simaihe@mail.shufe.edu.cn

© Springer Nature Switzerland AG 2019
D.-Z. Du et al. (eds.), *Nonlinear Combinatorial Optimization*,
Springer Optimization and Its Applications 147,
https://doi.org/10.1007/978-3-030-16194-1_4

81

to the sign of second order partial derivatives with respect to different dimensions of decision variable. Also, in revenue and inventory management, we are often interested in whether the decisions of two different products would influence each other. For example, different brands of smartphones are called "substitutable goods," since one can replace the function of the other. On the other hand, extra consumption of smartphones would boost sales number of the accessories, which we often define as "complementary goods." These properties can often be characterized by submodularity and supermodularity of customer utility function.

The objective of this chapter is to introduce some basic concepts, algorithms, and applications of discrete convex optimization. Discrete convex analysis is a deep research direction, and we aim at providing a quick survey of the classical results related to optimization problems applicable in operations management. Moreover, we emphasize on the motivations and intuitions behind the concepts and proofs, and we omit certain details of proofs due to page limit. Readers may refer to Topkis's book [29] for more detailed examples, discussions, and classical applications in supply chain management, as well as game theory related topics. Mutora [23] and his long list of research works provide a thorough survey of the theoretical foundation of discrete convex analysis, including the duality theory in discrete domain. And Vondrak's Ph.D. thesis [30] provides a survey of many crucial ideas in designing combinatorial algorithms by utilizing submodularity.

Section 2 introduces the basic concepts, e.g., **lattices, submodular function, and comparative statics**. Fundamental properties of submodular functions and lattice sets are introduced. Examples arisen from applications are given to illustrate how to model problems with submodularity and other discrete convex properties.

Section 3 focuses on classical results of submodular set function optimization. **Separation Lemma** and **convex extensions** are introduced, and the minimization algorithm over submodular functions is established based on convex extensions. Moreover, greedy approaches and multi-linear extension based smooth-greedy algorithms are introduced for the maximization problems.

Section 4 discusses online and dynamic algorithms utilizing submodularity. $L^{\natural}$-convexity plays a key role for dynamic inventory control problems, while the diminishing return property guarantees $1 - \frac{1}{e}$ approximation ratio of greedy algorithm in online bipartite matching.

2 Basic Definitions and Properties

This section introduces the basics of submodularity and lattice structure. Section 2.1 illustrates the intuition of developing such concepts. Section 2.2 defines the basic concepts and establishes the basic properties. Section 2.3 discusses a special application where only submodularity only holds locally near the optimum solution path.

2.1 Motivation

To begin with, we start with the following observations:

1. When a competitor lowers the price of its product, one often needs to also lower his/her own price.
2. When the inventory level of a product is low, retailers often raise the price.
3. In public spaces, one would naturally lower his/her voice, if the others are doing so.

To explain these observations and to further study the related problems, one needs to provide reasonable mathematical models:

Example 1 Suppose the sales quantity Q_i of retailer i is a function $Q_i(p_i, p_j)$ of the price p_i of retailer i, and its major competitor's price p_j, and the corresponding profit is $R_i(p_i, p_j) = (p_i - c_i)Q_i(p_i, p_j)$.

The simplest assumption is $Q_i(p_i, p_j) = (a_i + b_{ii}p_i + b_{ij}p_j)_+$ with $b_{ii} < 0, b_{ij} > 0$. The optimum price $p_i^* = \frac{a_i + b_{ij}p_j - b_{ii}c_i}{-2b_{ii}}$ for $\max\{R_i(p_i, p_j) \mid p_i \geq 0\}$ is indeed increasing with respect to p_j. Note that $b_{ij} = \frac{\partial^2}{\partial p_i \partial p_j}R_i(p_i, p_j) > 0$ is the crucial assumption, and can be generalized for other types of demand functions.

Naturally, one would like to extend the question to the following general **comparative statistics** question:

Problem 1 Given function $f(x, y) : \Re^2 \to \Re$, where x is the decision variable, and y is the input parameter (maybe the decision of another player). We consider the minimization problem $\min\{f(x, y) : (x, y) \in S\}$ within domain S. When would the optimum decision $x(y)$ be monotonically increasing with respect to input parameter y?

We analyze quadratic functions first:

Theorem 1 *If*

$$f(x, y) = \frac{1}{2}(x\ y) A \begin{pmatrix} x \\ y \end{pmatrix} + b^T \begin{pmatrix} x \\ y \end{pmatrix} + c$$

is a strongly convex function($A \succ 0$). The optimum solution of $\min\{f(x, y) \mid x \in \Re\}$ is defined as $x^*(y)$. Then $x^*(y)$ is monotonically increasing with respect to y when $A_{12} < 0$.

Proof Due to strong convexity, $A_{11} > 0$. By first order condition $A_{11}x^*(y) + A_{12}y + b_1 = 0$, the optimum solution is

$$x^*(y) = -\frac{A_{12}}{A_{11}}y - \frac{b_1}{A_{11}}.$$

Therefore, $x^*(y)$ is monotonically increasing with respect to y when $A_{12} < 0$.

Fig. 1 Idea of proof for general problem

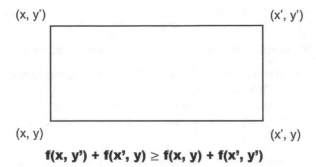

$$f(x, y') + f(x', y) \geq f(x, y) + f(x', y')$$

Next, we establish a more general result by dropping the quadratic assumption, by a proof with potential to be generalized in discrete domain:

Theorem 2 *If $f(x, y) : \Re^2 \to \Re$ is a strongly convex C^2 function. The optimum solution of $\min\{f(x, y) \mid x \in \Re\}$ is defined as $x^*(y)$. Then $x^*(y)$ is increasing with respect to y if $\frac{\partial^2}{\partial x \partial y} f(x, y) \leq 0$ for all $(x, y) \in \Re^2$.*

Proof Firstly, we note that for any $x \leq x', y \leq y'$,

$$f(x, y) + f(x', y') - f(x, y') - f(x', y) = \int_{s=x}^{x'} \int_{t=y}^{y'} \frac{\partial^2}{\partial s \partial t} f(s, t) ds dt \leq 0.$$

This condition is illustrated as in Figure 1.

Secondly, we prove the theorem by contradiction. Due to strong convexity, $x^*(y)$ is uniquely defined for each $y \in \Re$. If for $y < y'$ we have $x^*(y) > x^*(y')$, let's denote $x = x^*(y')$ and $x' = x^*(y) > x$. Then $f(x, y') = \min\{f(s, y') \mid s \in \Re\} \leq f(x', y')$ and $f(x', y) = \min\{f(s, y) \mid s \in \Re\} \leq f(x, y)$. Therefore,

$$0 \geq \int_{s=x}^{x'} \int_{t=y}^{y'} \frac{\partial^2}{\partial s \partial t} f(s, t) ds dt = [f(x, y) - f(x', y)] + [f(x', y') - f(x, y')] \geq 0.$$

Consequently, $f(x, y') = f(x', y')$ and $f(x', y) = f(x, y)$, which contradicts with the uniqueness of $x^*(y)$ and $x^*(y')$.

There are two crucial conditions in the above proof:

1. For any (x, y') and (x', y) in the domain S, if $x \leq x', y \leq y'$, then $(x, y), (x', y') \in S$.
2. For any $x \leq x', y \leq y', f(x, y) + f(x', y') - f(x, y') - f(x', y) \leq 0$.

In the following subsection, we generalize the first condition to the so-called Lattice structure, and the second condition to submodular property of functions.

2.2 Definition

In high dimensional discrete domain, the first condition in the above subsection is generalized as:

Definition 1 (Lattice)

1. Partial Order: $x \leq y$ if and only if $x_i \leq y_i$ for all indices i.
2. Maximization (or) Operation: $x \vee y$ defined as $(x \vee y)_i = \max\{x_i, y_i\}$.
3. Minimization (and) Operation: $x \wedge y$ defined as $(x \wedge y)_i = \min\{x_i, y_i\}$.
4. Lattice: $L \subseteq \mathfrak{R}^n$ is a lattice if and only if $x \vee y, x \wedge y \in L$ for any $x, y \in L$.
5. Sublattice: If L' is a subset of lattice L and $x \vee y, x \wedge y \in L'$ for any $x, y \in L'$, we call L' a sublattice of L.
6. For a set of points $\{x^j \in \mathfrak{R}^n : j \in S\}$, we can define $\vee_{j \in S} x^j$ as $\left(\vee_{j \in S} x^j\right)_i = \sup\{x_i^j \mid j \in S\}$ and $\left(\wedge_{j \in S} x^j\right)_i = \inf\{x_i^j \mid j \in S\}$. These are well defined when S is a finite set, or when $\{x^j : j \in S\}$ is within a bounded region.

Some important classes of lattices are listed as follows:

1. Any totally ordered set (e.g., single dimensional set) is a lattice!
2. Finite Cartesian product $L = \prod_{j \in S} L_j$ of lattices $L_j : j \in S$ is still a lattice when $|S|$ is finite.
3. Intersection $L = \bigcap_{j \in S} L_j$ of lattices $L_j : j \in S$ is still a lattice, regardless of the size of S.
4. Orthogonal projections and orthogonal slices of lattices are still lattices.
5. Linearly constrained set $\{(x, y) : ax - by \geq c\}$ with $a, b \geq 0$.

Theorem 3 *Suppose $L \subseteq \mathfrak{R}^N$ is a compact sublattice. Then there is a minimum element $\underline{x}$ and a maximum element $\overline{x}$ in L.*

Proof Because L is compact, its projection $L_i = \{y \mid \exists x \in L, x_i = y\}$ on i-th dimension is also compact. Define $\underline{x}_i = \inf\{x_i \mid x \in L\}$, which is well defined because L_i is compact, and will be reached by a certain point, which we denote as y^i, i.e., $y_i^i = \underline{x}_i$ and $y^i \in L$. Now we consider the point $\wedge_{i=1}^N y^i \in L$. It follows from definition that $\underline{x} \leq \wedge_{i=1}^N y^i$. Furthermore, $\left(\wedge_{i=1}^N y^i\right)_i \leq y_i^i = \underline{x}_i$, therefore $\wedge_{i=1}^N y^i \leq \underline{x}$. Consequently $\underline{x} = \wedge_{i=1}^N y^i \in L$. Similarly, we have $\overline{x} \in L$.

The second condition in the above subsection is extended to the concept of submodularity:

Definition 2 (Submodular Function)

1. A function $f(x) : L \to \mathfrak{R}$ defined on lattice L is called a submodular function if $f(x) + f(y) \geq f(x \vee y) + f(x \wedge y)$ for any $x, y \in L$.
2. Equivalent Definition (Decreasing Incremental):
   If $d, u \geq 0$ and $d^T u = 0$, then $f(x + d) - f(x) \geq f(x + u + d) - f(x + u)$.

3. Equivalent Definition (Local Condition):
   If f is defined on $\mathbf{Z}^n$, and $f(x + \mathbf{1}_i) - f(x) \geq f(x + \mathbf{1}_i + \mathbf{1}_j)) - f(x + \mathbf{1}_j)$
   for all indices $i \neq j$.
4. **Supermodular Function**: A function f is supermodular if and only if $-f$ is submodular.

Example 2 (Examples of Submodular Functions)

1. Quadratic Functions $\frac{1}{2}x^T A x + b^T x + c$ with $A_{ij} \leq 0$ for all $i \neq j$.
2. A C^2 function $f(x) : \Re^n \to \Re$ with $\frac{\partial^2}{\partial x_i \partial x_j} f(x) \leq 0$ for all $i \neq j$.
3. $g(x - y)$ with convex function $g(z) : \Re \to \Re$.
4. $g(\sum_{i=1}^{n} x_i)$ with concave function $g(z) : \Re \to \Re$.
5. Cobb–Douglas function $f(x) = \prod_i x_i^{\alpha_i}$ defined on $\Re_+^n$, with $\alpha \in \Re_+^n$.
6. $\|x - y\|_2^2 = \sum_i (x_i - y_i)^2$ and $\|x - y\|_1 = \sum_i |x_i - y_i|$.
7. Nonnegative linear combinations, expectations, and limitations of submodular functions are still submodular.
8. $g(f(x))$ is submodular, if $f : \Re^n \to \Re$ is submodular, $g : \Re \to \Re$ is concave and monotonically increasing.

A set function $f(S) : 2^N \to \Re$ is defined on the set 2^N of all subsets of N.

Definition 3 (Submodular Set Function)

1. A set function is called submodular set function, if $f(A) + f(B) \geq f(A \cup B) + f(A \cap B)$ for any $A, B \subseteq N$.
2. Equivalent Definition (Local Condition): For any set $A \subseteq N$, and two elements $i, j \in N$, $f(A \cup \{i\}) + f(A \cup \{j\}) \geq f(A) + f(A \cup \{i, j\})$.
3. Connection with Submodular Function: Define $F : \{0, 1\}^N \to \Re$ as $F(\mathbf{1}_S) = f(S)$, then f is a submodular set function if and only if F is a submodular function.

There is a special class of submodular function generalizing the concept of rank in linear algebra:

Definition 4 (Rank Function) A set function $F : 2^N \to \Re$ which satisfies $F(\emptyset) = 0$ (normalized), $F(A) \leq F(B)$ for all $A \subseteq B$ (monotonicity) and submodularity, is called a **rank function.**

One example of rank function $R(S)$ defined on set of vectors $S = \{v_i \in \Re^m : i \in K\}$ is the rank of the spanning space of S.

Now we extend the monotonicity result in Theorem 2 to discrete scenario:

Theorem 4 (Theorem 2.7.1 in Topkis [29]) *If $f(x) : L \to \Re$ is a submodular function defined on lattice domain L, then the optimum solution set $\mathrm{argmin}_{x \in X} f(x)$ is a sublattice.*

Proof We prove this by definition. Suppose both $u, v \in \mathrm{argmin}_{x \in X} f(x)$, then $f(u) = f(v) = \min_{x \in X} f(x)$. Therefore $f(u \vee v) \geq \min_{x \in X} f(x) = f(u)$ and $f(u \wedge v) \geq \min_{x \in X} f(x) = f(v)$. It follows that $f(u \vee v) + f(u \wedge v) \geq f(u) + f(v)$.

But by submodularity, $f(u \vee v) + f(u \wedge v) \leq f(u) + f(v)$. Combine the two above inequalities, $f(u \vee v) = f(u \wedge v) = f(u) = f(v) = \min_{x \in X} f(x)$, and both $u \vee v, u \wedge v \in \operatorname{argmin}_{x \in X} f(x)$.

Next, we establish the monotonicity of the optimum decision set, with respect to input parameters. For this purpose, we need to first define the set monotonicity, which basically is the monotonicity of both the largest and smallest elements of the sets, if they do exist.

Definition 5 (Set Monotonicity) Set S_t is called monotonically increasing with respect to t, if for any $t < s$, $x \in S_t$, and $y \in S_s$, there exist a $u \in S_s$ and $v \in S_t$ such that $u \geq x$ and $v \leq y$. This implies that the $\bar{x}(t) = \vee_{x \in S_t} x$ and $\underline{x}(t) = \wedge_{x \in S_t} x$ are both increasing in t.

An important fact is slices of lattice remains to be a lattice, which is illustrated in the following theorem. The proof of this theorem follows directly from the definition and is omitted here.

Theorem 5 (Monotonicity of Lattice Slices) *If $S \subseteq X \times T$ is a sublattice of $X \times T$ for lattices X and T, then $S_t = \{x \mid (x, t) \in S\}$ is increasing on t, when it's nonempty.*

Theorem 6 (Topkis, Theorem 2.8.2) *Suppose $f(x, t) : S \rightarrow \Re$ is a submodular function defined on sublattice $S \subseteq X \times T$, where both X and T are lattices. Then $X^*(t) = \operatorname{argmin}\{f(x, t) : (x, t) \in S\}$ is increasing with respect to t when it is nonempty, and the set $\{(u, t) \mid u \in X^*(t)\}$ is a sublattice.*

Proof We first prove that the set $L = \{(u, t) \mid u \in X^*(t)\}$ is a sublattice by definition. For any $(u, t), (v, s) \in L$, without losing generality we assume $t \leq s$. By definition, we have $\min\{f(x, s) : (x, s) \in S\} = f(v, s)$ and $\min\{f(x, t) : (x, t) \in S\} = f(u, t)$. And it follows from lattice structure that both $(u \vee v, s) = (u, t) \vee (v, s)$ and $(u \wedge v, t) = (u, t) \wedge (v, s)$ are in set S. Therefore, $f(u \vee v, s) \geq \min\{f(x, s) : (x, s) \in S\} = f(v, s)$ and $f(u \wedge v, t) \geq \min\{f(x, t) : (x, t) \in S\} = f(u, t)$. However, by submodularity of f we have

$$f(u \vee v, s) + f(u \wedge v, t) = f((u, t) \vee (v, s)) + f((u, t) \wedge (v, s)) \leq f(u, t) + f(v, s).$$

It could only hold when $f(u \vee v, s) = f(v, s)$ and $f(u \wedge v, t) = f(u, t)$. Therefore, $u \vee v \in X^*(s)$ and $u \wedge v \in X^*(t)$, and by definition we have $(u, t) \vee (v, s) = (u \vee v, s) \in L$ and $(u, t) \wedge (v, s) = (u \wedge v, t) \in L$.

By Theorem 5, set $X^*(t) = \{x \mid (x, t) \in L\}$ increases with respect to t.

Corollary 1 (Topkis, Corollary 2.8.1) *If $f(x)$ is a submodular function defined on lattice domain $X \subseteq \Re^n$, then $f(x) - y^T x$ is submodular on domain $X \times \Re^n$, and $\operatorname{argmin}_{x \in X} f(x) - y^T x$ increases with respect to y.*

Proof Function $-y^T x$ is submodular, so is $f(x) - y^T x$ on domain $X \times \Re^n$, applying Theorem 6 we obtain the monotonicity.

For submodular functions, another important characteristic which mimics the convexity in continuous domain is the classical preservation under minimization property:

Theorem 7 (Preservation of Submodularity) *Suppose both S and T are lattices and $X \subseteq S \times T$ is a sublattice. Function $f : X \to \Re$ is a submodular function. Then the function $g(y) = \min\{f(x, y) \mid (x, y) \in X\}$ is a submodular function defined on sublattice domain $Y = \{y \mid \exists(x, y) \in X\}$.*

Proof We first prove the lattice structure of Y by definition. For any $y, y' \in Y$, there exists $x, x' \in S$ such that $(x, y) \in X$ and $(x', y') \in X$. Since X is a lattice, $(x \vee x', y \vee y') = (x, y) \vee (x', y') \in X$ and $(x \wedge x', y \wedge y') = (x, y) \wedge (x', y') \in X$. Therefore, $y \vee y'$ and $y \wedge y'$ are both in Y.

Secondly, we establish the submodularity of g by constructive proof, which is very useful in establishing properties of discrete convexity. For $y, y' \in Y$, there exists $z, z' \in S$ such that both $(z, y), (z', y') \in X$, $f(z, y) = g(y)$, and $f(z', y') = g(y')$. Therefore,

$$g(y \vee y') + g(y \wedge y') \leq f(z \vee z', y \vee y') + f(z \wedge z', y \wedge y')$$
$$= f[(z, y) \vee (z', y')] + f[(z, y) \wedge (z', y')]$$
$$f[(z, y) \vee (z', y')] + f[(z, y) \wedge (z', y')] \leq f(z, y) + f(z', y') = g(y) + g(y'),$$

where the first inequality is due to definition of g, the second inequality is due to submodularity of f, and the last equality is due to definition of z and z'.

2.3 Local Submodularity

In practice, it is often difficult to guarantee the submodularity of a function over the whole domain. However, for the monotonicity of optimum solution we only need the submodularity in a small region, i.e., a neighborhood of the optimum solution set path. In the following example, we use local supermodularity to explain why one retailer's price should decrease, if its competitors' prices are dropping.

Example 3 (Discrete Choice Model) A popular model that captures customer choice between substitutable goods is the so-called random utility (discrete choice) model. In this model, customers have random utility $\xi_i(p_i)$ for goods i with price p_i, where $u_i(p_i) = E\xi_i(p_i)$ is the expected utility. A random customer would choose the goods which give him/her the best (realized) utility. When the random noises $\xi_i(p_i) - u_i(p_i)$ follow independent Gumbel distributions, the probability that a customer would choose goods i from a set S of goods is $P_i = \frac{u_i(p_i)}{1+\sum_{j \in S} u_j(p_j)}$, while the probability of not choosing anything is $P_0 = \frac{1}{1+\sum_{j \in S} u_j(p_j)}$. One thing to note that is, a popular choice in practice is to use the logistic model: $u_i(p_i) = e^{\alpha_i p_i + \beta_i}$. Retailer i's expected profit from a random customer is therefore, $R_i = (p_i - c_i)P_i$

if the cost per unit is c_i. We adopt the classical notation that all prices other than p_i are denoted as p_{-i}, and optimum solution is $p_i^*(p_{-i}) = \text{argmax}\{R_i(p_i, p_{-i}) \mid p_i \in \mathfrak{R}_+\}$. We assume $u_j'(p_j) < 0$ for each retailer j, which is intuitive as customer's utility would decrease with respect to the price of goods.

Lemma 1 *If each u_i is a C^2 function, then in an open neighborhood of optimum solution path $\{(p_i^*(p_{-i}), p_{-i}) \mid p_{-i} \in \mathfrak{R}_+^{n-1}\}$, we have $\frac{\partial^2}{\partial x_i \partial x_j} R_i(p) > 0$.*

Proof The profit is negative when $p_i = 0$, and tends to 0 when $p_i \to \infty$, by continuity of R_i the optimum solution exists. Since the function R_i is also a C^2 function, we only need to verify the condition $\frac{\partial^2}{\partial x_i \partial x_j} R_i(p) > 0$ for $p_i = p_i^*(p_{-i})$. By optimality condition at $p_i = p_i^*(p_{-i})$,

$$0 = \frac{\partial}{\partial x_i} R_i(p) = \frac{1}{(1 + \sum_{j \in S} u_j(p_j))^2} \left[u_i(p_i)(1 + \sum_{j \in S} u_j(p_j)) + (p_i - c_i)u_i'(p_i)(1 + \sum_{j \in S} u_j(p_j) - u_i(p_i)) \right],$$

and $u_i(p_i)(1 + \sum_{j \in S} u_j(p_j)) + (p_i - c_i)u_i'(p_i)(1 + \sum_{j \in S} u_j(p_j) - u_i(p_i)) = 0$.

$$\frac{\partial^2}{\partial x_i \partial x_j} R_i(p)$$
$$= \frac{-u_j'(p_j)}{(1 + \sum_{j \in S} u_j(p_j))^3} \left[u_i(p_i)(1 + \sum_{j \in S} u_j(p_j)) + (p_i - c_i)u_i'(p_i)(1 + \sum_{j \in S} u_j(p_j) - 2u_i(p_i)) \right]$$
$$= \frac{-u_j'(p_j)}{(1 + \sum_{j \in S} u_j(p_j))^3} (p_i - c_i)u_i'(p_i)(-2u_i(p_i)) > 0.$$

Theorem 8 *When $u_i(p_i) = e^{\alpha_i p_i + \beta_i}$, $p_i^*(p_{-i})$ is continuous, and it is monotonically increasing with respect to p_{-i}.*

Proof We first prove the strongly concavity of $\ln R_i(p_i, p_{-i})$ in p_i. Notice that

$$\frac{\partial}{\partial p_i} \ln R_i = \frac{1}{p_i - c_i} + \frac{u_i'(p_i)(1 + \sum_{j \in S \setminus \{i\}} u_j)}{u_i(p_i)(1 + \sum_{j \in S} u_j)} = \frac{1}{p_i - c_i} + \alpha_i \frac{1 + \sum_{j \in S \setminus \{i\}} u_j}{1 + \sum_{j \in S} u_j}.$$

Notice that $\alpha_i < 0$ and $u_i(p_i)$ is decreasing with respect to p_i, $\frac{\partial}{\partial p_i} \ln R_i$ is a decreasing function with respect to p_i, and $\ln R_i$ is a strongly concave function with respect to p_i. Since R_i is C^2, and strongly concave in p_i, $p_i^*(p_{-i})$ is continuous.

By the local supermodularity, there exists a small neighborhood $N_\epsilon = \{p \in \mathfrak{R}_+^n \mid \|p - (p_i^*(p_{-i}), p_{-i})\|_\infty \leq \epsilon\}$ of any point $(p_i^*(p_{-i}), p_{-i})$ on optimum solution path, inside which $\frac{\partial^2}{\partial x_i \partial x_j} R_i(p) > 0$. Therefore, by applying Theorem 6 in the box, for any $q_{-i} \in [p_{-i}, p_{-i} + \epsilon e]$ we have

$$x = \text{argmax}\{R_i(p_i, q_{-i}) \mid p_i \in [p_i^*(p_{-i}) - \epsilon, p_i^*(p_{-i}) + \epsilon]\} \geq p_i^*(p_{-i}).$$

By log-concavity of function R_i, local optimum x within region $[p_i^*(p_{-i}) - \epsilon, p_i^*(p_{-i}) + \epsilon]$ is on the same side of the point $p_i^*(p_{-i}) \in [p_i^*(p_{-i}) - \epsilon, p_i^*(p_{-i}) + \epsilon]$ with the global optimum point $p_i^*(q_{-i})$, therefore

$$p_i^*(q_{-i}) = \text{argmax}\{R_i(p_i, q_{-i}) \mid p_i \in \Re\} \geq p_i^*(p_{-i}).$$

3 Optimization with Submodular Set Functions

In this section, we introduce the classical results for optimization over submodular set functions. Section 3.1 introduces the Lovasz extension for submodular set function. Section 3.2 discusses the polymatroid optimization. In Section 3.3, Lovasz extension is utilized for minimization of submodular set functions, with a fast gradient projection based algorithm. In Section 3.4, we analyze greedy and double greedy approaches for monotone and nonmonotone submodular set functions maximization problem. In Section 3.5, the smooth-greedy approaches based on multi-linear extension are analyzed for submodular set functions maximization problem with matroid constraint.

3.1 Extensions of Submodular Set Function

We first recall two important definitions in convex optimization theory. The **convex hull** of a set X is defined as $\mathbf{Conv}(X) = \mathbf{Cl}\{\sum_i s_i x_i : \sum_i s_i = 1, s_i \geq 0, x_i \in X\}$, where **Cl** defines the closure of a set. **Epigraph** of a function f is defined as $\mathbf{epigraph}(f) = \{(x, t) : f(x) \leq t\}$. A classical fact in convex optimization theory is that: **A function is convex if and only if its epigraph is a convex set.**

For each given set function $f : 2^N \to \Re$, we can define $\tilde{f} : \{0, 1\}^N \to \Re$ as $\tilde{f}(\mathbf{1}_S) = f(s)$, where $\mathbf{1}_S$ is the characteristic vector defined as $x \in \Re^N$ with $x = 1$ if $i \in S$ and $x_i = 0$ if $i \notin S$. We treat extreme points $\{0, 1\}^N$ of a box as the set of subsets 2^N, where $\mathbf{1}_S$ is equivalent to set S.

Definition 6 (Convex Extension and Lovasz Extension)

1. Given function $f : X \to \Re$, we define its **convex hull** $f^- : \mathbf{Conv}(X) \to \Re$ as

$$f^-(x) = \inf\{\sum_i s_i^j f(x_i^j) : \lim_{j \to \infty} x^j = x, \sum_i s_i^j x_i^j = x^j, \sum_i s_i^j = 1, s_i^j \geq 0, x_i^j \in X\},$$

which is the largest convex function below f.

2. Given a set function $f : 2^N \to \Re$, the **Lovasz extension** $f^L(x) : [0, 1]^N \to \Re$ is defined as $f^L(x) = \sum_{j=1}^m s_j f(S_j)$, where $\{S_j\}$ is the unique decreasing series of sets $N = S_1 \supset S_2 \supset S_3 \supset \cdots \supset S_m = \emptyset$ such that $x = \sum_j s_j \mathbf{1}_{S_j}$ for $\sum_j s_j = 1, s_j \geq 0$.

3. Equivalent Definition of Lovasz Extension: Take uniform distribution $\xi \in [0, 1]$, then $f^L(x) = E_\xi f(\{i : x_i \geq \xi\})$.
4. For any $S \subseteq N$, $f^L(1_S) = f^-(1_S) = f(S)$. So both are extensions for set functions.

Theorem 9 *Convex hull f^- of any function f is convex, and it is the largest convex function below f.*

Proof We analyze the epigraph of f^-:

$$\mathbf{epigraph}(f^-) = Cl\{(x, t) : \exists \sum_i s_i = 1, s_i \geq 0, x_i \in X, \sum_i s_i x_i = x, \sum_i s_i f(x_i) \leq t\}$$
$$= Cl\{(x, t) : \exists \sum_i s_i = 1, s_i \geq 0, x_i \in X, \sum_i s_i x_i = x, \sum_i s_i t_i = t, f(x_i) \leq t_i\}$$
$$= \mathbf{Conv}(\bigcup_i \{(x_i, t_i) : f(x_i) \leq t_i\}),$$

which is a convex set. Therefore, by convex optimization theory $f^-(x)$ is convex.

Because f^- is an extension of f, it is below f. Next we prove that any convex function g below f is also below f^-. For any $s \in \Re_+^n$ and $x_i \in X, i = 1, 2, \cdots, N$ with $\sum_i s_i = 1, s_i \geq 0, \sum_i s_i x_i = x$, by convexity we have

$$g(x) \leq \sum_i s_i g(x_i) \leq \sum_i s_i f(x_i).$$

Therefore, it follows from definition that

$$f^-(x) = \inf\{\sum_i s_i^j f(x_i^j) : \lim_{j \to \infty} x^j = x, \sum_i s_i^j x_i^j = x^j, \sum_i s_i^j = 1, s_i^j \geq 0, x_i^j \in X\} \geq \inf\{g(x^j)\} \geq g(x).$$

Theorem 10 *If f is a submodular set function, then $f^-(x) = f^L(x)$ and f^L is convex. Reversely, if the Lovasz extension f^L of a set function f is convex, f has to be submodular.*

Proof We can formulate the convex extension as a linear programming problem:

$$f^-(x) = \min \sum_{S \subseteq N} \lambda_S f(S)$$
$$\text{s.t.} \quad \sum_{S:i \in S} \lambda_S = x_i \; \forall i \in N$$
$$\sum_S \lambda_S = 1$$
$$\lambda \geq 0,$$

whose dual is

$$f^-(x) = \max t + \sum_{i \in N} y_i x_i$$
$$\text{s.t.} \quad \sum_{i \in S} y_i \leq f(S) - t \; \forall S \subseteq N.$$

For any given $x \in [0, 1]^N$, there exists order π of indices such that $x_{\pi_1} \leq x_{\pi_2} \leq \cdots \leq x_{\pi_N}$. Define $S_j = \{\pi_j, \cdots, \pi_N\}$ for $j = 1, 2, \cdots, N$ and $S_{N+1} = \emptyset$. Define $\lambda_{S_j} = x_{\pi_j} - x_{\pi_{j-1}}$ with $x_{\pi_0} = 0$ and $x_{\pi_{N+1}} = 1$, and $\lambda_S = 0$ if else. Then $\lambda \geq 0$ and $\sum_{j=1}^{N+1} \lambda_j = 1$. Furthermore,

$$\sum_{S:i\in S}\lambda_S = \sum_{j:j\leq\pi^{-1}(i)}\lambda_{S_j} = \sum_{j:j\leq\pi^{-1}(i)}x_{\pi_j} - x_{\pi_{j-1}} = x_i.$$

Therefore, λ is a primal feasible solution with the given x.

For the dual problem, define $t = f(\emptyset)$ and $y_i = f(S_{\pi^{-1}(i)}) - f(S_{1+\pi^{-1}(i)}) = f(S_{\pi^{-1}(i)}) - f(S_{\pi^{-1}(i)}\setminus\{i\})$. For any set $S = \{\pi_{j_1}, \pi_{j_2}, \cdots, \pi_{j_m}\}$ with $j_1 < j_2 < \cdots < j_m$, denote $S^k = \{\pi_{j_1}, \pi_{j_2}, \cdots, \pi_{j_k}\}$. Then

$$\sum_{i\in S}y_i = \sum_{i\in S}f(S_{\pi^{-1}(i)}) - f(S_{\pi^{-1}(i)}\setminus\{i\}) \leq \sum_{k=1}^{m}f(S^k) - f(S^{k+1}) = f(S) - f(\emptyset) = f(S) - t.$$

Therefore, (t, y) is a dual feasible solution.

Next we establish the strong duality, that is,

$$\sum_{S\subseteq N}\lambda_S f(S) = \sum_{j=1}^{N+1}(x_{\pi_j} - x_{\pi_{j-1}})f(S_j) = \sum_{j=1}^{N}x_{\pi_j}[f(S_j) - f(S_{j+1})] + x_{\pi_{N+1}}f(S_{N+1}) - x_{\pi_0}f(S_1) = \sum_{i\in N}x_i y_i + t.$$

Take all j such that $\lambda_{S_j} > 0$, these (λ_{S_j}, S_j) define the Lovasz extension $f^L(x)$. Therefore $f^-(x) = \sum_{j:\lambda_{S_j}>0}\lambda_{S_j}f(S_j) = f^L(x)$. Because f^- is always convex, so is f^L.

If f^L is convex, then for any $S, T \subseteq N$, consider point $x = \frac{1_S + 1_T}{2} = \frac{1_{S\cap T} + 1_{S\cup T}}{2}$. By definition, $f^L(x) = \frac{f(S\cap T) + f(S\cup T)}{2}$. By convexity, $f^-(x) \leq \frac{f(S) + f(T)}{2}$. Therefore

$$f(S) + f(T) \geq 2f^-(x) = 2f^L(x) = f(S\cap T) + f(S\cup T).$$

In convex optimization theory, the **Separation Lemma** guarantees the existence of "dual certificate" of an optimum primal solution for a convex optimization problem, which is a big step towards strong duality. For submodular set functions, we have the following:

Theorem 11 (Frank's Discrete Separation Theorem) *If $f(S)$, $g(S)$ are submodular and supermodular set functions defined on sublattice domain $D \subseteq 2^N$, respectively, and $f(S) \geq g(S)$ for all $S \subseteq N$, then there exists a modular (linear) function $L(S) = c + \sum_{i\in S}l_i$ such that $f(S) \geq L(S) \geq g(S)$ for all $S \subseteq N$.*

Proof We prove for $D = 2^N$ first. Since both f and $-g$ are submodular set functions, their Lovasz extensions f^L and $(-g)^L$ are convex. Note that $f(S) + (-g)(S) \geq 0$ for all $S \subseteq N$, it follows from definition that $f^L(x) + (-g)^L(x) = (f + (-g))^L(x) \geq 0$ for all $x \in [0, 1]^N$. Due to Separation Lemma in convex optimization theory, there exists a linear function $L^-(x) : [0, 1]^N \rightarrow \Re$ such that $f^L(x) \geq L^-(x) \geq -(-g)^L(x)$ for all $x \in [0, 1]^N$. Constraint this L^- function in $\{0, 1\}^N$ we obtain the modular function

$$L(S) = L^-(1_S) = L^-(0) + \sum_{i \in S}[L^-(e_i) - L^-(0)],$$

which satisfies $f(x) \geq L(x) \geq g(x)$.

For $D \neq 2^N$, we can extend the function f to domain 2^N by defining $f(S) = +\infty$ for all $S \notin D$. Similarly, we extend g by defining $g(S) = -\infty$ for all $S \notin D$. The extended functions are still submodular and supermodular, and we can apply the proof for the full domain 2^N directly.

Optimum solution of a convex function can be verified by a tangent hyperplane which touches the epigraph of the convex function. Similarly, we have the following existence result for the certificate of optimum solution of submodular set function minimization problem:

Corollary 2 *If $f(S)$ is a submodular set functions defined on domain 2^N, and $L \subseteq 2^N$ is a sublattice. Then S^* is the optimum solution for $\min\{f(S) \mid S \in L\}$ if and only if there exists a modular set function $L : 2^N \to \Re$ such that $f(S^*) = l(S^*)$, $f(S) \geq l(S)$ for all $S \subseteq N$ and $L \subseteq \{S : l(S) \geq l(S^*)\}$.*

This is a direct application of Theorem 11, and the fact that $f(S) \geq f(S^*) \geq 2f(S^*) - f(S)$ for all $S \in L$.

3.2 Polymatroid Optimization

In the proof of Theorem 10, the dual formulation of f^- has been discussed:

$$f^-(x) = \max t + \sum_{i \in N} y_i x_i$$
$$\text{s.t.} \quad \sum_{i \in S} y_i \leq f(S) - t \; \forall S \subseteq N.$$

The optimum solution for the dual problem is $y_i = f(S_{\pi^{-1}(i)}) - f(S_{1+\pi^{-1}(i)}) = f(S_{\pi^{-1}(i)}) - f(S_{\pi^{-1}(i)}\setminus\{i\})$, where the order π corresponds to the increasing order of x_i: $x_{\pi_1} \leq x_{\pi_2} \leq \cdots \leq x_{\pi_N}$. For sets $S^k = \{\pi_{j_1}, \pi_{j_2}, \cdots, \pi_{j_k}\}$,

$$\sum_{i \in S^k} y_i = \sum_{i \in S^k} f(S_{\pi^{-1}(i)}) - f(S_{\pi^{-1}(i)}\setminus\{i\}) = \sum_{k=1}^{m} f(S^k) - f(S^{k+1}) = f(S^k) - t.$$

Therefore, S^k corresponds to the tight dual constraints, and the optimum solution can be obtained by the greedy process: rank the coefficients in the objective from highest (π_N) to lowest (π_1), find the maximum possible value y_j one by one.

We conclude this observation into the so-called polymatroid optimization framework:

Definition 7 (Polymatroid Optimization) Given a nonnegative set function r : $2^N \rightarrow \Re_+$, it induces a polytope (with exponentially many linear constraints)

$$\mathbf{P}(r, N) = \{x \in \Re_+^N \mid \sum_{j \in S} x_j \leq r(S) \; \forall S \subseteq N\}.$$

This polytope is called a **polymatroid** if r is a rank function.

Problem 2 How to maximize a linear objective function with a polymatroid constraint

$$max \left\{ \sum_{j \in N} c_j x_i \mid x \in \mathbf{P}(r, N) \right\}.$$

Algorithm 1: Greedy optimum

---

1 $S_0 = \emptyset$ Find the **decreasing** order of coefficients: $c_{\pi_1} \geq c_{\pi_2} \geq \cdots \geq c_{\pi_N}$;
2 Find the maximum possible value for x_{π_t} one by one, in increasing order of
   t: **for each** $t = 1, 2, \ldots, N$ **do**
3 $\quad \mid \quad S_t = \{\pi_1, \pi_2, \cdots, \pi_t\}, x_{\pi_t} = r(S_t) - r(S_{t-1})$;
4 **end**

Theorem 12 *The greedy Algorithm 1 is optimum for Problem 2.*

This theorem has been established in [8]. We can prove the theorem by constructing primal–dual solution with no duality gap, where the primal solution x is already constructed by the greedy algorithm, and the dual is exactly the same as the primal solution in Theorem 10.

Furthermore, in [15], He et al. established the following structural result of polymatroid optimization:

Theorem 13 (Preservation of Submodularity) *If $r : 2^N \rightarrow \Re$ is a rank function, the function*

$$F(c) = \max \left\{ \sum_{j \in N} c_j x_i \mid x \in \mathbf{P}(r, N) \right\}$$

is a submodular function, and the function

$$\hat{F}(S) = \max \left\{ \sum_{j \in S} c_j x_i \mid x \in \mathbf{P}(r, S) \right\}$$

is a rank function for given $c \in \Re^N_+$. Furthermore,

$$\hat{F}(S) = \max \left\{ \sum_{j \in S} f_j(x_j) \mid x \in \mathbf{P}(r, S) \right\}$$

is a rank function if the objective function is separable concave and $f_j(0) = 0$.

Proof Due to space limitation, we provide an abstract proof with the main ideas here. Firstly, because the objective function is continuous, and the domain is compact, the optimum value $F(c)$ is also continuous. Secondly, negative coefficient c_i would yield $x_i = 0$, so we only need to focus in the nonnegative domain $c \in \Re^N_+$. Lastly, we only need to prove that for any given $C \in \Re^N_+$ and two different indices $i, j \in N$, if u_i and v_j are nonnegative vectors with only positive values in index i and j, respectively, then $F(C + u_i) + F(C + v_j) \geq F(C) + F(C + u_i + v_j)$.

Now we can fix all but two dimensions i, j. We then segment the two dimensional space $(c_i, c_j) \in \Re^2_+$ into small grids by the values of other $C_k, k \neq i, j$. We only need to prove inside each grid since local submodularity implies global submodularity. Inside each small grid, the line $c_i = c_j$ cuts the grid into two pieces, and by Theorem 12 there is a uniform optimum solution in each piece, as illustrated in Figure 2. We note the optimum solution in the left piece $(c_i \leq c_j)$ as x_L, then $F(c) = x_L^T(c - C) + F(C)$; also the optimum solution in the right piece $(c_i \geq c_j)$ is noted as x_R, so $F(c) = x_R^T(c - C) + F(C)$ when $c_i \geq c_j$, inside this small grid.

Without losing generality, we set $F(C) = 0$, and assume $C_j \geq C_i$. Note $b = C + u_i$ and $a = C + v_j$, then $C = a \wedge b$ and $C + u_i + v_j = a \vee b$. If $C_i + |v_j| \leq C_j$, then a, b are not separate by the line, and $F(c)$ is the same linear function for $a, b, a \vee b, a \wedge b$, so the submodularity directly follows. If $C_i + |v_j| > C_j$, then a, b are in different piece, with $F(b) = x_R^T(b - C) \geq x_L^T(b - C)$ and $F(a) = x_L^T(a - C) \geq x_R^T(a - C)$. The line $c_i = c_j$ intersects line from $C = a \wedge b$ to b at $z = (C_j, C_j)$ as in Figure 2. Note that $a \wedge b = a \wedge z$, so we have

$$F(a) + F(z) = x_L^T(a - C) + x_L^T(z - C) = x_L^T(a \wedge z - C + a \vee z - C) = F(a \wedge b) + F(a \vee z).$$

Because $z = (a \vee z) \wedge b$ and $a \vee b = (a \vee z) \vee b$, we have

$$F(a \vee z) + F(b) \geq x_R^T(a \vee z - C) + x_R^T(b - C) = x_R^T[(a \vee z) \wedge b + a \vee b - 2C] = F(a \wedge b) + F((a \vee z) \vee b) = F(z) + F(a \vee b).$$

Adding these two inequalities up, we obtain

$$F(a) + F(b) \geq F(a \wedge b) + F(a \vee b).$$

Therefore $F(a)$ is submodular in $\Re^N$.

For set function $\hat{F}(S)$, note that $\hat{F}(S) = F(c \mid S)$, where $(c \mid S)_i = c_i$ if $i \in S$ and 0 if else. The submodularity of $\hat{F}$ then directly follows from submodularity of F. For the proof of separable objective functions, please refer to Theorem 3 in [15].

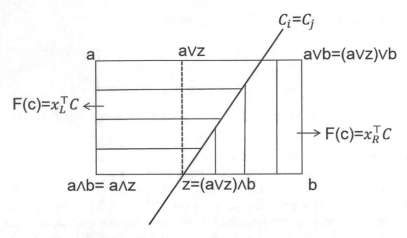

Fig. 2 Idea of proof for preservation of submodularity

3.3 Minimization of Submodular Set Function

In this subsection, we discuss how to solve submodular set function minimization problems. It relies on the fact that minimizer of the Lovasz extension can be reached at the extreme points of the polytope, which is a counter-intuitive result since this property holds mostly for concave functions instead of convex functions.

Theorem 14 (Minimization of Submodular Set Function) *If* $f : 2^N \to \Re$ *is a submodular set function, then the minimizer of its Lovasz extension in domain* $[0, 1]^N$ *can be obtained at vertex points:* $\min_{x \in [0,1]^N} f^L(x) = \min_{S \subseteq N} f(S)$.

Proof By submodularity and Theorem 10, $f^L = f^-$.

$$
\begin{aligned}
\min_{x \in [0,1]^N} f^L(x) = \min_{x \in [0,1]^N} \min \sum_{S \subseteq N} \lambda_S f(S) && = \min \sum_{S \subseteq N} \lambda_S f(S) \\
\text{s.t.} \ \sum_{S:i \in S} \lambda_S = x_i, \forall i && \text{s.t.} \ 0 \leq \sum_{S:i \in S} \lambda_S \leq 1, \forall i \\
\sum_S \lambda_S = 1 && \sum_S \lambda_S = 1 \\
\lambda_S \geq 0 \ \forall S && \lambda_S \geq 0 \ \forall S.
\end{aligned}
$$

Notice that $0 \leq \sum_{S:i \in S} \lambda_S \leq 1$ follows from the fact that $\sum_S \lambda_S = 1$ and $\lambda_S \geq 0$,

$$
\min_{x \in [0,1]^N} f^L(x) = \min\{\sum_{S \subseteq N} \lambda_S f(S) \mid \sum_S \lambda_S = 1, \lambda \geq 0\} = \min_{S \subseteq N} f(S).
$$

By Theorem 14, if we can find an optimum solution for $\min\{f^L(x) \mid x \in [0, 1]^N\}$, it corresponds to the optimum solution of the discrete problem $\min\{f(S) \mid S \subseteq N\}$. For convex optimization problem $\min\{f^L(x) \mid x \in [0, 1]^N\}$, we can evaluate the value and subgradient of $f^L(x)$ at x by the linear programming formulation and its dual in proof of Theorem 14. The exact algorithms for submodular function minimization are quite extensive, interested readers can refer to Section 10.2 of [23], or the research papers [5, 14, 17, 18, 25]. In particular, Schrijver's algorithm [25] achieves $O(n^5)$ iterations, with $O(n^7)$ function evaluation and $O(n^8)$ arithmetic operations (see Yvgen [31]), and the improved Iwata–Fleischeer–Fugishige's algorithm [17] can solve the problem within $O(n^7 \ln n)$ function evaluation and arithmetic operations.

In practice, speed of the algorithm is often an important factor, while the precision can be sacrificed for speed. Next, we introduce a fast algorithm based on subgradient method to optimize $f^L(x)$ within high precision. After obtaining a high quality solution $x \in [0, 1]^N$ for $f^L(x)$, by the definition of Lovasz extension, we can identify at most $N + 1$ set S_j such that $f^L(x)$ is the convex combination of $f(S_j)$. Therefore, $\min_j f(S_j) \leq f(x)$. We introduce the classical result of gradient projection method in the following theorem:

Theorem 15 (Gradient Projection Method) *Suppose $g : X \to \Re$ is a convex function defined on closed convex set X with diameter R. If we apply the gradient projection method: $x_{t+1} = (x_t - \alpha_t d_t) |_X$, where d_t is a subgradient of g at x_t whose length is uniformly upper bounded by G, $\alpha_t \geq 0$ is the step length, and $y |_X$ is the projection of y in convex set X defined as $y |_X = \mathrm{argmin}\{\|z - y\| \mid z \in X\}$. Then*

$$\min_{t \leq T}[g(x_t) - g(x^*)] \leq \frac{G^2(\sum_{t=1}^T \alpha_t^2) + R^2}{2 \sum_{t=1}^T \alpha_t}.$$

In this error bound estimation, taking $\alpha_t = \frac{R}{G} \frac{1}{\sqrt{T}}$ for fixed horizon T would yield upper bound $\frac{RG}{\sqrt{T}}$, and taking horizon independent step length $\alpha_t = \frac{R}{G} \frac{1}{\sqrt{t}}$ would yield upper bound $\frac{RG(1 + \ln T)}{2\sqrt{T}}$.

Proof Suppose $x^* \in X$ is the optimum solution, then

$$\|x_{t+1} - x^*\|^2$$
$$\leq \|x_t - \alpha_t d_t - x^*\|^2 \qquad \longleftarrow (x_t - \alpha_t d_t - x_{t+1})^T (y - x_{t+1}) \leq 0 \forall y \in X$$
$$\leq \|x_t - x^*\|^2 - 2\alpha_t d_t^T (x_t - x^*) + \alpha_t^2 G^2 \qquad \longleftarrow \|d_t\| \leq G$$
$$\leq \|x_t - x^*\|^2 - 2\alpha_t[g(x_t) - g(x^*)] + \alpha_t^2 G^2 \longleftarrow \text{ by convexity } g(x^*) \geq g(x_t) + d_t^T (x^* - x_t).$$

Therefore,

$$2\alpha_t[g(x_t) - g(x^*)] \leq \alpha_t^2 G^2 + \|x_t - x^*\|^2 - \|x_{t+1} - x^*\|^2.$$

Sum these inequalities up, we have

$$\left(2 \sum_{t=1}^T \alpha_t\right) \min_{t \leq T}[g(x_t) - g(x^*)] \leq \sum_{t=1}^T 2\alpha_t[g(x_t) - g(x^*)] \leq G^2 \left(\sum_{t=1}^T \alpha_t^2\right) + R^2.$$

More general cases have been studied by Haucbaum et al. [16].

Problem 3

$$\min \ f(S)$$
$$\textbf{s.t. } a_{ij}x_i + b_{ij}x_j \geq c_{ij} \quad \text{for all } (i, j) \in A$$
$$S \subseteq N,$$

where $x = 1_S$ is the characteristic vector of S, A is a set of pairs (allowing multiple copies of the same pair), and $f : 2^N \rightarrow \Re$ is a submodular set function.

The following result has been established by Haucbaum et al. [16]:

Theorem 16 *If f is submodular and constraints are monotone (feasible set is a lattice), then it's (strongly) polynomial-time solvable. If f is nonnegative submodular, and the constraints satisfy round up property or f is monotone, then it's 2-approximable in polynomial time.*

Proof Firstly, we preprocess all the constraints. Note that all variables are $\{0, 1\}$ variables. We first remove all redundant constraints. If a constraint $a_{ij}x_i + b_{ij}x_j \geq c_{ij}$ implies x_i or x_j equals to a certain value, then we can replace this constraint by two single dimensional constraints, which are either redundant or can be removed by fixing the variable. Repeatedly simplifying all such constraints, the left over constraints with two variables would all be of the form $x_i \geq x_j$, $x_i + x_j \leq 1$, or $x_i + x_j \geq 1$. Furthermore, constraints of type $a_{ij}x_i - b_{ij}x_j \geq c_{ij}$, where $a_{ij}, b_{ij} \geq 0$, would be reduced to simple single dimensional constraints, or the constraints of type $x_i \geq x_j$ (or $x_i \leq x_j$), constraints of type $a_{ij}x_i - b_{ij}x_j \geq c_{ij}$, where $a_{ij} \geq 0, b_{ij} \leq 0$, would be reduced to simple single dimensional constraints, or the constraints of type $x_i + x_j \leq 1$, or $x_i + x_j \geq 1$. If there is a group of cyclic constraints $x_{i_1} \geq x_{i_2} \geq \cdots x_{i_n} \geq x_{i_1}$, we further simplify it by replacing all x_{i_j} with a single variable.

When all the constraints are monotone, the constraints after simplification would all be the form $x_i \geq x_j$ for directed pairs $(i, j) \in E$. The problem now reduces to submodular minimization over a ring, which is solvable in (strong) polynomial time in the size of the underlying graph. A simple explanation is that, we can reform the problem into minimization of another submodular set function over set 2^E. For this purpose, for constraint $x_i \geq x_j$ we define variable $y_{ij} = 1$ if $x_i = 1$ and $x_j = 0$, and $y_{ij} = 0$ if else. And define base set B of indices as those indices never appear in the left side of $\geq$ constraints, and we define $y_i = x_i$ for all $i \in B$. Now, each x_i can be defined by $\vee_{k \in S_k} y_k$ for certain set $S_i \subseteq E \cup B$ (basically, in the ordered graph, S_i is the set children edges of i, as well as the leaves grow from node i). It can be easily proved that for a monotone set function F, and set $T \subseteq E \cup B$, define set $S(T) = \{i \mid S_i \cap T \neq \emptyset\}$, then function $G(T) = F(S(T))$ is also submodular for submodular function f. And the constraint for original variables is embedded in the transformation $S(T)$, so the constraint for function G becomes $T \subseteq E \cup B$.

For general cases, suppose the problem after simplification is of form:

$$\min \ f(x)$$
$$\text{s.t. } x_i \geq x_j, \forall (i, j) \in E,$$
$$x_i + x_j \geq 1, \forall (i, j) \in U,$$
$$x_i + x_j \leq 1, \forall (i, j) \in V.$$

We introduce two copies of original variables $x^+ = x \in \{0, 1\}^n, x^- = -x \in \{0, -1\}^n$. Then the original problem can be reformulated by

$$\min \ \frac{f(x^+) + f(-x^-)}{2}$$
$$\text{s.t. } x_i^+ \geq x_j^+, x_i^- \leq x_j^- \forall (i, j) \in E,$$
$$x_i^+ - x_j^- \geq 1, -x_i^- + x_j^+ \geq 1, \forall (i, j) \in U,$$
$$x_i^+ - x_j^- \leq 1, -x_i^- + x_j^+ \leq 1, \forall (i, j) \in V,$$
$$x_i^+ + x_i^- = 0, \forall i$$
$$x_i^+ \in \{0, 1\}, x_i^- \in \{0, -1\}, \forall i.$$

Dropping the only nonmonotone constraints $x_i^+ + x_i^- = 0$, we obtain a relaxed problem with only monotone constraints, which can be solved exactly. Suppose optimum solution is (x^+, x^-) with objective value $V^* \leq \mathscr{O}PT$. However, notice that $y = \lceil \frac{x^+ - x^-}{2} \rceil$ and $z = \lfloor \frac{x^+ - x^-}{2} \rfloor$ are both feasible for the original problem. However, $y = x^+ \vee (-x^-)$ and $z = x^+ \wedge (-x^-)$. By submodularity, $f(y) + f(z) \leq f(x^+) + f(-x^-)$. Because f is nonnegative, $f(y) \leq f(x^+) + f(-x^-) = 2V^* \leq 2\mathscr{O}PT$.

3.4   Maximization of Submodular Set Function

There are many scenarios where one needs to maximize a submodular set function. For example, consider a social network where people's decision is influenced by their friends. When a company needs to place a number of individual advertisements, e.g., via phone calls, a crucial problem is which group of people should they reach to maximize the total effect, within given budget constraint. A simplified model is the so-called Max-k-Cover problem:

Problem 4 (Max-k-Cover) Given a set of sets $\{S_j \subseteq N \mid j \in A\}$, find k sets which covers the most number of elements.

We can also assume that each element (customer) covered has a different value:

Problem 5 (The Maximum Coverage Problem) Given a set of $S_1, S_2, \cdots, S_m \subseteq N$. For each element $i \in N$, it has a value $v_i \geq 0$, and for each set $S \subseteq N$ the value function is defined as $V(S) = \sum_{i \in S} v_j$. We need to select K sets $\{S_i \mid j \in A\}$, and to maximize the maximum value $V(\bigcup_{j \in A} S_j)$.

One of the most important characteristic of these problem is the submodularity of objective function, with respect to the selected set of sets. The proof is straightforward and is omitted here.

Proposition 1 *We define* $U(A) = V(\bigcup_{j \in A} S_j)$, *then U is a submodular set function:*

$$V(\cup_{i \in A} S_j) + V(|\cup_{i \in B} S_j|) \geq V(|\cup_{i \in A \cap B} S_j|) + V(|\cup_{i \in A \cup B} S_j|) \text{ for any } A, B \subseteq \{1, 2, \cdots, m\}.$$

A related problem arises from application is **assort optimization**, where one needs to place advertisements of goods on the front-page of its website for maximum sales effect.

Problem 6 (Assortment Optimization) There are K advertisement slots of a webpage, which we need to select from a set N of goods from a certain category. The goods are substitutable to each other, that is, increasing sales from one product would hurt (or has no effect to) sales of the other product, so the more the goods placed on the webpage, the lesser the contribution from the advertisement of the next goods added. In some classical literatures, e.g., [6], the total sales revenue $V(S)$ from displacement of set S of goods on the webpage is assumed to be increasing and submodular. And we aim to solve the cardinality constrained maximization problem:

$$\max\{V(S) : |S| \leq K, S \subseteq N\}.$$

Because Max-Cut problem is well-known to be NP-hard, and the cut weight $V(S) = \sum_{i \in S, j \notin S} w_{ij}$ is submodular in S, submodular set function maximization with cardinality constraint is also NP-Hard. The hardness to approximate result has been established by Feige [11]:

Theorem 17 (Max-Hardness) *Consider cardinality constrained submodular maximization problem* $\max\{f(S), |S| \leq K, S \subseteq N\}$ *for rank function (submodular, normalized, and increasing)* $f : 2^N \to \mathfrak{R}$. *Unless $P = NP$, there is no polynomial-time algorithm which achieves approximation ratio strictly better than $1 - \frac{1}{e}$ in general (for general setting of K).*

There is a simple greedy algorithm which can achieve the best possible approximation ratio $1 - \frac{1}{e}$ for cardinality constrained maximization problem of rank functions.

Algorithm 2: Greedy algorithm: rank function maximization

---

1 Initialization $t = 0$, $S_t = \emptyset$;
2 **foreach** $t = 1, 2, \ldots, K$ **do**
3 $\quad$ Find the element $i \notin S_t$ with maximum improvement for function value:
$\quad\quad i_t = \mathrm{argmax}\{\mathrm{f}(S_{t-1} \cup \{i\})\}$;
4 $\quad$ Define $S_t = S_{t-1} \cup \{i\}$;
5 **end**
6 Return S_K

Theorem 18 (Greedy for Rank Function) *The greedy algorithm above achieves approximation ratio* $1 - (1 - \frac{1}{K})^K \geq 1 - \frac{1}{e}$ *for* $\max\{f(S), |S| \leq K, S \subseteq N\}$, *if the function f is a rank function.*

Proof Define the optimum solution as S^*, and optimum value $\mathscr{O}PT = f(S^*)$. For any set $S \subseteq N$, note that elements in $S^* \backslash S$ as $\{j_1, j_2, \cdots, j_m\}$, and $S^k = S \cup \{j_1, j_2, \cdots, j_k\}$. Then

$$\sum_{i \in S^*}[f(S+\{i\}) - f(S)] = \sum_{k=1}^{m}[f(S+\{j_k\}) - f(S)] \geq \sum_{k=1}^{m}[f(S^k)] - f(S^{k-1})] = f(S \cup S^*) - f(S).$$

Due to monotonicity of f, we have $f(S \cup S^*) \geq f(S^*) = \mathscr{O}PT$. Consequently,

$$\max\{f(S + \{i\}) - f(S) : i \in S^*\} \geq \frac{1}{K}(\mathscr{O}PT - f(S)).$$

Therefore, for any t and set S_t, the greedy algorithm outputs set S_{t+1}:

$$\mathscr{O}PT - f(S_{t+1}) \leq \mathscr{O}PT - f(S_t) - [f(S_{t+1}) - f(S_t)] \leq \mathscr{O}PT - f(S_t) - \frac{1}{K}[\mathscr{O}PT - f(S)] \leq \left(1 - \frac{1}{K}\right)[\mathscr{O}PT - f(S_t)].$$

This implies that

$$\mathscr{O}PT - f(S_K) \leq \left(1 - \frac{1}{K}\right)^K [\mathscr{O}PT - f(S_0)] \leq \left(1 - \frac{1}{K}\right)^K \mathscr{O}PT,$$

and

$$f(S_K) \geq \left[1 - \left(1 - \frac{1}{K}\right)^K\right] \mathscr{O}PT \geq \left(1 - \frac{1}{e}\right)\mathscr{O}PT.$$

The $1 - \frac{1}{e}$-approximation is tight for rank function due to Theorems 17 and 18. In the remainder of this subsection, we discuss a more general case by relaxing the monotonicity assumption of objective function.

Problem 7 (Nonmonotone Submodular Function Maximization) Given a non-negative submodular set function $f : 2^N \rightarrow \Re_+$, suppose we can evaluate $f(S)$ for any $S \subseteq N$. How should we solve the problem:

$$\max\{f(S) : S \subseteq N\}.$$

The hardness to approximate result is established in [12]:

Theorem 19 (Hardness for Nonmonotone Submodular Function Maximization) *Suppose $f : 2^N \rightarrow \Re_+$ is a submodular set function, which we can evaluate the function value on each $S \subseteq N$. Then for any $\epsilon > 0$, an algorithm which can approximate the general maximization problem of approximation ratio $\frac{1}{2} + \epsilon$ needs to call the valuation oracle exponentially many times. This is also true even if f is known to be symmetric, i.e., $f(S) = f(N \backslash S)$.*

Buchbinder et al. [3] recently established the tight approximation algorithm, based on the idea of forward–backward greedy search:

Algorithm 3: $1/2$-Randomized approximation algorithm

---

1 Initialization $t = 0$, $A_0 = \emptyset$, $B_0 = N$;
2 Given random order $u_1, u_2, \cdots, u_N$ of $1, 2, \cdots, N$;
3 foreach $t = 1, 2, \ldots, N$ **do**
4   Define
         $a_t = [f(A_{t-1} \cup \{u_t\}) - f(A_{t-1})]_+, b_t = [f(B_{t-1} \backslash \{u_t\}) - f(B_{t-1})]_+;$
5   With probability $p_t = \frac{a_t}{a_t + b_t}$, we add u_t to A_{t-1}: $A_t = A_{t-1} \cup \{u_t\}$,
         $B_t = B_{t-1}$;
6   Else (with probability $1 - p_t = \frac{b_t}{a_t + b_t}$), remove u_t from B_{t-1}:
         $A_t = A_{t-1}, B_t = B_{t-1} \backslash \{u_t\}$;
7 end
8 Return $A_N = B_N$. Note Define $p_t = 0$ if both $a_t = b_t = 0$.

This algorithm maintains increasing random series of sets $\{A_t\}$ and decreasing series of sets $\{B_t\}$, by gradually deciding whether an element should be added to A_t, or removed from B_t, based on whether its potential is improving the function value. It stops at $A_N = B_N$. Next, we define a series of sets S_t to assist our analysis of the algorithm. Suppose the optimum solution of $\max\{f(S) : S \subseteq N\}$ is S^*, with the optimum value noted as $\mathscr{O}PT = f(S^*)$. We define the random set $S_t = (S^* \cup A_t) \cap B_t$ and value $V_t = E[f(S_t)]$. Then we have for all t, $A_t \subseteq S_t \subseteq B_t$, $S_0 = S^*$, $f(S_0) = \mathscr{O}PT$, and $S_N = A_N = B_N$.

To prove the approximation result, we quantify the potential loss of function value from S_{t-1} to S_t by the following technical lemma from [30]:

Lemma 2 *For any t, the algorithm outputs*

$$E[f(S_{t-1}) - f(S_t)] \leq \frac{1}{2}[f(A_t) - f(A_{t-1}) + f(B_t) - f(B_{t-1})].$$

Proof By definition we have $B_t - A_t = \{u_{t+1}, u_{t+2}, \cdots, u_N\}$, and $u_t \in B_{t-1} \backslash A_{t-1}$. If $a_t = b_t = 0$, then by definition $f(A_{t-1} \cup \{u_t\}) - f(A_{t-1}) \leq 0$, $f(B_{t-1} \backslash \{u_t\}) - f(B_{t-1}) \leq 0$. Note that $A_{t-1} \subseteq B_{t-1} \backslash \{u_t\}$, it follows from submodularity we have

$$0 \geq f(A_{t-1} \cup \{u_t\}) - f(A_{t-1}) \geq f(B_{t-1}) - f(B_{t-1} \backslash \{u_t\}) \geq 0.$$

Notice the algorithm outputs $A_t = A_{t-1} \cup \{u_t\}$, $B_t = B_{t-1}$, so $f(A_t) - f(A_{t-1}) = f(B_{t-1}) - f(B_t) = 0$. If $u_t \in S_{t-1}$, we have $S_t = S_{t-1}$ and $f(S_{t-1}) - f(S_t) = 0$. If $u_t \notin S_{t-1}$, then the algorithm outputs $S_t = S_{t-1} \cup \{u_t\}$, consequently $A_{t-1} \subseteq S_{t-1} \subseteq B_{t-1} \backslash \{u_t\}$. By submodularity we have

$$0 \geq f(A_{t-1} \cup \{u_t\}) - f(A_{t-1}) \geq f(S_{t-1} \cup \{u_t\}) - f(S_{t-1}) \geq f(B_{t-1}) - f(B_{t-1} \backslash \{u_t\}) \geq 0,$$

which implies that $f(S_{t-1}) - f(S_t) = f(A_t) - f(A_{t-1}) = f(B_{t-1}) - f(B_t) = 0$.

Now we consider the case $a_t + b_t > 0$. If $u_t \in S^*$, then $u_t \in S_{t-1}$, $S_t = S_{t-1}$ with probability $p_t = \frac{a_t}{a_t + b_t}$, and $S_t = S_{t-1} \backslash \{u_t\}$ with probability $1 - p_t$. Note that $A_{t-1} \subseteq S_{t-1} \backslash \{u_t\}$, by submodularity $f(S_{t-1}) - f(S_{t-1} \backslash \{u_t\}) \leq f(A_{t-1} \cup \{u_t\}) - f(A_{t-1}) = a_t$. Therefore

$$E[f(S_{t-1}) - f(S_t)] \leq (1 - p_t)a_t = \frac{a_t b_t}{a_t + b_t}.$$

If $u_t \notin S^*$, then $u_t \notin S_{t-1}$, $S_t = S_{t-1} \cup \{u_t\}$ with probability $p_t = \frac{a_t}{a_t + b_t}$, and $S_t = S_{t-1}$ with probability $1 - p_t$. Because $S_{t-1} \subseteq B_{t-1} \backslash \{u_t\}$, it follows from submodularity that $f(S_{t-1}) - f(S_{t-1} \cup \{u_t\}) \leq f(B_{t-1} \backslash \{u_t\}) - f(B_{t-1}) = b_t$. Therefore we also have

$$E[f(S_{t-1}) - f(S_t)] \leq p_t b_t = \frac{a_t b_t}{a_t + b_t}.$$

Note that $p_t = 0$ if $a_t = 0, b_t > 0$, $p_t = 1$ if $a_t > 0, b_t = 0$, so whenever $a_t + b_t > 0$,

$$f(A_t) - f(A_{t-1}) + f(B_t) - f(B_{t-1})$$
$$= p_t[f(A_{t-1} \cup \{u_t\}) - f(A_{t-1})] + (1 - p_t)[f(B_{t-1} \backslash \{u_t\}) - f(B_{t-1})]$$
$$= p_t[f(A_{t-1} \cup \{u_t\}) - f(A_{t-1})]_+ + (1 - p_t)[f(B_{t-1} \backslash \{u_t\}) - f(B_{t-1})]_+$$
$$= p_t a_t + (1 - p_t)b_t.$$

Therefore,

$$\frac{a_t b_t}{a_t + b_t} \leq \frac{a_t^2 + b_t^2}{2(a_t + b_t)} = \frac{p_t a_t + (1 - p_t) b_t}{2}$$

$$= \frac{1}{2}[f(A_t) - f(A_{t-1}) + f(B_t) - f(B_{t-1})].$$

Theorem 20 ($\frac{1}{2}$-Approximation) *If the function $f : 2^N \to \Re_+$ is a nonnegative submodular set function, then Algorithm 3 achieves $\frac{1}{2}$-approximation ratio, i.e.,*

$$E[f(A_N)] \geq \frac{1}{2}\mathcal{O}PT.$$

Proof Adding the inequalities in Lemma 2 for all $t = 1, 2, \cdots, N$, we obtain

$$E[f(S_0)] - E[f(S_N)] = \sum_{t=1}^{N} E[f(S_{t-1}) - f(S_t)]$$

$$\leq \frac{1}{2}[f(A_N) - f(A_0) + f(B_N) - f(B_0)].$$

It then follows from $S_N = A_N = B_N$, and $f(A_0), f(B_0) \geq 0$ that

$$E[f(S_0)] - E[f(S_N)] \leq \frac{1}{2}[f(A_N) + f(B_N)] = f(S_N).$$

Because $S_0 = S^*$, $f(A_N) = f(S_N) \geq \frac{1}{2}f(S_0) = \frac{1}{2}\mathcal{O}PT$.

3.5 Multi-Linear Relaxation and Submodular Function Maximization

In this section, we introduce another line of approach to deal with submodular function maximization problems, which utilize the so-called **multi-linear relaxation**.

Definition 8 (Multi-Linear Relaxation) Given a set function $f : 2^N \to \Re$, we define its multi-linear relaxation by rounding a continuous point $x \in [0, 1]^N$ to $\{0, 1\}^N$: $F(x) = E[f(\xi(x))]$, where $\xi(x) \in \Re^N$ takes value $\xi(x)_i = 1$ with probability x_i, and $\xi(x)_i = 0$ with probability $1 - x_i$ independently.

Because the multi-linear relaxation is defined via expectation, it is straightforward to see:

Theorem 21

$$\max\{F(x) \mid x \in [0, 1]^N\} = \max\{f(x) \mid x \in \{0, 1\}^N\}.$$

In the remainder of the section, we introduce variations of submodular maximization problem, and how to utilize the multi-linear relaxation for solving these problems. We start with the general matroid constrained problem:

Definition 9 (Matroid) A matroid $\mathbf{M} = (\mathbf{X}, \mathbf{I})$ consists of the ground set $\mathbf{X}$ and the independent set $\mathbf{I} \subseteq 2^{\mathbf{X}}$ which is a set of subsets of $\mathbf{X}$, if it satisfies the following:

1. For any $A \subseteq B$ and $B \in \mathbf{I}$, it has to be $A \in \mathbf{I}$.
2. For any $A, B \in \mathbf{I}$ and $|A| < |B|$, there exists $x \in B \setminus A$ such that $A \cup \{x\} \in \mathbf{I}$.

Matroids are discrete sets, whose convex hull are actually polymatroids. In the following, we first illustrate how matroid induces a rank function, and how this rank function defines a polymatroid which is the convex hull of the matroid.

Theorem 22 (Matroid Rank) *Define* $r(S) = \max\{|X| \mid X \subseteq S, X \in \mathbf{I}\}$, *if* $\mathbf{M}$ *is a matroid, then* $r(S)$ *is a rank function.*

Proof It follows from definition that $r(S)$ is monotonically increasing and $r(\emptyset) = 0$, so we only need to verify the submodularity. For any $i, j \notin S$, if $r(S \cup \{i, j\}) = r(S \cup \{i\})$ or $r(S \cup \{i, j\}) = r(S \cup \{j\})$, it follows from monotonicity that $r(S) + r(S \cup \{i, j\}) \leq r(S \cup \{i\}) + r(S \cup \{j\})$. If else, then $r(S \cup \{i, j\}) > r(S \cup \{i\})$ and $r(S \cup \{i, j\}) > r(S \cup \{j\})$. Define $A \doteq \mathrm{argmax}\{|X| \mid X \subseteq S \cup \{i, j\}, X \in \mathbf{I}\}$. Note that $r(S \cup \{i, j\}) = |A|$ and $A \setminus j \subseteq S \cup \{i\}$, by the definition of independent set $A \setminus j \in \mathbf{I}$, so $r(S \cup \{i\}) \geq |A| - 1 = r(S \cup \{i, j\}) - 1$. Because $r(S \cup \{i\}) < r(S \cup \{i, j\})$, we have $r(S \cup \{i\}) = r(S \cup \{i, j\}) - 1$. Similarly, $r(S \cup \{j\}) = r(S \cup \{i, j\}) - 1$.

Define $B \doteq \mathrm{argmax}\{|X| \mid X \subseteq S, X \in \mathbf{I}\}$. Because $|B| = r(S) \leq r(S \cup \{i\}) < r(S \cup \{i, j\}) = |A|$, it follows from definition of independent set that there exists $x \in A \setminus B$ with $B \cup \{x\} \in \mathbf{I}$. By the definition of B, $x \notin S$ because otherwise $B \cup \{x\} \subseteq S$ is a larger independent set in S. Therefore, $x \in (S \cup \{i, j\}) \setminus S = \{i, j\}$, so $x = i$ or $x = j$. If $x = i$, then $r(S \cup \{i\}) \geq |B + i| = |B| + 1 = r(S) + 1$. Similarly, if $x = j$, we also have $r(S \cup \{j\}) \geq r(S) + 1$ if $x = j$. Therefore, we always have

$$r(S \cup \{i\}) + r(S \cup \{j\}) \geq r(S) + 1 + r(S \cup \{i, j\}) - 1 = r(S) + r(S \cup \{i, j\}).$$

In linear algebra, the set of linearly independent vectors forms an independent set for ground set of all vectors in $\Re^N$. The rank function induced by this independent set is exactly the rank of the spanning space of a set of vectors.

Theorem 23 (Matroid to Polymatroid) *For a matroid* $\mathbf{M}$ *with induced rank function* r, *define polytope* $\mathbf{P}(\mathbf{M}) = \mathbf{Conv}\{\mathbf{1}_S \mid S \in \mathbf{I}\}$, *then*

$$\mathbf{P}(\mathbf{M}) = \mathbf{P}(r, X) = \left\{ x \in \Re_+^X \mid \sum_{j \in S} x_j \leq r(S) \ \forall S \subseteq X \right\}$$

and it is a polymatroid.

Proof For any independent set $A \in \mathbf{I}$ and $S \subseteq X$, it follows from $A \cap S \in \mathbf{I}$ that $r(S) \geq |A \cap S| = \sum_{i \in S} (\mathbf{1}_A)_i$. Therefore $\mathbf{M} \subseteq \mathbf{P}(\mathbf{M})$.

Reversely, in the proof of Theorem 23 we showed that $r(S \cup \{i\}) - r(S) = 0$ or 1. By the optimum solution structure in Theorem 12, all the vertices of polytope $\mathbf{P}(\mathbf{M})$ are 0/1 vector. Suppose one vertex is $v = \mathbf{1}_A$, which corresponds to set A. If A is not an independent set, then by definition $r(A) \leq |A| - 1$. However, $\sum_{i \in A} v_i = \sum_{i \in A} 1 = |A| > r(A)$, which contradicts the constraint in the definition of $\mathbf{P}(\mathbf{M})$. Therefore, any vertices of the polytope are an element in $\mathbf{M}$.

For the matroid constrained rank function maximization problem:

$$\max\{f(S) : S \in \mathbf{I}\},$$

where $f : 2^N \rightarrow \Re_+$ is a rank function and $\mathbf{M} = (\mathbf{N}, \mathbf{I})$ is a matroid, we introduce the algorithm in [30]. Firstly, they use the smooth-greedy algorithm to obtain solution x such that $F(x) \geq \left(1 - \frac{1}{e} - o(1)\right) \mathcal{O}PT$, then they apply **pipage rounding** to gradually round each indices to 0 or 1. Since the multi-linear extension is defined by expectation form, rounding (or even greedy) would naturally yield integer solution with better quality. To start with, we consider the smooth process:

Algorithm 4: Smooth differential equation

---
1 Initialization: set $\delta = \frac{1}{m^2}$, $t = 0$, $y_{ij}(t) = 0$;

2 For any $y \in [0,1]^N$, define $I(y) = \max\{\sum_{j \in S} \frac{\partial}{\partial y_j} F(y) \mid S \in \mathbf{I}\}$;

3 Define $y(t)$ by differential equation $y(0) = 0$, $\frac{d}{dt} y(t) = \mathbf{1}_{I(y)}$;

4 Output $y(1)$;

The step 2 of solving $I(y)$ is doable because it is equivalent to polymatroid optimization problem as in Section 3.2, which can be solved by simple greedy process.

Theorem 24 (Smooth Process) *For the problem* $\max\{f(S) : S \in \mathbf{I}\}$ *with rank function* r *and polymatroid* $\mathbf{M} = (\mathbf{N}, \mathbf{I})$, *the smooth process outputs*

$$F(y(1)) \geq \left(1 - \frac{1}{e}\right) \mathcal{O}PT.$$

Proof Firstly, because $0 \leq \frac{\partial}{\partial t} y_j(t)) \leq 1$ for any index $j \in N$, $y(t)$ is always feasible for $t \in [0, 1]$. Define the optimum solution as $S^* = \text{argmax}\{f(S) : S \in \mathbf{I}\}$, then $f(S^*) = \mathcal{O}PT$. For any $y \in [0, 1]^N$, define random set R_y by independently randomly selecting index $i \in N$ with probability y_i, and not selecting i with probability $1 - y_i$.

For any two sets $S, T \subseteq N$, we define $f_S(T) = f(S + T) - f(S)$, and $f_S(j) = f_S(\{j\})$. By submodularity, for any set $S \subseteq N$ we have:

$$OPT = f(S^*) \leq f(S \cup S^*) \leq f(S) + \sum_{j \in S^*} f_S(j).$$

Define $f_{R_y}(j) = E_{S \sim R_y} f_S(j)$ and notice that $F(y) = E_{S \sim R_y} f(S)$, then

$$OPT \leq E_{S \sim R_y}\left[f(S) + \sum_{j \in S^*} f_S(j) \right]$$

$$= F(y) + \sum_{j \in S^*} f_{R_y}(j) \leq F(y) + \max_{S \in \mathbf{I}} \sum_{j \in S} f_{R_y}(j),$$

where the last inequality follows from the fact that $S^* \in \mathbf{I}$. Note that

$$F(y) = \sum_{S \subseteq N} f(S) \prod_{i \in S} y_i \prod_{i \notin S} (1 - y_i),$$

which implies that

$$\frac{\partial}{\partial y_j} F(y) = F(y \mid y_j = 1) - F(y \mid y_j = 0)$$

$$= E\left[f(R_y \cup \{j\}) - f(R_y \setminus \{j\}) \right] \geq f_{R_y}(j).$$

Therefore, the differential equation process satisfies

$$\frac{d}{dt} F(y(t)) = \sum_{j \in I(y)} \frac{\partial}{\partial y_j} F(y(t))$$

$$= \max_{S \in \mathbf{I}} \sum_{j \in S} \frac{\partial}{\partial y_j} F(y(t)) \geq \max_{S \in \mathbf{I}} \sum_{j \in S} f_{R_y}(j) \geq OPT - F(y(t)).$$

Combine with the fact that $F(y(0)) \geq 0 = (1 - e^{-0}) \mathcal{O}PT$, we have for any $t \in [0, 1]$,

$$F(y(t)) \geq (1 - e^{-t}) \mathcal{O}PT.$$

Since the smooth solution can't be obtained exactly, one can apply the following algorithm for $1 - \frac{1}{e} - o(1)$ approximation ratio:

Algorithm 5: Smooth greedy

---

1 Set $\delta = \frac{1}{M}$, $M \geq N^2$, $t = 0$, $y(0) = 0$;
2 **foreach** $t = 0, 1, 2, \ldots, M - 1$ **do**
3   Define $\omega_j(t) \sim f_{R_{y(t\delta)}}(j)$, which can be obtained within any required error by sampling algorithm;
4   Define $I(t) = \mathrm{argmax}\{\mathrm{sum}_{j \in S}\omega_j(t) \mid S \in \mathbf{I}\}$;
5   Take $y((t+1)\delta) = y(t\delta) + \delta \mathbf{1}_{I(t)}$
6 **end**
7 Output $y(1)$.

Many well-known combinatorial problems can be reformulated into matroid constrained rank function maximization:

Problem 8 (The Submodular Social Welfare Problem) There are a set P (m many) of players and a set N of resources. Player i's utility function is $w_i(S_i)$ if receiving set S_i of resources, which is assumed to be a rank function. How should we distribute resources among a group of people, to maximize the social utility $\sum_{i=1}^{m} w_i(S_i)$? Without losing generality, we assume that each resource is of single unbreakable unit, and this assumption can be relaxed to multi-units without altering the following process as well as its analysis.

By making m copies (i, j) of each item j, and an allocation $\{S_1, S_2, \cdots, S_m\}$ uniquely corresponds to set $S = \bigcup_{i=1}^{m}\{(i, j) \mid j \in S_i\}$. We obtain a matroid $\mathbf{M}$ is defined by the ground set $\mathbf{X} = P \times N$, the independent set

$$\mathbf{I} = \{S \subseteq \mathbf{X} \mid |S \cap \{P \times \{j\}\}| \leq 1 \text{ for all } j \in N\}.$$

Then the problem is reduced to classical matroid constraints rank function maximization.

When each player also faces the bin packing problem, the problem becomes the general assignment problem.

Problem 9 (General Assignment Problem) There is a set P of players, and a set N of items. Each player i has only 1 unit of capacity which can't be exceeded. Receiving the item j would yield utility v_{ij}, but also consumes capacity c_{ij} of the player i.

Note that each player has a feasible set $\mathscr{F}_i \subseteq 2^N$ of possible choices for each player i, we can construct the matroid $\mathbf{X} = (\mathbf{X}, \mathbf{I})$ by ground set $\mathbf{X} = \{(i, S_i) \mid S_i \in \mathscr{F}_i, i \in P\}$, and

$$\mathbf{I} = \{\mathbf{S} \subseteq \mathbf{X} \mid \text{At most one set } S_i \text{ assigned to each player } i\}.$$

To avoid assigning one item to multiple players, the objective function is changed to

$$f(\mathbf{S}) = \sum_{j \in N} \max\{v_{ij} : j \in S_i, (i, S_i) \in \mathbf{S}\}.$$

The GAP can also be solved via the so-called configuration LP approach as in [13], which plays significant role in combinatorial optimization.

Algorithm 6: Configuration LP+greedy rounding

---

1 Define $V_i(S) = \sum_{j \in S} v_{ij}$;
2 Solve the configuration LP problem

$$\max \sum_{i \in P} \sum_{S \in \mathscr{F}_i} y_{i,S} V_i(S)$$
$$\text{s.t. } \sum_{S \in \mathscr{F}_i} y_{i,S} \leq 1 \qquad \forall i \in P$$
$$\sum_{(i,S):j \in S, S \in \mathscr{F}_i} y_{i,S} \leq 1 \ \forall j \in N$$
$$y_{i,S} \geq 0 \qquad \forall i \in P, S \in \mathscr{F}_i$$

to obtain the fractional optimum solution $\{y_{i,S}\}$;
3 For each player i, independently select one $S_i = S$ with probability $y_{i,S}$, which is doable because $\sum_{S \in \mathscr{F}_i} y_{i,S} \leq 1$;
4 For each item j, allocate it to the player with the best value v_{ij}.

Note that for general assignment problem, step 2 of the above algorithm can be solved by reformulating with a linear programming problem by assignment variables x_{ij} for (continuous) amount of item j assigned to player i. Fleischer et al. [13] showed that this greedy rounding algorithm yields $1 - \frac{1}{e}$ approximation ratio:

Theorem 25 *The configuration LP can be solved exactly, and the greedy rounding yields $1 - \frac{1}{e}$ approximation ratio with respect to the fractional solution.*

Problem 10 (Budget Constrained Maximization) Given a monotone submodular function $f : 2^N \to \Re_+$, suppose we can evaluate $f(S)$ for any $S \subseteq N$. And for each item $i \in N$ it consumes nonnegative budget of c_i. How should we solve budget constrained problem:

$$\max \left\{ f(S) : \sum_{i \in S} c_i \leq B, S \subseteq N \right\}.$$

The first $1 - \frac{1}{e} - o(1)$ approximation algorithm for the budget constrained maximization problem was achieved by Sviridenko [28], later improved by Badanidiyuru and Vondrák [2] and Ene and Nguyen [9]. The detailed algorithms are quite involved and lengthy, readers may refer to the listed research papers for reference.

All algorithms split the items into two groups, those with large value and those with small value. The large valued ones are of small number, which can be guessed, or decided with the help of multi-linear extension form. For the small valued ones, missing one small valued items due to capacity would lose at most ϵ ratio. Therefore one can simply apply cost-efficiency greedy approach to fill the capacity.

4 Discrete Convexity in Dynamic Programming

Submodularity and other discrete convex properties are also very useful in dynamic and online decision problems. In Section 4.1, we present the concept of $L^\sharp$-convexity. Applications in dynamic inventory control problems are discussed in Section 4.2. In Section 4.3, online matching problems are introduced.

4.1 $L^\sharp$-Convexity

All the discussions of submodular function optimization in Section 3 are focused on set functions. In most practical problems, one needs to deal with decision variables in broader domain. Since the submodular function minimization relies heavily on Lovasz extension, a natural question is, when would the Lovasz extension of a function coincide with its convex hull in common discrete domain?

Definition 10 ($L^\sharp$-Convex Set) A set $D \subseteq \mathbf{Z}^N$ is called $L^\sharp$-convex, if $\{(x, t) \mid x - te \in D, t \in \mathbf{Z}_+\}$ is a sublattice, i.e.,

$$(x + te) \wedge y \in D \text{ and } x \vee (y - te) \in D \text{ for all } x, y \in D, t \in \mathbf{Z}_+,$$

where $e \in \Re^N$ is the all one vector.

Definition 11 ($L^\sharp$-Convex Function) For $L^\sharp$-convex domain D, we call a function $f : L \to \Re$ a $L^\sharp$-convex if the function $g(x, t) = f(x - te)$ is a submodular function on sublattice domain $\{(x, t) \mid x - te \in D, t \in \mathbf{Z}_+\}$.

Theorem 26 *The condition of $L^\sharp$-convexity is equivalent to: (Condition A) $f(x) + f(y) \geq f((x + te) \wedge y) + f(x \vee (y - te))$ for any $x, y \in D, t \in \mathbf{Z}_+$. When $D = \mathbf{Z}^N$, the next two conditions are also equivalent conditions for $L^\sharp$-convexity:*

1. *(Condition B) $f(x) + f(y) \geq f(\lfloor \frac{x+y}{2} \rfloor) + f(\lceil \frac{x+y}{2} \rceil)$ for any $x, y \in D$.*
2. *(Condition C) If we define the Lovasz extension $f^L(x)$ within each integer grid, and merge them together, it is well defined and coincides with the convex hull: $f^L = f^-$.*

Note: when the function $f : \Re^N \to \Re$ is a C^2 function defined on continuous domain, the condition is equivalent to the Hessian of $M = \nabla^2 f(x)$ that is always a diagonal dominated M-matrix for any x, i.e., $M_{ij} \leq 0$ for all $i \neq j$.

Proof Firstly, for any $x, y \in D, t \in \mathbf{Z}_+$, we denote $z = x + te$. Then $(z, t) \vee (y, 0) = (z \vee y, t), (z, t) \wedge (y, 0) = (z \wedge y, 0), z \vee y - te = x \vee (y - te)$. Notice that

$$f(x) + f(y) - f((x + te) \wedge y) - f(x \vee (y - te))$$
$$= g(z, t) + g(y, 0) - g((z, t) \wedge (y, 0)) - g((z, t) \vee (y, 0)).$$

So the condition (A) is equivalent to the submodularity of g.

Secondly, we show that condition (B) implies submodularity of g when the domain $D = \mathbf{Z}^N$, and vice versa. Note that we only need to verify the submodularity locally, i.e.:

1. $g(x, t) + g(x + e_i + e_j, t) \leq g(x + e_i, t) + g(x + e_j, t)$ for all $i \neq j$, where e_i is unit length vector which only takes value of 1 at index i,
2. $g(x, t) + g(x + e_i, t + 1) \leq g(x + e_i, t) + g(x, t + 1)$ for all $x \in D, i \in N$, and $t \in \mathbf{Z}_+$.

The first inequality follows from

$$f(x + e_i - te) + f(x + e_j - te) \geq f(x - te + \lfloor \frac{e_i + e_j}{2} \rfloor) + f(x - te + \lceil \frac{e_i + e_j}{2} \rceil)$$
$$= f(x - te) + f(x - te + e_i + e_j).$$

The second inequality follows from

$$f(x + e_i - te) + f(x - (t + 1)e) \geq f(x - te + \lfloor \frac{e_i - e}{2} \rfloor) + f(x - te + \lceil \frac{e_i - e}{2} \rceil)$$
$$= f(x + e_i - (t + 1)e) + f(x - te).$$

Reversely, when g is submodular, we start with the case $|x_i - y_i| \leq 1$ for all $i \in N, \lfloor \frac{x+y}{2} \rfloor = x \wedge y$, and $\lceil \frac{x+y}{2} \rceil = x \vee y$. Therefore

$$f\left(\lfloor \frac{x+y}{2} \rfloor\right) + f\left(\lceil \frac{x+y}{2} \rceil\right) = g(x \wedge y, 0) + g(x \vee y, 0)$$

$$\leq g(x, 0) + g(y, 0) = f(x) + f(y).$$

Now we prove that condition (A) woud imply condition (B). If condition (B) is violated by some pair of (x, y), we define (x^*, y^*) as the minimal pair which violates the condition (B), i.e., solution for

$$\min\{\|x - y\|_1 \mid f(x) + f(y) < f\left(\lfloor \frac{x+y}{2} \rfloor\right) + f\left(\lceil \frac{x+y}{2} \rceil\right), x, y \in D'\},$$

where we can constraint D is a finite box neighborhood D' of a violation pair. For the inequality to hold, there exists at least one index k such that $|x_k^* - y_k^*| \geq 2$. Without losing generality, we assume $x_k^* \leq y_k^* - 2$. Note that for any $x_i, y_i \in \mathbf{Z}$, if $x_i \leq y_i - 1$, then $\min\{x_i + 1, y_i\} = x_i + 1$ and $\min\{x_i, y_i - 1\} = y_i - 1$, if $x_i \geq y_i - 1$, then $\min\{x_i + 1, y_i\} = y_i$ and $\min\{x_i, y_i - 1\} = x_i$. Therefore $(x^* + e) \wedge y^* + x^* \vee (y^* - e) = x^* + y^*$, and $|((x^* + e) \wedge y^*)_i - (x^* \vee (y^* - e))_i| \leq |x_i^* - y_i^*|$ for all index i. Furthermore,

$$|((x^* + e) \wedge y^*)_k - (x^* \vee (y^* - e))_k| \leq |y_i^* - x_i^* - 2| = y_k^* - x_k^* - 2 < y_k^* - x_k^*,$$

which implies that $\|(x^* + e) \wedge y^* - x^* \vee (y^* - e)\|_1 < \|x^* - y^*\|_1$. Because (x^*, y^*) is the minimal pair which violates the condition, and the fact that $(x^* + e) \wedge y^* + x^* \vee (y^* - e) = x^* + y^*$, we have

$$f((x^* + e) \wedge y^*)) + f(x^* \vee (y^* - e)) \geq f(\lfloor \frac{x^* + y^*}{2} \rfloor) + f(\lceil \frac{x^* + y^*}{2} \rceil).$$

However, it follows from condition (A) that

$$f(x^*) + f(y^*) \geq f((x^* + e) \wedge y^*)) + f(x^* \vee (y^* - e)).$$

These two inequalities contradict the definition of (x^*, y^*), so we proved that there is no pair $x, y \in \mathbf{Z}^N$ which can violate condition (B).

Thirdly, we establish the equivalence of $L^\sharp$-convexity with the convexity of Lovasz extension. When f is $L^\sharp$-convex, it has been established that f is submodular in each small grid; therefore, we can define f^L in each grid. Next we prove this definition coincides with the convex extension, by showing that for convex combinations of $x = \sum_{z \in D} \alpha_z z$, the minimum combination of function values can be achieved in the smallest grid near x, for any x with no integer value. For those x with integer value, i.e., within intersection of multiple small grids, we can apply continuity argument.

For each given finite convex combination $x = \sum_{z \in D} \lambda_z z$ with value $V = \sum_{z \in D} \alpha_z f(z)$, the support $\{z \mid \alpha_z > 0\}$ is a finite set and can be assumed to be contained in a finite box $B = [-M, M]^N$. Consider all convex combinations of x in B with better value, i.e., $\Lambda = \{\lambda \mid \sum_{z \in B} \lambda_z f(z) \leq V, \sum_{z \in B} \lambda_z = 1, \lambda \geq 0\}$ which is nonempty because $\alpha \in \Lambda$. Define the potential function $P(\lambda) = \sum_{z \in B} \lambda_z \|z\|_2^2$ for convex combination λ defined on B. And define β as the solution for $\min\{P(\lambda) \mid \lambda \in \Lambda\}$, with support $\mathbf{Supp}_\beta = \{z \mid \beta_z > 0\}$. If it is not contained in the smallest box, then there exists $u, v \in \mathbf{Supp}_\beta$ and index i such that $v_i - u_i \geq 2$. It follows from the condition (A) that we can find $w = (u + e) \wedge v, y = u \vee (v - e) \in D$, which satisfies $f(u) + f(v) \geq f(w) + f(y), u + v = w + y$, and $w, y \in B$. Furthermore, for each index j, note that if $u_j \geq v_j - 1$, then $w_j = v_j$ and $y_j = u_j$, and if $u_j \leq v_j - 2$, then $w_j = u_j + 1$ and $y_j = v_j - 1$, therefore $u_j^2 + v_j^2 \geq w_j^2 + y_j^2$ for all $j \in N$, and $u_i^2 + v_i^2 - 2 \geq w_i^2 + y_i^2$. Therefore, denoting $\delta = \min\{\beta_u, \beta_v\} > 0$, the convex combination $\hat{\beta}$ defined as

$$\hat{\beta}_z = \begin{cases} \beta_z, & \text{if } z \in B\backslash\{u, v, w, y\} \\ \beta_z - \delta, & \text{if } z \in \{u, v\} \\ \beta_z + \delta, & \text{if } z \in \{w, y\} \end{cases}$$

satisfies

$$\sum_{z \in B} \hat{\beta}_z z = \sum_{z \in B} \beta_z z + \delta(w + y - u - v) = \sum_{z \in B} \beta_z z = x$$
$$\sum_{z \in D} \hat{\beta}_z f(z) = \sum_{z \in D} \beta_z f(z) + \delta(f(w) + f(y) - f(u) - f(v)) \leq \sum_{z \in D} \beta_z f(z) = VP(\hat{\beta}) = \sum_{z \in D} \hat{\beta}_z \|z\|^2 = \sum_{z \in D} \beta_z \|z\|^2 + \delta(\|w\|^2 + \|y\|^2 - \|u\|^2 - \|v\|^2)$$
$$\leq \sum_{z \in D} \beta_z \|z\|^2 - 2\delta \leq P(\beta) - 2\delta,$$

which contradicts with the minimum of potential function in the definition of β. Therefore, we showed that for the minimum convex combination β in the definition of $f^-(x)$, $z_j = \lfloor x_j \rfloor$, or $\lceil x_j \rceil$ for any $z \in \mathbf{Supp}_\beta$ and index $j \in N$. Therefore f^L coincides with the f^-, which is well defined and convex.

Reversely, if $f^L = f^-$, for any $x, y \in D$ we have

$$f(x) + f(y) \geq 2f^-(\frac{x+y}{2}) = 2f^L(\frac{x+y}{2}) = f(\lfloor \frac{x+y}{2} \rfloor) + f(\lceil \frac{x+y}{2} \rceil).$$

For C^2 function f defined on continuous domain, note that for any $x \in \Re^N$ and $t \in \Re_+$,

$$f(x + te_i) + f(x - te) - f(x + te_i - te) + f(x)$$

$$= \int_{s=0}^t \int_{r=0}^t \sum_{j=1}^N \frac{\partial^2}{\partial x_i \partial x_j} f(x + se_i + re - te) ds dr,$$

the submodularity of g across x_i and t is equivalent to diagonal dominance of $\nabla^2 f(x)$ on index i. Also, for any $i \neq j \in N$,

$$f(x + te_i) + f(x + te_j) - f(x) - f(x + te_i + te_j)$$

$$= \int_{s=0}^t \int_{r=0}^t \sum_{j=1}^N \frac{\partial^2}{\partial x_i \partial x_j} f(x + se_i + re_j) ds dr,$$

so the submodularity of g across x_i and x_j is equivalent to the off-diagonal (i, j)-th element of symmetric matrix $\nabla^2 f(x)$ that is non-positive.

Theorem 27 *For a $L^\sharp$-convex function $f : D \to \Re$ defined on $L^\sharp$-convex domain $D = \mathbf{Z}^N$, any local minimum solution x, i.e., $f(x) \leq f(y)$ for all $y \in D$ such that $\|y - x\|_\infty \leq 1$, is also a global minimum solution.*

Proof This result directly follows from the fact that x is local minimum for convex function f^L. Alternatively, we can establish the result via combinatorial approach:

Define set $S = \{y \mid f(z) < f(x), z \in D\}$. If S is nonempty, there exists $y \in S$ (may be not unique) which is closest to x in L^1 distance. Because x is local optimum and $f(x) > f(y)$, $\|y - x\|_\infty \geq 2$. It follows $2f(x) > f(x) + f(y) \geq f(\lfloor \frac{x+y}{2} \rfloor) + f(\lceil \frac{x+y}{2} \rceil)$ that $\min\{f(\lfloor \frac{x+y}{2} \rfloor), f(\lceil \frac{x+y}{2} \rceil)\} < f(x)$. But when $\|y - x\|_\infty \geq 2$, both $\lfloor \frac{x+y}{2} \rfloor$ and $\lceil \frac{x+y}{2} \rceil$ are strictly closer to x than y. Therefore both $\lfloor \frac{x+y}{2} \rfloor$ and $\lceil \frac{x+y}{2} \rceil$ are not in S, which contradicts with the fact that $\min\{f(\lfloor \frac{x+y}{2} \rfloor), f(\lceil \frac{x+y}{2} \rceil)\} < f(x)$.

Because local minimum is global minimum, minimization of $L^\natural$-convex function can be achieved by local search of improving directions, readers may refer to [26] for details. When precision can be traded for speed, the gradient projection approach in Theorem 15 should be applied, if the effective domain is compact.

4.2 $L^\natural$-Convexity in Dynamic Inventory System

In this subsection, we introduce some important applications of submodular function optimization. In particular, many inventory related problems arise in supply chain management, where one needs to handle complex inventory system dynamically. Because inventory of different goods may substitute for each other, and inventory for perishable goods (e.g., fresh fruit, fresh milk, etc.) with different expiration dates is also substitutable for each other, we often model the problem by submodularity or other related properties, and utilize these properties for obtaining better inventory strategy.

There are many applications of $L^\natural$-convex function in dynamic inventory management. We start with a simple example. Suppose there is a retailer who has n classes of goods, while class $i - 1$ goods can be updated to class i ($i = 2, 3, \cdots, n$) with upgrading cost c_i per unit overnight. The retailer can also purchase from supplier for class 1 goods with cost c_1 per unit overnight. There are random demand D_i^t of type i goods at day t, and unsatisfied demand will be backlogged (booked for future sales) with penalty cost b_i per unit day. Unsold class i goods will incur holding cost h_i per unit day. The retailer needs to decide the amount q_1^t of class 1 goods to purchase from supplier, as well as the amount q_i^t upgraded for class i goods from class $i - 1$, $i = 2, \cdots, n$. Denote the inventory of class i goods at the beginning of day t as I_i^t, then $I_i^{t+1} = I_i^t + q_i^t - D_i^t - q_{i+1}^t$. Therefore the decision corresponds to dynamic programming:

$$V_t(I^t) = E_{D_t} \min \left\{ \sum_{i=1}^n c_i q_i^t + \sum_{i=1}^n f_i(I_i^t - D_i^t) + V_{t+1}(I^{t+1}) : 0 \leq q_i^t \right\},$$

where $f_i(x) = -b_i x$ if $x \leq 0$ and $f_i(x) = h_i x$ if $x \geq 0$, $V_{T+1} \equiv 0$, and the overnight decision q^t depends on the realized demand D^t during the day. It we take the transformation of $S_i^t = \sum_{j \geq i} I_i^t$, then $S_i^{t+1} = S_i^t + q_i^t - \sum_{i \geq s} D_i^t$, and $I_i^t =$

$S_i^t - S_{i+1}^t$. We define $C_t(S^t) = V_t(I^t)$, which satisfies the dynamic programming:

$$C_t(S^t) = E_{D_t} \min \left\{ \sum_{i=1}^{n} c_i \left(S_i^{t+1} - S_i^t - \sum_{i \geq s} D_i^t \right) + \sum_{i=1}^{n} f_i(S_i^{t+1} - S_{i+1}^{t+1}) \right.$$

$$\left. + V_{t+1}(I^{t+1}) : S_i^{t+1} \geq S_i^t + \sum_{j \geq i} D_i^t \right\}.$$

Next, we show that the function C is always $L^\natural$-convex, and the function V is a so-called multimodular function. For this purpose, we need to establish that $L^\natural$-convexity can be preserved under minimization, similar to the preservation of submodular property in Theorem 7. We refer to a theorem in [33]:

Theorem 28 (Preservation of $L^\natural$-Convexity) *Suppose $S \subseteq X \times Y$ is a $L^\natural$-convex set, and $f : S \to \Re$ is a $L^\natural$-convex function. Then the function $g(y) = \min\{f(x, y) \mid (x, y) \in S\}$ is a $L^\natural$-convex function defined on $L^\natural$-convex set $T = \{y \mid \exists (x, y) \in S\}$*

Proof We first prove the $L^\natural$-convexity of T by definition. For any given $y - te, y' - t'e \in T$ with $t, t' \in \mathbf{Z}_+$, by definition there exists $x, x' \in X$ such that $(x, y - te), (x', y' - t'e) \in S$. Define $z = x + te$ and $z' = x' + te$, then $(z, y) - te = (x, y - te)$ and $(z,', y') - t'e = (x', y' - t'e)$. By $L^\natural$-convexity of S,

$$(z \vee z' - (t \vee t')e, y \vee y' - (t \vee t')e) = (z, y) \vee (z', y') - (t \vee t')e \in S$$

and

$$(z \wedge z' - (t \wedge t')e, y \wedge y' - (t \wedge t')e) = (z, y) \wedge (z', y') - (t \wedge t')e \in S.$$

It follows that $y \vee y' - (t \vee t')e \in T$ and $y \wedge y' - (t \wedge t')e \in T$.

Next, we establish the $L^\natural$-convexity of function g by constructive approach. Define $h(x, y, t) = f((x, y) - te)$ and $\hat{h}(y, t) = g(y - te)$. For any given $y - te, y' - t'e \in T$ with $t, t' \in \mathbf{Z}_+$, by definition there exists $x, x' \in X$ such that $f(x, y - te) = g(y - te)$ and $f(x', y' - t'e) = g(y' - t'e)$. Define $z = x + te$ and $z' = x' + te$, then $g(y - te) = f((z, y) - te) = h(z, y, t)$ and $g(y' - t'e) = f((z', y') - t'e) = h(z', y', t')$. By definition of $L^\natural$-convexity, the function h is submodular, therefore

$$h(z, y, t) + h(z', y', t') \geq h(z \vee z', y \vee y', t \vee t') + h(z \wedge z', y \wedge y', t \wedge t').$$

It follows from $(z, y) - te = (x, y - te) \in S$ and $(z', y') - t'e = (x', y' - t'e) \in S$ that $\left(z \vee z' - (t \vee t')e, y \vee y' - (t \vee t')e \right) \in S$, therefore

$$h(z \vee z', y \vee y', t \vee t') = f(z \vee z' - (t \vee t')e, y \vee y' - (t \vee t')e) \geq g(y \vee y' - (t \vee t')e).$$

Similarly,

$$h(z \wedge z', y \wedge y', t \wedge t') = f(z \wedge z' - (t \wedge t')e, y \wedge y' - (t \wedge t')e) \geq g(y \wedge y' - (t \wedge t')e).$$

Combine these inequalities together, we obtain

$$g(y - te) + g(y' - t'e) \geq g(y \vee y' - (t \vee t')e) + g(y \wedge y' - (t \wedge t')e),$$

which implies the $L^\sharp$-convexity of function g.

Furthermore, for the dynamic inventory control problem we notice the following two facts:

1. The set $(x, y) \in \Re^2 \mid x - y \leq c\}$ is $L^\sharp$-convex.
2. For convex function $f : \Re \to \Re$, $f(x - y)$ is always $L^\sharp$-convex.

Therefore, if C_{t+1} is $L^\sharp$-convex, for each given D_t, so is the function $F_{D^t}(S^t) = \min \left\{ \sum_{i=1}^n c_i(S_i^{t+1} - S_i^t - \sum_{i \geq s} D_i^t) + \sum_{i=1}^n f_i(S_i^{t+1} - S_{i+1}^{t+1}) + V_{t+1}(I^{t+1}) : S_i^{t+1} \geq S_i^t + \sum_{j \geq i} D_i^t \right\}$. By linearity in definition of $L^\sharp$-convexity, we know that $C_{t+1}(S^t) = E_{D^t} F_{D^t}(S^t)$ is also $L^\sharp$-convex. So we can inductively establish the $L^\sharp$-convex property for C_t:

Theorem 29 *The function C_t is $L^\sharp$-convex, and the original function V_t is multimodular.*

Definition 12 (Multimodular Function) A function $f : X \to \Re$ is defined on $S \subseteq \Re^N$, which is $X = \{x : a_j^T x \leq b, j = 1, 2, \cdots, m\}$, where each a_j is of form $\sum_{i=K}^L e_i$, i.e., vector with value 1 on consecutive indices, and 0 if else, or $\sum_{i=K}^L e_i$, for different $1 \leq K \leq L \leq N$. If $\Phi(x) = f(x_1 - y, x_2 - x_1, \cdots, x_N - x_{N-1})$ is submodular on $S = \{(x, y) \in \Re^{N+1} \mid (x_1 - y, x_2 - x_1, \cdots, x_N - x_{N-1}) \in X\}$ is a submodular function, we say f is a multimodular function.

Multimodular function is essentially $L^\sharp$-convex function under a linear transformation, which was established in [24]:

Theorem 30 *Suppose we define set $Z = \{z \in \Re^N \mid (z_1, z_1 - z_2, z_2 - z_3, \cdots, z_n - z_{n-1}) \in X\} = \{z \mid (z, 0) \in S\}$ and function $g(z) = f(z_1, z_1 - z_2, z_2 - z_3, \cdots, z_n - z_{n-1})$. Then $f : X \to \Re$ is a multimodular function, is equivalent to $g : Z \to \Re$, and is a $L^\sharp$-convex function.*

Multimodularity, or equivalently under transformation, $L^\sharp$-convexity are used to characterize the dynamic decision systems and the corresponding optimum solutions for inventory management of perishable goods [4] and [21], as well as the queueing system [1].

For optimizing the $L^\sharp$-convex functions in the dynamic system, one could not simply apply the greedy local search algorithm, because the state space is exponentially large which makes it impossible to recursively solve for function values at all states as the classical dynamic programming approach does. Therefore, one can only hope to solve the problem approximately in dynamic system by adaptive approximation approach, which uses classes of simple functions to approximate each V_t, and recursively find the best approximation for each state function V_t. In particular, [22] uses linear functions for inventory problem of perishable goods. In Sun et al. [27], a class of quadratic functions has been used, according to the following Lemma they established based on Murota's characterization of quadratic $L^\sharp$ functions:

Lemma 3 *A quadratic function $f : \Re^N \to \Re$ is multimodular if and only it can be expressed as $f(x) = \sum_{i=1}^{N} \sum_{j=1}^{i} Q^{ij}(\sum_{l=j}^{i} x_l)$, where $Q^{ij}(x) = a_{ij}x^2 + b_{ij} + c_{ij}$ with $a_{ij} \geq 0$.*

Recently, Chen et al. [4] develop the basis function approach which approximates the original function by a linear combination $\hat{F}(x) = \sum_{i=1}^{N} \sum_{j=1}^{i} B^{ij}(\sum_{l=i}^{j} x_l)$ of basis functions B^{ij}, which can be recursively constructed by solving single dimensional optimization problems. This approach is much more flexible by allowing a much broader class of functions to be used to approximate the original function, and does achieve significant improvement in practice.

4.3   Online/Dynamic Matching

Submodularity also has important applications in dynamic matching. To begin with, we analyze the static matching. Consider the bipartite matching problem with two sets (A and B) of nodes, and edges $(i, j) \in E \subseteq A \times B$. Each edge $(i, j) \in E$ is associated with a weight w_{ij}. The objective is to match the nodes to maximize the total matching weight, constraint to that each node can be matched to at most one node. This problem can be modeled by:

$$W(A, B) = \max \sum_{i \in A, j \in B} w_{ij} x_{ij}$$
$$\text{s.t.} \sum_{j \in B} x_{ij} \leq 1 \forall i \in A$$
$$\sum_{i \in A} x_{ij} \leq 1 \forall j \in B$$
$$x_{ij} = 0 \text{ or } 1 \forall i \in A, j \in B.$$

Since the constraint matrix is unimodal, there is no integrality gap, we can replace the integer constraints by $x_{ij} \geq 0$.

We can even consider a more general formulation:

$$U(a, b) = \max \sum_{i \in A, j \in B} w_{ij} x_{ij}$$
$$\text{s.t.} \sum_{j \in B} x_{ij} \leq a_i \forall i \in A$$
$$\sum_{i \in A} x_{ij} \leq b_j \forall j \in B$$
$$x_{ij} \geq 0,$$

which satisfies $W(A, B) = U(\mathbf{1}_A, \mathbf{1}_B)$.

Intuitively, we can view a and b as resources, and resources in a (or b) are substitutes to each other, so adding more and more resources in a (or b) has diminishing return. However, resources in a are complementary to those in b, so adding resource in a would boost the values of resources in b, and vice versa. Next, we show the function $U(a, b)$ is submodular within a, b, but supermodular across them:

Theorem 31 *The function $U(a, -b)$ is submodular! By setting $a = \mathbf{1}_A$ and $b = -\mathbf{1}_B$), we know that $W(A, B)$ is submodular in A for fixed B.*

Proof Taking the dual, we have

$$U(a, -b)$$

$$= \min \left\{ \sum_{i \in A} a_i y_i - \sum_{j \in B} b_j z_j \mid y \in \mathfrak{R}_+^A, z \in \mathfrak{R}_+^B, y_i + z_j \geq w_{ij}, \forall i \in A, j \in B \right\}.$$

By defining variable $v = -y$, it becomes

$$U(a, -b)$$

$$= \min \left\{ -\sum_{i \in A} a_i v_i - \sum_{j \in B} b_j z_j \mid v \in \mathfrak{R}_-^A, z \in \mathfrak{R}_+^B, -v_i + z_j \geq w_{ij}, \forall i \in A, j \in B \right\}.$$

Note that the objective function $-a^T v - b^T z$ is submodular in (a, b, v, z), and the feasible domain is a lattice, by Theorem 7 the function $U(a, -b)$ is submodular.

For online matching problems, submodularity or equivalently the diminishing return property plays a crucial role. We present a more general online matching case in [19].

Problem 11 There is a fixed group A of players, and a group of items arrive stochastically. The items are of different types $j \in T$, at each time t the type j_t of the item arrives following an i.i.d distribution. At the end (time N), player i receives S_i of items, with submodular utility $V_i(S_i)$. We need to match each item at the time it arrives, and aim at maximizing the total matching score at the end. One thing to note that is, by setting $V_i(S_i) = \max_{j \in S_i} w_{ij}$, we can reduce this problem to an online bipartite matching problem.

Consider the following greedy allocation rule:

Algorithm 7: Greedy online matching

---

1 Initialize with $S_i^0 = \emptyset$ for all $i \in A$. **foreach** $t = 1, 2, \ldots, N$ **do**

2 | Match the item arrives (of type j_t) to player i_t with maximum matching weight (or equivalently, maximum improvement) ;

3 |

$$i_t = \text{argmax}\{V_i(S_i^{t-1} \cup \{j_t\}) - V_i(S_i^{t-1}) \mid i \in S_{t-1}\}$$

$S_{i_t}^t = S_{i_t}^{t-1} \cup \{i_t\}, S_i^t = S_i^{t-1}$ for all $i \neq i_t$;

4 end

Theorem 32 *The greedy algorithm achieves $1 - \frac{1}{e}$ approximation ratio.*

Proof By submodularity of V_i and two sets X and Y of items, we have $V_i(X \cup Y) - V_i(X) \leq \sum_{j \in Y} c_j(Y)[V_i(X \cup \{j\}) - V_i(X)]$. Denote p_j as the arrival probability of type j items, and $c_j(S)$ as the type j items used in set S, then the optimum offline matching value with expected number of arrival is bounded by:

$$\mathcal{O}PT \leq \max \sum_{i \in A, S} V_i(S) x_{i,S}$$
$$\text{s.t.} \sum_{i \in A} \sum_{S: j \in S} x_{i,S} c_j(S) \leq p_j N, \forall j \in T$$
$$\sum_S x_{i,S} \leq 1, \forall i \in A$$
$$x_{i,S} \geq 0, \forall i \in A \text{ and set } S.$$

Denote $y_{ij} = \sum_{S: j \in S} x_{i,S} c_j(S)$ and $z_{ij} = \frac{y_{ij}}{p_j N}$, then $\sum_{i \in A} z_{ij} \leq 1$ and $z \geq 0$. When item of type j arrives at time t, consider the random allocation which assigns this item to player i with probability z_{ij}. Then the expected gain is

$$\sum_{i,j} p_j z_{ij}[V_i(S_i^{t-1} \cup \{j\}) - V_i(S_i^{t-1})] = \frac{1}{N} \sum_{i,j} y_{ij}[V_i(S_i^{t-1} \cup \{j\}) - V_i(S_i^{t-1})].$$

The greedy has at least the expected gain, denote actual matching value at stage t as V^t. Therefore,

$$E[V^t] - V^{t-1} \geq \frac{1}{N} \sum_{i,S} x_{i,S} \left[\sum_{j \in S} c_j(S)[V_i(S_i^{t-1} \cup \{j\}) - V_i(S_i^{t-1})] \right]$$
$$\geq \frac{1}{N} \sum_{i,S} x_{i,S}[V_i(S_i^{t-1} \cup S) - V_i(S_i^{t-1})].$$

Therefore

$$E[V^t] - V^{t-1} \geq \frac{1}{N}\mathscr{O}PT - \frac{1}{N}\sum_{i,S} x_{i,S} V_i \left(S_i^{t-1} \right)$$

$$\geq \frac{1}{N}\left[\mathscr{O}PT - \sum_i V_i(S_i^{t-1}) \right] = \frac{1}{N}\left[\mathscr{O}PT - V^{t-1} \right].$$

This approximation ratio is tight in online stochastic matching scenario, and similar algorithms and analyses have been established for other online matching and online allocation problems [10, 20, 32], etc.

The diminishing return effect from submodularity can be applied to quantify the matching efficiency, when we combine matching stages with other type of operations, in algorithm design. In a recent work, He et al. [7] studied the matching problem in kidney exchange, which matches donated kidneys from non-directed donors, as well as kidneys from relatives of patients who does not match with own targeted relative, to other patients. The matching process has been divided into two stages, in the first stage random walk mechanism has been applied to achieve efficient chains in difficult patients, and in the second stage bipartite matching algorithms are applied to further reduce number of unmatched patients. Submodularity of second stage matching score, with respect to the available (unmatched) difficult patients for matching at beginning of stage two, is utilized to transfer the analysis in stage one, to an analysis of the full mechanism. By this approach, the first non-asymptotic bound on matching efficiency has been established for medium size random graphs.

References

1. Altman, E., Gaujal, B., Hordijk, A.: Multimodularity, convexity, and optimization properties. Math. Oper. Res. **25**(2), 324–347 (2000)
2. Badanidiyuru, A., Vondrák, J.: Fast algorithms for maximizing submodular functions. In: Proceedings of the Twenty-Fifth Annual ACM-SIAM Symposium on Discrete Algorithms, pp. 1497–1514 (2014)
3. Buchbinder, N., Feldman, M., Seffi, J., Schwartz, R.: A tight linear time (1/2)-approximation for unconstrained submodular maximization. SIAM J. Comput. **44**(5), 1384–1402 (2015)
4. Chen, S., Li, Y., Yang, Y., Zhou, W.: Managing Perishable inventory systems with age-differentiated demand. In: SSRN Scholae Papers, Society Science Research Network (2018)
5. Cunningham, W.H.: On submodular function minimization. Combinatorica **5**(3), 185–192 (1985)
6. Davis, J.M., Topaloglu, H. and Williamson, D.P.: Assortment optimization over time. Oper. Res. Lett. **43**(6), 608–611 (2015)
7. Ding, Y., Ge, D., He, S., Ryan, C.T.: A nonasymptotic approach to analyzing kidney exchange graphs. Oper. Res. **66**(4), 918–935 (2018)
8. Edmonds, J.: Submodular functions, matroids, and certain polyhedra. In: Combinatorial Structure and Their Applications, pp. 69–87. Gordon and Breach, New York (1970)

9. Ene, A., Nguyen, H.L.: A Nearly-linear Time Algorithm for Submodular Maximization with a Knapsack Constraint. Preprint. arXiv:1709.09767 (2017)
10. Esfandiari, H., Korula, N., Mirrokni, V.: Bi-objective online matching and submodular allocations. In: Advances in Neural Information Processing Systems, pp. 2739–2747 (2016)
11. Feige, U.: A threshold of ln n for approximating set cover. J. ACM (JACM) **45**(4), 634–652 (1998)
12. Feige, U., Mirrokni, V.S., Vondrák, J.: Maximizing non-monotone submodular functions. SIAM J. Comput. **40**(4), 1133–1153 (July 2011)
13. Fleischer, L., Goemans, M.X., Mirrokni, V.S., Sviridenko, M.: Tight approximation algorithms for maximum general assignment problems. In: Proceedings of the Seventeenth Annual ACM-SIAM Symposium on Discrete Algorithm, pp. 611–620 (2006)
14. Grötschel, M., Lovász, L., Schrijver, A.: The ellipsoid method and its consequences in combinatorial optimization. Combinatorica **1**(2), 169–197 (1981)
15. He, S., Zhang, J., Zhang, S.: Polymatroid optimization, submodularity, and joint replenishment games. Oper. Res. **60**(1), 128–137 (2012)
16. Hochbaum, D.S.: Submodular problems-approximations and algorithms. arXiv:1010.1945 [cs] (2010)
17. Iwata, S.: A faster scaling algorithm for minimizing submodular functions. SIAM J. Comput. **32**(4), 833–840 (2003)
18. Iwata, S., Fleischer, L., Fujishige, S.: A combinatorial strongly polynomial algorithm for minimizing submodular functions. J. ACM **48**(4), 761–777 (2001)
19. Kapralov, M., Post, I., Vondrák, J.: Online submodular welfare maximization: greedy is optimal. In: SIAM, pp. 1216–1225 (2013)
20. Karp, R.M., Vazirani, U.V., Vazirani, V.V.: An optimal algorithm for on-line bipartite matching. In: Proceedings of the Twenty-Second Annual ACM Symposium on Theory of Computing, pp. 352–358 (1990)
21. Li, Q., Yu, P.: Multimodularity and its applications in three stochastic dynamic inventory problems. Manuf. Serv. Oper. Manag. **16**(3), 455–463 (2014)
22. Li, Q., Yu, P., Wu, X.: Managing perishable inventories in retailing: replenishment, clearance sales, and segregation. Oper. Res. **64**(6), 1270–1284 (2016)
23. Murota, K.: Discrete Convex Analysis. SIAM, Philadelphia (2003)
24. Murota, K.: Note on multimodularity and L-convexity. Math. Oper. Res. **30**(3), 658–661 (2005)
25. Schrijver, A.: A combinatorial algorithm minimizing submodular functions in strongly polynomial time. J. Combin. Theory Ser. B **80**(2), 346–355 (2000)
26. Shioura, A.: Algorithms for L-convex function minimization: connection between discrete convex analysis and other research fields. J. Oper. Res. Soc. Jpn. **60**(3), 216–243 (2017)
27. Sun, P., Wang, K., Zipkin, P.: Quadratic approximation of cost functions in lost sales and perishable inventory control problems, Working paper (2015)
28. Sviridenko, M.: A note on maximizing a submodular set function subject to a knapsack constraint. Oper. Res. Lett. **32**(1), 41–43 (2004)
29. Topkis, D.M.: Supermodularity and Complementarity. Princeton University, Princeton (2011)
30. Vondrák, J.: Submodularity in combinatorial optimization. PhD thesis, Charles University, Charles (2007)
31. Vygen, J.: A note on Schrijver's submodular function minimization algorithm. J. Combin. Theory Ser. B **88**(2), 399–402 (2003)
32. Wang, Y., Wong, S.C.-W.: Matroid online bipartite matching and vertex cover. In: Proceedings of the 2016 ACM Conference on Economics and Computation, pp. 437–454 (2016)
33. Zipkin, P.: On the Structure of Lost-Sales Inventory Models. Oper. Res. **56**(4), 937–944 (2008)

Thresholding Methods for Streaming Submodular Maximization with a Cardinality Constraint and Its Variants

Ruiqi Yang, Dachuan Xu, Min Li, and Yicheng Xu

Abstract Constrained submodular maximization (CSM) is widely used in numerous data mining and machine learning applications such as data summarization, network monitoring, exemplar-clustering, and nonparametric learning. The CSM can be described as: Given a ground set, a specified constraint, and a submodular set function defined on the power set of the ground set, the goal is to select a subset that satisfies the constraint such that the function value is maximized. Generally, the CSM is NP-hard, and cardinality constrained submodular maximization is well researched. The greedy algorithm and its variants have good performance guarantees for constrained submodular maximization. When dealing with large input scenario, it is usually formulated as streaming constrained submodular maximization (SCSM), and the classical greedy algorithm is usually inapplicable. The streaming model uses a limited memory to extract a small fraction of items at any given point of time such that the specified constraint is satisfied, and good performance guarantees are also maintained. In this chapter, we list the up-to-date popular algorithms for streaming submodular maximization with cardinality constraint and its variants, and summarize some problems in streaming submodular maximization that are still open.

R. Yang · D. Xu
Department of Operations Research and Scientific Computing, Beijing University of Technology, Beijing, People's Republic of China
e-mail: yangruiqi@emails.bjut.edu.cn; xudc@bjut.edu.cn

M. Li (✉)
School of Mathematics and Statistics, Shandong Normal University, Jinan, People's Republic of China
e-mail: liminemily@sdnu.edu.cn

Y. Xu
Shenzhen Institutes of Advanced Technology, Chinese Academy of Sciences, Shenzhen, People's Republic of China
e-mail: yc.xu@siat.ac.cn

© Springer Nature Switzerland AG 2019
D.-Z. Du et al. (eds.), *Nonlinear Combinatorial Optimization*,
Springer Optimization and Its Applications 147,
https://doi.org/10.1007/978-3-030-16194-1_5

123

1 Introduction

Submodular maximization (SM) is widely used in machine learning, such as data summarization [2], budget allocation [1, 23], recommender systems [8, 19], and nonparametric learning [13]. Submodular maximization with constraints like cardinality, knapsack, and matroid gives rise to a lot of interest [11, 17, 21, 22]. In constrained submodular maximization (CSM), we are given a ground set V of size n, a specified constraint ξ, and a submodular set function $f : 2^V \rightarrow R_+$. The goal is to choose a set $S \in \xi$ such that $f(S)$ is maximized, i.e.,

$$\max_{S \subseteq V, S \in \xi} f(S).$$

Let ξ be a cardinality constraint, that is, to say, choose a subset $S \subseteq V$ with $|S| \leq k$ such that $f(S)$ is maximized, where k is an given integer. The submodular maximization with a cardinality constraint is generally NP-hard proofed by Feige [11]. Under the assumption of monotonicity and non-negativity, Nemhauser et al. [21] give a simple and effective greedy algorithm by choosing a marginal gain maximum item with respect to current solution in each iteration. The greedy algorithm obtains a $(1 - e^{-1})$-approximation, and the performance guarantee is tight in several classes of monotone submodular functions which is showed by Feige [11] and Nemhauser and Wolsey [20]. The marginal gain can be defined as follows. In streaming submodular maximization with cardinality constraint (SSMCC), either the items arrive in a stream fashion or the ground set cannot fit in the main memory, which leads to great loss of efficacy. Badanidiyuru et al. [2] first measure the performance of streaming algorithms by four parameters as follows:

- the #passes, the number of the algorithm called over the stream;
- the memory, bounded by $\max_t |M_t|$, where M_t is the main memory at time t;
- the running time, defined as the number of oracle queries make;
- the approximation ratio, the ratio of the final solution set value over the optimum.

The rest of this chapter is organized as follows.

In Section 2, we introduce the sieve-streaming algorithms presented by Badanidiyuru et al. [2], who give the first effective streaming algorithm by carefully choosing threshold value for SSMCC.

In Section 3, we investigate the variants of SSMCC. In particular, we review the algorithms introduced by Epasto et al. [10], who consider a more generalized model that defined as submodular maximization with a cardinality constraint over sliding windows (SMCCSW) in Section 3.1. In Section 3.2, following from the work of Mitrović et al. [19], who consider a streaming robust submodular maximization (SRSM), we recommend the partitioned threshold approach. In Section 3.3, we present the streaming algorithms for dynamic deletion-robust submodular maximization (DDRSM), which was defined by Mirzasoleiman et al. [18]. Motivated by the "right to be forgotten," in their scenario, the item set generates at a fast pace and in real time, at any time point, there is an arrived subset and a deletion subset. The goal is to decide a series of solution sets over time.

In Section 4, we firstly introduce the elegant guarantees on the performance of greedy for maximizing nonsubmodular function which are measured by submodular ratio and generalized curvature combined by Bian et al. [3]. Secondly, We restate a streaming algorithm for maximizing weakly submodular introduced by Elenberg et al. [9].

In Section 5, we restate some other applications of thresholding methods for streaming constrained submodular maximization including streaming submodular maximization with a knapsack [14], d-knapsack [25], and p-matchoid constraint [4, 12], just to name a few.

2 Thresholding Methods for Streaming Submodular Maximization with Cardinality Constraint

SSMCC can be formally defined as follows. A ground streaming set $V = \{e_1, \ldots, e_n\}$ is ordered in an arbitrary manner, nonnegative monotone submodular function $f : 2^V \to R+$, at each iteration t, any algorithm may derive a candidate set $M_t \subseteq V$, and be ready to output a solution set $S_t \subseteq M_t$ with $|S_t| \leq k$. For any $S \subseteq V, e \in V$, define $\Delta_f(e|S) = f(S \cup e) - f(S)$ as the *marginal gain* of item e to set S. Badanidiyuru et al. [2] present their sieve-streaming algorithms by three parts which depend on if we know any information of the optimal solution.

2.1 Sieve-Streaming in Case the Optimum is Known

Let *opt* be the optimum, and assume we know *opt* in advance. Set v is an α-approximation of optimum with $v \in [\alpha opt, opt]$, where $\alpha \in [0, 1]$ is a constant. Let $T = \frac{\beta v - f(S_{t-1})}{k - |S_{t-1}|}$ be the threshold value, where $\beta(> 0)$ is a parameter and S_{t-1} is the solution set just before encountering e_t in the stream. Then the streaming algorithm can be restated as Algorithm 1.

Algorithm 1 Sieve-streaming-know-opt

Input: $v \in [\alpha opt, opt]$, $\beta \in (0, 1)$
1: $S_0 \leftarrow \emptyset, t = 1$
2: **if** $\Delta_f(e_t|S_{t-1}) \geq T$ and $|S_t| \leq k$ **then**
3:      $S_t \leftarrow S_{t-1} + e_t$
4:      $t \leftarrow t + 1$
5: **end if**
6: return S_n

Lemma 1 ([2]) *For any iteration t of Algorithm 1, we have*

$$f(S_t) \geq \frac{\beta v |S_t|}{k}.$$

Proof We show the proof by induction.

- Base case, the lemma obviously holds for $t = 0$, i.e., $S_t = \emptyset$.
- Inductions step. Assume the lemma holds for S_{t-1}, i.e.,

$$f(S_{t-1}) \geq \frac{\beta v |S_{t-1}|}{k}.$$

At time t, item e_t arrives, and is added to S_{t-1}, we have

$$f(S_{t-1} \cup e_t) - f(S_{t-1}) \geq \frac{\beta v - f(S_{t-1})}{k - |S_{t-1}|},$$

where the inequality follows from Algorithm 1. Then

$$
\begin{aligned}
f(S_t) = f(S_{t-1} \cup e_t) &\geq (1 - \frac{1}{k - |S_{t-1}|}) f(S_{t-1}) + \frac{\beta v}{k - |S_{t-1}|} \\
&\geq (1 - \frac{1}{k - |S_{t-1}|}) \frac{\beta v |S_{t-1}|}{k} + \frac{\beta v}{k - |S_{t-1}|} \\
&= \frac{\beta v (|S_{t-1} + 1|)}{k} \\
&= \frac{\beta v (|S_t|)}{k},
\end{aligned}
$$

where the second inequality is obtained by the assumption in induction step.

Theorem 1 ([2]) *At the end of Algorithm 1, we have*

$$f(S_n) \geq \max\{\alpha\beta, 1 - \beta\}opt.$$

Proof

- Case 1. If $S_n = k$, we have

$$f(S_n) \geq \alpha\beta opt,$$

which follows from Lemma 1.
- Case 2. If $S_n < k$, we assume $OPT \setminus S_n = \{o_1, \ldots, o_l\}$, let $OPT_i = \{o_1, \ldots, o_i\}$ for any $i \in [l]$, and set S_i be the current solution set just encountering o_i for any $i \in [l]$. By Algorithm 1,

$$\Delta_f(o_i | S_i) < \frac{\beta v - f(S_i)}{k - |S_i|} \leq \frac{\beta v}{k}.$$

Then

$$opt - f(S_n) \leq f(S_n \cup OPT) - f(S_n)$$

$$= \sum_{i=1}^{l} (f(S_n \cup OPT_i) - f(S_n \cup OPT_{i-1}))$$

$$\leq \sum_{i=1}^{l} (f(S_i \cup o_i) - f(S_i))$$

$$\leq \frac{l\beta v}{k}$$

$$\leq \beta opt,$$

where the first inequality follows from the monotonicity. Thus,

$$f(S_n) \geq (1 - \beta)opt.$$

Corollary 1 ([2]) *For any $\epsilon > 0$, $\alpha = 1 - \epsilon$, and $\beta = 0.5$, Algorithm 1 makes one pass over the stream, and keeps $(0.5 - \epsilon)$-approximation, while the memory is bounded by $O(k)$ and the running time is $O(n)$, for SSMCC.*

2.2 Sieve-Streaming in Case We Know the Maximum

If we know $m = \max_{e \in V} f(e)$, we can guess the *opt* by geometric series in candidate set $O = \{(1 + \epsilon)^i | i \in Z, (1 + \epsilon)^i \in [m, km]\}$, while guessed value just sacrifices a small factor. For any candidate v, let $T = \frac{\beta v - f(S_{t-1}^v)}{k - |S_{t-1}^v|}$, where S_{t-1}^v be the solution set just before encountering e_t according to v. The algorithm can be redescribed as follows (see Algorithm 2).

Algorithm 2 Sieve-streaming-know-maximum

Input: $m = \max_{e \in V} f(e)$;
1: $O \leftarrow \{(1 + \epsilon)^i | i \in Z, (1 + \epsilon)^i \in [m, km]\}$;
2: $S_0^v \leftarrow \emptyset, t \leftarrow 0$;
3: **for** each $v \in O$ **do**
4:     **if** $\Delta_f(e_t | S_{t-1}^v) \geq \frac{\beta v - f(S_{t-1}^v)}{k - |S_{t-1}^v|}$ and $|S_t^v| \leq k$ **then**
5:         $S_t^v \leftarrow S_{t-1}^v + e_t$;
6:         $t \leftarrow t + 1$;
7:     **end if**
8: **end for**
9: return $\arg \max_{v \in O} f(S_t^v)$.

Theorem 2 ([2]) *If the maximum singleton value is known in advance, Algorithm 2 keeps the same approximation ratio, while the memory increases to $O(\epsilon^{-1} k \log k\epsilon)$, and the running time increases to $O(\epsilon^{-1} n \log k)$.*

2.3 Sieve-Streaming in Case Nothing Is Known

In fact, there is a simple two passes presented by Algorithm 2, because one can obtain the maximum singleton value m by making one pass over the data set. But, they give an estimate of the maximum singleton value m on the fly, within one pass over the data set. The algorithm can be restated as Algorithm 3.

Theorem 3 ([2]) *Algorithm 3 makes one pass over the stream, and keeps $(0.5 - \epsilon)$-approximation, while the memory is bounded by $O(\epsilon^{-1} k \log k)$, and the running time is at most $O(\epsilon^{-1} n \log k)$.*

Algorithm 3 Sieve-streaming

Input: $O \leftarrow \{(1 + \epsilon)^i | i \in Z\}$;
 1: **for** each $v \in O$ **do**
 2:     $S_0^v \leftarrow \emptyset, t \leftarrow 0, m \leftarrow 0$;
 3: **end for**
 4: $m = \max\{m, f(e_t)\}$;
 5: $O_t \leftarrow \{(1 + \epsilon)^i | (1 + \epsilon)^i | \in [m, 2km]\}$;
 6: Delete all S_t^v such that $v \notin O_t$;
 7: **for** each $v \in O_t$ **do**
 8:     **if** $\Delta_f(e_t | S_{t-1}^v) \geq \frac{\beta v - f(S_{t-1}^v)}{k - |S_{t-1}^v|}$ and $|S_t^v| \leq k$ **then**
 9:         $S_t^v \leftarrow S_{t-1}^v + e_t$;
10:         $t \leftarrow t + 1$;
11:     **end if**
12: **end for**
13: return arg $\max_{v \in O_n} f(S_t^v)$.

---

3 Thresholding Algorithms for Variants of Streaming Submodular Maximization with Cardinality Constraint

3.1 An $(0.333 - \epsilon)$ Approximation Algorithm for Submodular Maximization over Sliding Windows

Epasto et al. [10] consider a submodular maximization with cardinality constraint over the last W items in the stream which is formally defined as the *submodular maximization with cardinality constraint over sliding window* (SMCCSW). In this model, let $W \in Z$ be the size of the sliding window. For any time t, define *active* window A_t, which contains the last W items in the stream. The goal is to choose a subset $S_t \subseteq A_t$ that satisfy constraint ξ such that $f(S_t)$ is as possible as close to $f(OPT(A_t, \xi))$, where $OPT(A_t, \xi)$ is an optimal solution over active window A_t. They firstly introduce a streaming algorithm (SA) for sub-stream by enumerating thresholds. Secondly they give a submodular smooth histograms algorithm (SSHA) to deal with the sliding setting. They present the first non-trivial submodular smooth histograms algorithm which keeps a constant approximation of the optimum, while the memory is bounded by sublinear in the size of the window for SMCCSW. First, we restate their streaming algorithm as follows.

3.1.1 Stream Algorithm for Streaming Submodular Maximization with Cardinality Constraint

---

Algorithm 4 SA(ϵ, η)

Input: $\epsilon \in (0, 1), \eta = \lfloor \frac{\log_{1+\epsilon} \frac{2km}{f(e_1)}}{} \rfloor$;

1: $O \leftarrow \{\frac{f(e_1)}{2k}, \frac{(1+\epsilon)f(e_1)}{2k}, \ldots, \frac{(1+\epsilon)^\eta f(e_1)}{2k}\}$;
2: $S_0^T \leftarrow \emptyset, t \leftarrow 0$;
3: **for** each $T \in O$ **do**
4:     **if** $\Delta_f(e_t | S_{t-1}^T) \geq T$ and $|S_t^T| \leq k$ **then**
5:         $S_t^T \leftarrow S_{t-1}^T + e_t$;
6:         $t \leftarrow t + 1$;
7:     **end if**
8: **end for**
9: return $\arg\max_{T \in O} f(S_n^T)$.

Theorem 4 ([10]) *For any $k' \in [k]$, we have*

$$f(ALG(V, k)) \geq \frac{(1-\epsilon)k}{k+k'} f(OPT(V, k')),$$

where ALG(V, k) is the returned set by SA on stream V.

Proof By submodularity, we have

$$f(OPT(V, k')) \in [m, k'm].$$

Then

$$\frac{f(OPT(V, k'))}{k + k'} \in \left[\frac{f(e_1)}{2k}, \frac{m}{2}\right].$$

Thus, we have a threshold value T_0 such that

$$T_0 \in \left[\frac{(1-\epsilon)f(OPT(V,k'))}{k+k'}, \frac{f(OPT(V,k'))}{k+k'} \right]. \qquad (1)$$

- Case 1. If $|S_n^{T_0}| = k$, then

$$f(ALG(V,k)) \geq f(S_n^{T_0}) \geq kT_0 \geq \frac{(1-\epsilon)k}{k+k'} f(OPT(V,k')).$$

- Case 2. If $|S_n^{T_0}| < k$, consider an item e_t that was not chosen in $S_n^{T_0}$. Then

$$\Delta_f(e_t|S_{t-1}^{T_0}) < T_0,$$

where $S_{t-1}^{T_0}$ is the solution set of S^{T_0} just before encountering e_t. Then

$$f(OPT(V,k')) - f(S_n^{T_0}) \leq \sum_{e \in OPT(V,k') \backslash S_n^{T_0}} \Delta_f(e|S_n^{T_0})$$

$$\leq \sum_{e \in OPT(V,k') \backslash S_n^{T_0}} \Delta_f(e|S_{t-1}^{T_0})$$

$$< |OPT(V,k') \backslash S_{t-1}^{T_0}|T_0$$

$$\leq \frac{k'}{k+k'} f(OPT(V,k')),$$

where the first two inequalities follow from submodularity, and the last inequality is obtained by inequality (1).

Then

$$f(ALG(V,k)) \geq f(S_n^{T_0}) \geq \frac{k}{k+k'} f(OPT(V,k')).$$

3.1.2 A Submodular Smooth Histograms Algorithm for Streaming Submodular Maximization over Sliding Windows

Following from Algorithm 5, they have one important observation as following lemma.

Lemma 2 ([10, 25]) *For any time t and $i \in [s]$, one of the following two properties always holds:*

1. $x_{i+1} = x_i + 1$;
2. there exists some $t' \le t$, *such that*

$$f(ALG([x_{i+2}, t'], k) \ge (1 - \theta)f(ALG([x_i, t'], k).$$

The main result is introduced as the following theorem.

Theorem 5 ([10]) *For any* $\epsilon > 0$ *and* $\theta = O(\epsilon)$, *Algorithm 5 keeps* $(0.333 - \epsilon)$-*approximation which uses* $O(\epsilon^{-2}k\log^2(k\Phi))$ *memory and has* $O(\epsilon^{-2}\log^2(k\Phi))$ *update time per element, for SMCCSW, where* Φ *is the ratio of maximum to minimum singleton values.*

Algorithm 5 Submodular smooth histograms algorithm for SMSW

Input: $k \in Z^+$, parameter $\beta \in (0, 1)$, window size W;
1: Initialize $s \leftarrow 0$;
2: **for** t=1,... **do**
3:    $s \leftarrow s + 1$;
4:    $x_s \leftarrow t$;
5:    **if** $x_2 < t - W + 1$ **then**
6:       Delete index x_1, and shift other indexes accordingly;
7:       $s \leftarrow s - 1$;
8:    **end if**
9:    **for** $i = 1, \ldots, s - 1$ **do**
10:       Process from x_i to x_s by Algorithm 4;
11:    **end for**
12:    **while** there exist $i \in [s - 1]$, $f(ALG([x_{i+2}, x_s], k) \ge (1 - \theta)f(ALG([x_i, x_s], k)$ **do**
13:       Delete x_{i+1}, and shift the remaining indexes accordingly;
14:       $s \leftarrow s - 1$;
15:    **end while**
16:    **if** $x_1 = \max\{1, t - W + 1\}$ **then**
17:       $S_1 = ALG([x_1, x_s], k)$;
18:    **else**
19:       $S_2 = ALG([x_2, x_s], k)$.
20:    **end if**
21: **end for**

3.2 A Partitioned Thresholding Approach for Streaming Robust Submodular Maximization (SRSW)

Mitrović et al. [19] consider a robust-streaming submodular maximization subject to a cardinality constraint (RSSMCC). In their scenario, the items arrive in a stream fashion, and τ items may be deleted from the algorithm's memory after the stream finished. The goal is to choose a *robust* subset form the stream. We redefined the robust set as follows.

Definition 1 ([19]) A set $S \subseteq V$, $D \subseteq V$ with $|D| \leq \tau$, S is *robust* for a parameter τ if there exists a subset $Z \subseteq S \setminus D$ of size at most k such that

$$f(Z) \geq cf(OPT(V \setminus D, k)),$$

where $OPT(V \setminus D, k)$ is an optimal solution of size k of $V \setminus D$ and $c > 0$ is a constant.

They propose a two-stage procedure for RSSMCC. They call the first stage as (STAR-T), restated as Algorithm 6. Let ρ is an π-approximation of $f(OPT(V \setminus D, k))$. In the second stage (STAR-T-GREEDY), they run the greedy algorithm presented by Nemhauser et al. [21] (see Algorithm 7). To simplify analysis, we say bucket $B_{i,j}$ is *full* if $|B_{i,j}| = \min\{k, 2^i\}$ for any partition part i. There are three important observations formally restated by the following lemmas.

Lemma 3 ([19]) *If there is a partition part $i \in S$ such that at least half of its buckets are full, i.e., $|\{B_{i,j} || B_{i,j}| = \min\{k, 2^i\}\}| \geq \frac{wk}{2^{i+1}}$, then*

Algorithm 6 STAR-T for RSSM

Input: Stream $V = \{e_1, e_2 \ldots, \}, k, \omega, \rho$;
1: Initialize $B_{i,j} \leftarrow \emptyset, \forall i \in [0, \lceil logk \rceil], j \in [1, \omega \lceil \frac{k}{2^i} \rceil], t \leftarrow 1$;
2: **for** $i = 0, \ldots, \lceil logk \rceil$ **do**
3:    **for** $j = 1, \ldots, \omega \lceil \frac{k}{2^i} \rceil$ **do**
4:       **if** $|B_{i,j}| < \min\{2^i, k\} \cap \Delta_f(e_t | B_{i,j}) \geq \frac{\rho}{\min\{2^i, k\}}$ **then**
5:          $B_{i,j} \leftarrow B_{i,j} \cup \{e_t\}$;
6:          break: $t \leftarrow t + 1$;
7:       **end if**
8:    **end for**
9: **end for**
10: return $S \leftarrow \cup_{i,j} B_{i,j}$.

Algorithm 7 STAR-T-GREEDY

Input: Set$S$, any query set $D \subseteq S$ with $|D| \leq \tau$
1: $Z_D^V \leftarrow GREEDY(k, S \setminus E)$
2: **return** Z_D^V

$$f(Z_D^V) \geq (1 - e^{-1})(1 - \frac{4\tau}{wk})\rho.$$

Let $B_i = arg \min_{B_{i,j}:|B_{i,j}| < \min\{k, 2^i\}} |B_{i,j} \cap D|$. We have the following lemma.

Lemma 4 ([19]) *If all partition parts in S no more than half of its buckets are full, then*

$$f(Z_D^V) \geq (1 - e^{-1/3})(f(B_{\lceil logk \rceil}) - \frac{4\tau}{wk}\rho).$$

There is another observation for the second case presented by the following lemma.

Lemma 5 ([19]) *If all partition parts in S no more than half of its buckets are full, then*

$$f(Z_D^V) \geq (1 - e^{-1})(f(OPT(V \setminus D, k)) - f(B_{\lceil logk \rceil}) - \rho),$$

where $B_{\lceil logk \rceil}$ is any not full bucket in the last partition part.

Theorem 6 ([19]) *Let $w \geq \lceil 4\tau \lceil \log k \rceil k^{-1} \rceil$, $\pi = \frac{1-e^{-1/3}}{2(1-e^{-1/3})+(1-e^{-1})(1-\lceil logk \rceil^{-1})}$, STAR-T makes one pass over the stream and gets a set S of size at most $O((k + \tau \log k) \log k)$ items. Furthermore, for any given set D with $|D| \leq \tau$, STAR-T-GREEDY returns a set $Z_D^V \in S \setminus D$ with $|Z_D^V| \leq k$ such that*

$$f(Z_D^V) \geq 0.149(1 - \lceil logk \rceil^{-1})f(OPT(V \setminus D, k)).$$

Proof Obviously, the size of S is bounded by $O((k + \tau logk)logk)$. Combining with Lemmas 4 and 5, we have

$$f(Z_D^V) \geq \frac{(1 - e^{-1/3})(1 - e^{-1})}{2 - e^{-1/3} - e^{-1}}(f(OPT(V \setminus D, k)) - (1 + \frac{4\tau}{wk})\rho),$$

where equality holds for $f(B_{\lceil logk \rceil}) = \frac{(1-e^{-1})f(OPT(V \setminus D,k))-((1-e^{-1})-(1-e^{-1/3})\frac{4\tau}{wk})\rho}{2-e^{-1/3}-e^{-1}}$.
Following Lemma 3 and the last inequality, we have

$$f(Z_D^V) \geq \frac{1}{\frac{2}{(1-e^{-1})(1-\frac{4\tau}{wk})} + \frac{1}{1-e^{-1/3}}} f(OPT(V \setminus D, k)).$$

For $w \geq \lceil 4\tau \lceil logk \rceil k^{-1} \rceil$, we have

$$f(Z_D^V) \geq \frac{(1 - e^{-1/3})(1 - e^{-1})(1 - \lceil logk \rceil^{-1})}{2(1 - e^{-1/3}) + (1 - e^{-1})}$$

$$\geq 0.149(1 - \lceil logk \rceil^{-1})f(OPT(V \setminus D, k)).$$

3.3 A Robust-Streaming Algorithm for Deletion-Robust Submodular Maximization

Mirzasoleiman et al. [18] introduce the dynamic deletion-robust submodular maximization problem (DDRSW), in which item set V is generated at a fast pace and real time. At any given time t, a subset $V_t \subseteq V$ is arrived, and a subset D_t is deleted. The model can be restated as follows:

$$OPT_t = \max_{A_t \in I_t} f(A_t)$$

$$s.t. I_t = \{A : A \subseteq V_t \setminus D_t \cap |A| \le k\}.$$

We are given a collection of subsets $\mathscr{S} = \{S_1, \ldots, S_n\}$, where $S_i \subseteq V = \{1, \ldots, n\}$, and utility function $f(A) = |\cup_{i \in A} S_i|, A \subseteq V$. Assume the stream arrives in order $S_1 = \{1, \ldots, n\}$, and then $S_i = \{i\}$ for $i \in [2, n]$. Streaming algorithms based on choosing items by marginal gains will suffer arbitrary badly performance. In the above example, any streaming algorithm may select the item S_1 firstly. Then the other items cannot be chosen since that the corresponding marginal gains are equal to zero. Once the first item is in the deletion set, the stream algorithms fail. To deal with the bad examples, they derive a series of non-overlapping solution sets such that if we suffer a deletion, there is only one solution set gets affected by constructing a cascading chain. They first give two operation sub-procedures as Algorithms 8 and 9. The main robust-streaming algorithm is restated as Algorithm 10.

Theorem 7 *The robust-streaming algorithm can keep the same approximation ratio, while the memory are bounded by $O(rM)$, and the average update time is at most $O(rT)$, where $M(T)$ is the memory (update time) bound of streaming algorithm $\mathscr{A}$ and r is the length of chain.*

Algorithm 8 Add(i, R) procedure
___
Input: Set $i \in [r]$, $R = \{e_1, \ldots, e_{|R|}\}$, Streaming algorithm $\mathscr{A}$;
 1: Initialize $l \leftarrow 1$;
 2: $[R_t^i, M_t^i, S_t^i] \leftarrow \mathscr{A}(e_l)$;
 3: **if** $R_t^i \ne \emptyset$ and $i < r$ **then**
 4:     Add$(i + 1, R_t^i)$;
 5: **end if**
 6: $l \leftarrow l + 1$.

Algorithm 9 Delete(e) procedure
___
 1: Initialize $i \leftarrow 1$;
 2: **if** $e \in M_t^i$ **then**
 3:     $R_t^i \leftarrow M_t^i \setminus e$;
 4:     $M_t^i \leftarrow$ null;
 5:     Add$(i + 1, R_t^i)$;
 6: **end if**
 7: $i \leftarrow i + 1$.

Algorithm 10 The robust-streaming for DDRSW

Input: Stream V_t, deletion set D_t, $r \leq m + 1$;
  1: Initialize $t \leftarrow 1$, $M_t^i \leftarrow 0$, $S_t^i \leftarrow \emptyset$, $\forall i \in [r]$;
  2: **while** $|(V_t \setminus V_{t-1}) \cup (D_t \setminus D_{t-1})| \neq 0$ **do**
  3:   **if** $|D_t \setminus D_{t-1}| \neq 0$ **then**
  4:     $e \leftarrow D_t \setminus D_{t-1}$;
  5:     Delete(e);
  6:   **else**
  7:     $e_t \leftarrow V_t \setminus V_{t-1}$;
  8:     Add($1, e_t$);
  9:   **end if**
 10:   $t \leftarrow t + 1$;
 11:   $S_t \leftarrow \{S_t^i | i = \min\{j \in [r], M_t^j \neq null\}\}$.
 12: **end while**

---

4 Thresholding Algorithms for Nonsubmodular Maximization

There is another important class of nonsubmodular optimization problems including subset selection [7], sparse approximation [6, 16], sparse m-estimation [15], just to name a few. Das and Kemple [7] first introduce *submodularity ratio* to characterize how close a set function is to being submodular. We redefined as follows.

Definition 2 ([7]) For any nonnegative set function f, the *submodularity ratio* is the largest scalar γ such that

$$\gamma \leq \min_{\forall S, T \subseteq V} \frac{\sum_{e \in T \setminus S} \Delta_f(e|S)}{\Delta_f(T|S)}.$$

There is also a definition of curvature, which quantifies how close a set submodular function is to being modular.

Definition 3 ([3, 5, 24]) For any nonnegative set function f, the *curvature* is the smallest scalar λ such that

$$\lambda = 1 - \min_{\forall S, T \subseteq V, e \in S \setminus T} \frac{\Delta_f(e|S \setminus \{e\} \cup T)}{\Delta_f(e|S \setminus \{e\})}.$$

Bian et al. [3] show the performance of the classical greedy algorithm by combining with the curvature and the submodularity ratio. By their technical analysis, there are strong theoretical guarantees on the performance of greedy algorithm for nonsubmodular maximization. The main results can be restated as follows.

Theorem 8 ([3]) *For any nonnegative nondecreasing set function f, its submodularity ratio is $\gamma \in [0, 1]$ and curvature is $\lambda \in [0, 1]$. Then the classical greedy is a $\lambda^{-1}(1 - e^{\lambda\gamma})$-approximation.*

Elenberg et al. [9] introduce a threshold-greedy algorithm for streaming weakly submodular maximization (SWSM). We first redefine the weakly submodular function.

Definition 4 ([9]) Any given monotone set function f is γ'-weakly submodular for an integer ν if

$$\gamma' \leq \min_{S,T \subseteq V : |S| \leq \nu, |T \setminus S| \leq \nu} \frac{\sum_{e \in T \setminus S} \Delta_f(e|S)}{\Delta_f(T|S)}.$$

Let $\rho' \in [0, \frac{\gamma'(\sqrt{2 - e^{-\gamma'/2}} - 1)}{2} f(OPT(V, k))]$, where $f(OPT(V, k))$ is assumed known in advance. The threshold-greedy algorithm for streaming γ'-weakly submodular maximization (γ'-$SWSM$) is restated as follows (see Algorithm 11).

Theorem 9 ([9]) *The streak algorithm gives at least $((1 - \epsilon)\frac{\gamma'}{2}(3 - e^{\gamma'/2} - 2\sqrt{2 - e^{-\gamma'/2}}))$-approximation, while the memory is bounded by $O(\epsilon^{-1}k\log k)$.*

Algorithm 11 Threshold-greedy (ρ') for γ'-$SWSM$

Input: Stream $V \leftarrow \{e_1, \ldots, e_n\}$, an integer k, ρ';
1: Initialize $t \leftarrow 1$, $S_0 \leftarrow \emptyset$;
2: **if** $\Delta_f(e_t|S_{t-1}) \geq \frac{\rho'}{k}$ and $|S_{t-1}| < k$ **then**
3: $\quad S_t \leftarrow S_{t-1} + e_t$;
4: $\quad t \leftarrow t + 1$;
5: **end if**
6: return S_t.

Following from that the $OPT(V, k)$ is not known in advance, they present a streak algorithm which does not depend on ρ'. The main results can be restated as the following theorem.

Algorithm 12 Streak for γ'-$SWSM$

Input: Stream $V \leftarrow \{e_1, \ldots, e_n\}$, an integer k, $\epsilon > 0$;
1: Initialize $t \leftarrow 1$, $S_0 \leftarrow \emptyset$, $m \leftarrow 0$;
2: $m \leftarrow \max\{m, f(e_t)\}$;
3: $O^t \leftarrow \{(1 - \epsilon)^i | i \in Z, (1 - \epsilon)^i \in [\frac{m(1-\epsilon)}{9k^2}, mk]\}$;
4: **for** each $\rho' \in O^t$ **do**
5: $\quad$ process Threshold-greedy (ρ');
6: **end for**
7: $t \leftarrow t + 1$;
8: return S_t.

---

5 Summarizations of Thresholding Methods for Other Submodular Maximization Applications

Huang et al. [14] consider the streaming submodular maximization with a knapsack constraint (SMKC). In their model, item arrives in streaming type, and each item e has size $c(e)$, and the goal is to choose a subset $S \subseteq V$ with $\sum_{e \in S} c(e) \leq B$, where B is a budget given in advance. Huang et al. give a simple marginal ratio threshold $(0.333 - \epsilon)$-approximation. In the end, they improve the ratio to 0.363 by the technical analysis. Simultaneously, the memory is bounded by $O(K poly(\epsilon^{-1}) polylogk)$. Wang et al. [25] consider a more general streaming submodular maximization with d-knapsack constraint (SSMd-KC), and they present a knapsack-stream algorithm, which is a $((1 + d)^{-1} - \epsilon)$-approximation, for SSM$d$-KC.

Theorem 10 ([12]) *The sample-streaming is a $(4p)^{-1}$-approximation, while the memory is bounded by $O(k)$, and values and independence oracle queries are at most $O(p^{-1}nkq)$, in expectation, where k is the maximum independence set size.*

Feldman et al. [12] consider streaming submodular maximization with a p-matchoid (SSMp-M). They first apply the subsampling technique to SSMp-M. We firstly formally redefined the p-matchoid as follows.

Definition 5 ([4, 12]) A set system (V, I) is *p-matchoid* for an integer p, if there exist q matroids $(V_1, I_1), \ldots, (V_q, I_q)$ with $V = \cup_{i=1}^{l} V_i$, $I = \{S \subseteq 2^V | S \cap V_i \in I_i, \forall i \in [l]\}$ such that each item of V appears at most p out of the above matroids.

Considering the ahead thresholding methods, most of their works assume the streams are in an arbitrary fixed order. Whether one item is added to solution set only relies on its marginal gain. However, the solution sets are obviously dependent on the streaming order. Feldman et al. [12] introduce *marginal contribution* of item e_t to the part set S. Formally we redefine the marginal contribution as follows.

Definition 6 ([12]) Let $e_1, \ldots, e_t$ be the items of V that are ordered by their arrival. For any set $S \subseteq V$, we define $f(e_t : S) = \Delta_f(e_t | S \cap \{e_1, \ldots, e_{t-1}\})$ as marginal contribution of item e_t to part set of S.

For any $T \subseteq V$, let $f(T : S) = \sum_{e \in T} f(e : S)$. The main algorithm has an exchange-candidate as sub-procedure introduced as Algorithm 14. The main sample-streaming algorithm is restated as follows (see Algorithm 15). The main results can be restated by the following theorem.

Algorithm 13 Knapsack-stream algorithm for streaming submodular maximization with d-knapsack

Input: Stream $V \leftarrow \{e_1, \ldots, e_n\}$, an integer $W, \epsilon > 0$;
1: Initialize $t \leftarrow 1, m \leftarrow \emptyset, M \leftarrow 0$;
2: $m \leftarrow \max\{m, f(e_t)\}$;
3: **if** $M < \frac{f(e_t)}{\min_{j \in [d]} c_j(e)}$ **then**
4:   $M \leftarrow \frac{f(e_t)}{\min_{j \in [d]} c_j(e)}, L \leftarrow f(e_t)$;
5: **end if**
6: $O^t \leftarrow \{(1+\epsilon)^i | i \in Z, (1+\epsilon)^i \in [L, M(1+d)]\}$;
7: **for** each $\rho'' \in O^t$ **do**
8:   **if** $\frac{\Delta_f(e_t | S_{\rho''})}{c_j(e_t)} \geq \frac{\rho''}{1+d}$ and $c(S_{\rho''}) + c(e_t) \leq W$ **then**
9:     $S_{\rho''} \leftarrow S_{\rho''} + e_t$;
10:   **end if**
11: **end for**
12: $S \leftarrow arg \max_{\rho'' \in O^t} f(S_{\rho''})$;
13: **if** $f(S) \geq f(e_m)$ **then**
14:   $S_t \leftarrow S$;
15:   $t \leftarrow t + 1$;
16: **else**
17:   $S_t \leftarrow \{e_m\}$;
18:   $t \leftarrow t + 1$.
19: **end if**

Algorithm 14 Exchange-candidate

Input: S_{t-1}, e_t;
1: **for** l=1:q **do**
2:   **if** $S_{t-1} \cap V_l \notin I_l$ **then**
3:     $C_l \leftarrow \{e \in S_{t-1} | (S_{t-1} - e + e_t) \cap V_l \in I_l\}$;
4:     $c_l \leftarrow arg \min_{e \in C_l} f(e : S_{t-1})$;
5:     $U_t \leftarrow U_{t-1} + c_l$;
6:   **end if**
7: **end for**
8: return U_t.

Algorithm 15 Sample-streaming for $SSMM_p$

1: $S_0 \leftarrow \emptyset, t \leftarrow 1$;
2: with probability $\frac{1}{2p+1}$;
3: $U_t \leftarrow$ Exchange-Candidate (S_{t-1}, e_t);
4: **if** $\Delta_f(e_t | S_{t-1}) \geq 2f(U_t : S_{t-1})$ **then**
5:   $S_t \leftarrow S_{t-1} \setminus U_t + e_t$;
6: **end if**
7: $t \leftarrow t + 1$;
8: return S_n.

Acknowledgements The first two authors are supported by National Natural Science Foundation of China (Nos. 11531014, 11871081). The third author is supported by the Higher Educational Science and Technology Program of Shandong Province (No. J17KA171). The fourth author is supported by China Postdoctoral Science Foundation funded project (No. 2018M643233) and National Natural Science Foundation of China (Nos.61433012, U1435215).

References

1. Alon, N., Gamzu, I., Tennenholtz, M.: Optimizing budget allocation among channels and influencers. In: Proceedings of the 21st International Conference on World Wide Web, pp. 381–388. ACM, New York (2012)
2. Badanidiyuru, A., Mirzasoleiman, B., Karbasi, A., Krause, A.: Streaming submodular maximization: massive data summarization on the fly. In: Proceedings of the 20th ACM SIGKDD International Conference on Knowledge Discovery and Data Mining, pp. 671–680. ACM, New York (2014)
3. Bian, A.A., Buhmann, J.M., Krause, A., Tschiatschek, S.: Guarantees for greedy maximization of non-submodular functions with applications. In: Proceedings of the 34th International Conference on Machine Learning, pp. 498–507. ACM, New York (2017)
4. Chakrabarti, A., Kale, S.: Submodular maximization meets streaming: matchings, matroids, and more. Math. Program. **154**(1–2), 225–247 (2015)
5. Conforti, M., Cornuéjols, G.: Submodular set functions, matroids and the greedy algorithm: tight worst-case bounds and some generalizations of the Rado-Edmonds theorem. Discrete Appl. Math. **7**(3), 251–274 (1984)
6. Das, A., Kempe, D.: Algorithms for subset selection in linear regression. In: Proceedings of the 40th Annual ACM Symposium on Theory of Computing, pp. 45–54. ACM, New York (2008)
7. Das, A., Kempe, D.: Submodular meets spectral: greedy algorithms for subset selection, sparse approximation and dictionary selection. In: Proceedings of the 28th International Conference on Machine Learning, pp. 1057–1064. ACM, New York (2011)
8. El-Arini, K., Guestrin, C.: Beyond keyword search: discovering relevant scientific literature. In: Proceedings of the 17th ACM SIGKDD International Conference on Knowledge Discovery and Data Mining, pp. 439–447. ACM, New York (2011)
9. Elenberg, E., Dimakis, A.G., Feldman, M., Karbasi, A.: Streaming weak submodularity: interpreting neural networks on the fly. In: Advances in Neural Information Processing Systems, pp. 4044–4054. MIT Press, Cambridge (2017)
10. Epasto, A., Lattanzi, S., Vassilvitskii, S., Zadimoghaddam, M.: Submodular optimization over sliding windows. In: Proceedings of the 26th International Conference on World Wide Web, pp. 421–430. ACM, New York (2017)
11. Feige, U.: A threshold of ln n for approximating set cover. J. ACM **25**(4), 634–652 (1998)
12. Feldman, M., Karbasi, A., Kazemi, E.: Do less, get more: streaming submodular maximization with subsampling (2018). Preprint. arXiv:1802.07098
13. Gomes, R., Krause, A.: Budgeted nonparametric learning from data streams. In: Proceedings of the 27th International Conference on Machine Learning, pp. 391–398. ACM, New York (2010)
14. Huang, C.C., Kakimura, N., Yoshida, Y.: Streaming algorithms for maximizing monotone submodular functions under a knapsack constraint. In: The 20th International Workshop on Approximation Algorithms for Combinatorial Optimization Problems, APPROX and the 21st International Workshop on Randomization and Computation, RANDOM (2017)
15. Jain, P., Tewari, A., Kar, P.: On iterative hard thresholding methods for high-dimensional m-estimation. In: Advances in Neural Information Processing Systems, pp. 685–693. MIT Press, Cambridge (2014)

16. Krause, A., Cevher, V.: Submodular dictionary selection for sparse representation. In: Proceedings of the 27th International Conference on Machine Learning, pp. 567–574. ACM, New York (2010)
17. Lin, H., Bilmes, J.: Multi-document summarization via budgeted maximization of submodular functions. In: Proceedings of the 2010 Annual Conference of the North American Chapter of the Association for Computational Linguistics, pp. 912–920. ACM, New York (2011)
18. Mirzasoleiman, B., Karbasi, A., Krause, A.: Deletion-robust submodular maximization: Data summarization with "the right to be forgotten". In: Proceedings of the 34th International Conference on Machine Learning, pp. 2449–2458. ACM, New York (2017)
19. Mitrović, S., Bogunovic, I., Norouzi-Fard, A., Tarnawski, J.: Streaming robust submodular maximization: a partitioned thresholding approach. In: Advances in Neural Information Processing Systems, pp. 4560–4569. MIT Press, Cambridge (2017)
20. Nemhauser, G.L., Wolsey, L.A.: Best algorithms for approximating the maximum of a submodular set function. Math. Oper. Res. 3(3), 177–188 (1978)
21. Nemhauser, G.L., Wolsey, L.A., Fisher, M.L.: An analysis of approximations for maximizing submodular set functions—I. Math. Program. 14(1), 265–294 (1978)
22. Segui-Gasco, P., Shin, H.S.: Fast non-monotone submodular maximisation subject to a matroid constraint (2017). Preprint. arXiv:1703.06053
23. Soma, T., Kakimura, N., Inaba, K., Kawarabayashi, K.: Optimal budget allocation: theoretical guarantee and efficient algorithm. In: Proceedings of the 31st International Conference on International Conference on Machine Learning, pp. 351–359. ACM, New York (2014)
24. Vondrák, J.: Submodularity and curvature: the optimal algorithm (combinatorial optimization and discrete algorithms). RIMS Kokyuroku Bessatsu (2010)
25. Wang, Y., Li, Y., Tan, K.L.: Efficient streaming algorithms for submodular maximization with multi-knapsack constraints (2017). Preprint. arXiv:1706.04764

Nonsubmodular Optimization

Weili Wu, Zhao Zhang, and Ding-Zhu Du

Abstract The nonsubmodular optimization is a hot research topic in the study of nonlinear combinatorial optimizations. We discuss several approaches to deal with such optimization problems, including supermodular degree, curvature, algorithms based on DS decomposition, and sandwich method.

1 Introduction

For any set function $f : 2^X \to R$, f is *submodular* if

$$f(A) + f(B) \geq f(A \cup B) + f(A \cap B).$$

f is *monotone nondecreasing* if

$$A \subset B \text{ implies } f(A) \leq f(B).$$

In the literature, there are many beautiful results on monotone nondecreasing submodular optimizations [3, 18, 24, 28, 29, 31, 34, 40] and nonmonotone submodular optimizations [14, 17, 23, 35]. However, in recent development of computer technology, many nonsubmodular optimization problems appear, such as sentiment analysis [2], cloud computing [12], machine learning [42], and social networks (such as viral marketing for complementary products [20], composed influence [45], and misinformation blocking [36]). Therefore, the study of nonsubmodular optimization becomes a hot research subject.

W. Wu · D.-Z. Du (✉)
Department of Computer Science, The University of Texas at Dallas, Richardson, TX, USA
e-mail: weiliwu@utdallas.edu; dzdu@utdallas.edu

Z. Zhang
Department of Computer Science, Zhejiang Normal University, Jinhua, Zhejiang, China

© Springer Nature Switzerland AG 2019   141
D.-Z. Du et al. (eds.), *Nonlinear Combinatorial Optimization*,
Springer Optimization and Its Applications 147,
https://doi.org/10.1007/978-3-030-16194-1_6

There are two classes of approaches to deal with nonsubmodular optimization problems. The first class consists of the traditional ones. Since the nonsubmodular optimization is, in nature, hard to deal with, we cannot find an efficient algorithm with satisfied guaranteed performance. In this class, the algorithm is often analyzed with some artificial parameter, such as the supermodular degree [13, 15, 16] or curvature [1, 8, 22, 38], with which some beautiful results on the approximation performance ratio may be established. However, those parameters are usually hard to be estimated and in many specific problems, they do not have a significant value.

Due to the above, one may give up the approximation performance ratio and establish other standards to evaluate the performance of algorithms. The second class of algorithms is designed based on this point. We may name two groups here.

The first group consists of local optimal algorithms. They are usually asked to terminate at a solution satisfying certain necessary optimality condition, such as local optimality conditions. We may find two local optimality conditions in the literature, which give two subgroups of algorithms for DS function optimizations [21, 27] and discrete DC function optimization [26, 43], respectively.

The second group consists of data-dependent approximation algorithms, which are evaluated by a new type of performance ratio. Those algorithms are also called sandwich methods [4, 25, 39, 45].

In this article, we briefly introduce the above-mentioned methods.

2 Supermodular Degree

Consider a set function $f : 2^X \to R$. The supermodular degree of an element $u \in X$ by a function f is defined to be $|\mathscr{D}^+(u)|$ where

$$\mathscr{D}_f^+(u) = \{v \in X \mid \exists A \subseteq X : \Delta_u f(A \cup \{v\}) > \Delta_u f(A)\}.$$

The supermodular degree of function f is defined by

$$\mathscr{D}_f^+ = \max_{u \in X} |\mathscr{D}^+(u)|.$$

When only function f is studied on supermodular degree, we may simply write $\mathscr{D}^+ = \mathscr{D}_f^+$.

With the supermodular degree, several theoretical results have been obtained [13, 15] in the literature.

The first one is for the monotone nonsubmodular maximization with matroid constraints as follows:

$$\max \ f(A)$$

$$\text{subject to} \ \ A \in \mathscr{C}_i \text{ for } i = 1, 2, \ldots, k,$$

where f is nonnegative and monotone nondecreasing and $(X, \mathscr{C}_i)$ is a matroid for $i = 1, 2, \ldots, k$. Consider the following greedy algorithm:

Greedy Algorithm 2.1

input:A monotone nondecreasing nonnegative set function $f : 2^X \to R$
    and k matroids $(X, \mathscr{C}_i)$;
$S_0 \leftarrow \emptyset$;
$i \leftarrow 0$;
while S_i is not a base **do**
    $i \leftarrow i + 1$;
    choose $u_i \in X \setminus S_{i-1}$ and $D_{i-1} \in \mathscr{D}^+(u_i)$ to maximize
    $f(D_i \cup \{u_i\} \cup S_{i-1}) - f(S_{i-1})$ subject to $D_i \cup \{u_i\} \cup S_{i-1} \in \cap_{i=1}^k \mathscr{C}_i$;
    set $S_i \leftarrow D_i \cup \{u_i\} \cup S_{i-1}$;
output S_i.

For this algorithm, there is the following theorem.

Theorem 1 (Feldman and Izsak [15]) *Greedy Algorithm 2.1 produces a* $\frac{1}{k(\mathscr{D}^+ +1)+1}$-*approximation solution for maximization of monotone nondecreasing nonnegative set function with k matroid constraints.*

Let $\mathscr{C} = \{A \mid |A| \leq k\}$. Then $(X, \mathscr{C})$ is a matroid. This means that the size constraint is a specific matroid constraint. With this constraint, the monotone nonsubmodular maximization has a better approximation solution.

Consider a maximization problem and a greedy algorithm as follows:

$$\max \ f(A)$$

$$\text{subject to} \ |A| \leq k,$$

where $f : 2^X \to R$ is nonnegative and monotone nondecreasing.

Greedy Algorithm 2.2

input a monotone nondecreasing nonnegative function $f : 2^X \to R$;
$S \leftarrow \emptyset$;
for $d = 1$ **to** $\mathscr{D}^+$, $v \in X$ and C with $|C| = k \mod (d + 1)$ **do**
    $S_0 \leftarrow C$;
    $i \leftarrow 0$;
    while S_i is not a base **do**
        $i \leftarrow i + 1$;
        choose $u_i \in X \setminus S_{i-1}$ and $D_i \subseteq \mathscr{D}^+(u_i)$ to maximize
        $f(D_i \cup \{u_i\} \cup S_{i-1}) - f(S_{i-1})$ subject to $|D_I \cup \{u_i\} \cup S_{i-1}| \leq k$
        and $|D_i| \leq d$;
        set $S_i \leftarrow D_i \cup \{u_i\} \cup S_{i-1}$;
    Sargmax$(f(S), f(S_i))$;
output S.

This algorithm produces a better approximation.

Theorem 2 (Feldman and Izsak [15]) *Greedy Algorithm 2.2 produces a* $(1 - e^{-1/(\mathscr{D}^+ +1)})$*-approximation solution for maximization of monotone nondecreasing nonnegative set function with a size constraint.*

The supermodular degree has been successfully applied to a few optimization problems, such as [16].

3 Curvature

There are several different definitions for curvature in the literature for study of different optimization problems, such as submodular minimization with submodular cover constraints [8, 38], submodular maximization with submodular knapsack constraint [22], and BP functions optimization [1]. In this section, we focus on one of them, the submodular minimization with submodular cover constraint.

Actually, this problem is closely related to the nonsubmodular minimization problem. By DS decomposition theorem (see Theorem 4), every set function f can be represented as $f = g - h$, where g and h are monotone nondecreasing submodular functions. From this representation, it is easy to know that min f can be solved through solving a sequence of problems $\min\{g - c \mid h \geq c\}$ for every discrete constant c.

Specifically, consider the following problem:

$$\min \ g(A)$$

$$\text{subject to} \ \ h(A) \geq b,$$

where g and h are monotone nondecreasing submodular functions on 2^X with $g(\emptyset) = h(\emptyset) = 0$, called *polymatroid functions*, and b is a constant. Define

$$\Delta_x g(A) = g(A \cup \{x\}) - g(A) \ \text{and} \ \Delta_x h(A) = h(A \cup \{x\}) - h(A).$$

Consider the following greedy algorithm.

Greedy Algorithm 3.1
input polymatroid functions g and h on 2^X, and a constant b.
$S \leftarrow \emptyset$;
while $h(S) < h$ **do**
      choose x to maximize $\frac{\Delta_x h(S)}{\Delta_x g(S)}$
      and set $S \leftarrow S \cup \{x\}$;
output S.

Define the curvature of g by

$$\chi(g) = \min_{A \subseteq X} \frac{\sum_{x \in A} g(\{x\})}{g(A)}.$$

Then the following holds.

Theorem 3 (Wan et al. [38]) *Greedy Algorithm 3.1 produces an approximation solution with performance ratio at most $\chi(g)H(\gamma)$, where*

$$\gamma = \max_{x \in X} h(\{x\})$$

and $H(\gamma) = \sum_{i=1}^{\gamma} 1/i$.

An interesting application of this result was given in [8]. Consider a wireless sensor network. Each sensor has a communication disk and a sensing disk with itself as common center. If sensor s_1 lies in the communication disk of s_2, then s_1 can receive message from s_2. When all sensors have the same size of communication disks and the same size of sensing disks, they are said to be homogeneous. In a homogeneous wireless sensor system, the communication network is a undirected graph, in which a virtual backbone is a connected dominating set, that is, it is a node subset such that every node is either in the subset or adjacent to the subset. Construction of the virtual backbone is an important issue in the study of wireless sensor networks [5, 19, 30, 32, 33, 37, 41]. Motivated from reducing routing cost and improving load balancing, routing-cost constraint was introduced in the construction [6]. Thus, many efforts have been made on routing-cost constrained construction problems [7, 9–11].

Let $G = (V, E)$ be a homogeneous wireless sensor network and D a subset of nodes. For each pair of nodes u and v, let $d(u, v)$ denote the shortest distance between u and v in G, i.e., the minimum number of edges on a path between u and v, and $d_D(u, v)$ the shortest distance between u and v in the subgraph induced by $D \cup \{u, v\}$. Let us study the following problem.

Routing-cost Constrained CDS: Given a homogeneous wireless sensor network $G = (V, E)$, find a minimum connected dominating set D such that

$$d_D(u, v) \leq \alpha d(u, v), \forall u, v \in V, \tag{1}$$

where α is a constant.

First, we would like to indicate that this problem can be formulated into a generalized hitting set problem as follows.

Generalized Hitting Set: Given m nonempty collections $\mathscr{C}_1, \mathscr{C}_2, \ldots, \mathscr{C}_m$ of subsets of a finite set X, find the minimum subset A of X such that every C_i has a member $S \subseteq A$.

Ding et al. [6] proved that to satisfy constraint (1), it is sufficient to satisfy

$$d_D(u, v) \leq \alpha + 1, \forall u, v \in X \text{ with } d(u, v) = 2. \tag{2}$$

For every pair of nodes u and v, let $\mathscr{C}_{uv}$ denote the collections of node subsets each of which is the set of intermediate nodes on a path between u and v with distance at most $\alpha + 1$. Let D be a node set hitting every $\mathscr{C}_{uv}$. Then D would satisfy constraint (2) and hence (1). This would imply that D is a connected dominating set. Thus, the routing-cost constrained CDS problem is equivalent to the generalized hitting set problem with input collections $\mathscr{C}_{uv}$.

We next reduce the generalized hitting set problem to a problem of submodular minimization with submodular cover constraint. To do so, let $\mathscr{C} = \cup_{u,v:d(u,v)=2} \mathscr{C}$. For every subcollection $\mathscr{A} \subseteq \mathscr{C}$, define

$$g(\mathscr{A}) = |\cup_{A \in \mathscr{A}} A|$$

and

$$h(\mathscr{A}) = |\{\mathscr{C}_{uv} \mid \mathscr{A} \cap \mathscr{C}_{u,v} \neq \emptyset\}|.$$

It is not hard to prove that g and h are monotone nondecreasing submodular functions with $g(\emptyset) = h(\emptyset) = 0$. Moreover, the generalized hitting set problem is equivalent to following

$$\min \ g(A)$$
$$\text{subject to} \ \ h(A) \geq f(\mathscr{C}),$$
$$\mathscr{A} \subseteq \mathscr{C}.$$

This equivalence means that $\mathscr{A}$ is a minimum solution of this problem if and only if $\cup_{A \in \mathscr{A}} A$ is the minimum solution of the generalized hitting set problem. It can be proved as follows.

Suppose $\mathscr{A}$ is the minimum solution of the above problem of submodular minimization with submodular cover constraint. For contradiction, suppose $\cup_{A \in \mathscr{A}} A$ is not a minimum generalized hitting set. Consider a minimum generalized hitting set D. Then $|D| < |\cup_{A \in \mathscr{A}} A|$. For each $\mathscr{C}_{uv}$, let C_{uv} be a subset of D, contained in $\mathscr{C}_{uv}$. Denote

$$\mathscr{C}_D = \{C_{uv} \mid u, v \in V \text{ with } d(u,v) = 2\}.$$

Then $h(\mathscr{C}_D) = h(\mathscr{C})$ and $g(\mathscr{C}_D) \leq |D| < g(\mathscr{A})$, a contradiction.

Conversely, suppose $\cup_{A \in \mathscr{A}} A$ is a minimum generalized hitting set. For contradiction, suppose $\mathscr{A}$ is not a minimum solution for the above problem of submodular minimization with submodular cover constraint. Consider a minimum solution $\mathscr{B}$ for it. Then $g(\mathscr{B}) < g(\mathscr{A})$. By the above argument, $\cup_{B \in \mathscr{B}} B$ is a generalized hitting set such that

$$|\cup_{B \in \mathscr{B}} B| = g(\mathscr{B}) < g(\mathscr{A}) = |\cup_{A \in \mathscr{A}} A|,$$

a contradiction.

Du et al. [8] proved that in this example, $\chi(g)$ and γ are constants and hence Greedy Algorithm 3.1 gives a constant-approximation for the routing-cost constrained CDS problem.

4 Local Optimality

There are two necessary conditions for minimality:

1. Let f be a set function on 2^X. Suppose A is a minimum solution of f in 2^X. Then $f(A) \leq f(A \setminus \{x\})$ and $f(A) \leq f(A \cup \{x\})$ for any $x \in X$.
2. Let $f = g - h$ be a set function and g and h submodular functions on the subsets of X. If set A is a minimum solution for $\min_{Y \subseteq X} f(Y)$, then $\partial h(A) \subseteq \partial g(A)$.

Condition 1 is obvious. Condition 2 needs a little explanation. First, let us explain what is the notation $\partial h(A)$. $\partial h(A)$ is the subgradient of function h at set A, defined as

$$\partial h(A) = \{c \in R^X \mid h(Y) \geq h(A) + \; <c, Y - A>\}.$$

Actually, for a submodular set function $h : 2^X \to R$, the subgradient at set A consists of all linear functions $c : X \to R$ satisfying $h(Y) \geq h(A) + c(Y) - c(A)$, where $c(Y) = \sum_{y \in Y} c(y)$. Each linear function c can also be seen as a vector in R^X, i.e., a vector c with components labeled by elements in X. The characteristic vector of each subset Y of X is a vector in $\{0, 1\}^X$ such that the component with label $x \in X$ is equal to 1 if and only if $x \in Y$. Here, for simplicity of notation, we use the same notation Y to represent the set Y and its characteristic vector.

To see condition 2, note that since A is a minimum solution for $\min_{Y \subseteq X} f(Y)$, we have $f(A) \leq f(Y)$ and hence $g(Y) - g(A) \geq h(Y) - h(A)$ for any $Y \subseteq X$. Therefore, for any $c \in \partial h(A)$, $g(Y) - g(A) \geq h(Y) - h(A) \geq c(Y) - c(A)$. This means that $\partial h(A) \subseteq \partial g(A)$.

Condition 2 requires that the set function f can be represented as a difference of two submodular functions. Actually, such a representation is available for any set function.

Theorem 4 (DS Decomposition [21, 27]) *Every set function $f : 2^X \to R$ can be expressed as the difference of two monotone nondecreasing submodular functions g snd h, i.e., $f = g - h$, where X is a finite set.*

The proof of this theorem is constructive [21], however not in polynomial-time. It is still an open problem whether there exists or not a polynomial-time algorithm to find a pair of submodular functions h and g for any given set function f such that $f = h - g$.

An important result follows from Theorem 4 is the following.

Theorem 5 (Sandwich Theorem [44]) *For any set function $f : 2^X \to R$ and any set $Y \subseteq X$, there are two modular functions $m_u : 2^X \to R$ and $m_l : 2^X \to R$ such that $m_u \geq f \geq m_l$ and $m_u(Y) = f(Y) = m_l(Y)$.*

There exist several algorithms in the literature, based on DS decomposition, such as the submodular–supermodular algorithm [27], the modular–modular algorithm [21], and the iterated sandwich algorithm [44]. Following is an example.

Submodular–Supermodular Algorithm for min $f(A)$:

- **Input** a set function $f : 2^X \to R$.
- Initially, compute a DS decomposition $f = g - h$ and choose an arbitrary set $A \subseteq X$.
- At each iteration, carry out following

    – Compute a modular lower bound m_{hl} for h such that $h(A) = m_{hl}(A)$.
    – Compute a minimum solution A^+ for $g - m_{hl}$.
    – If $f(A^+) = f(A)$, then stop iteration and go to **output**; else set $A \leftarrow A^+$ and start a new iteration.

- **Output** A.

In the above algorithm, how to compute modular lower bound m_{hl} for h? Following is the answer.

Lemma 1 (Iyer and Bilmes [27]) *For any submodular function $f : 2^X \to R$ and any set $Y \subseteq X$, there exists a modular function $m_l : 2^X \to R$ such that $f \geq m_l$ and $f(Y) = m_l(Y)$.*

Proof Put all elements of X into an ordering $X = \{x_1, x_2, \ldots, x_n\}$ such that $Y = \{x_1, x_2, \ldots, x_{|Y|}\}$. Denote $S_i = \{x_1, x_2, \ldots, x_i\}$. Define $m_l(\emptyset) = f(\emptyset)$ and for $\emptyset \neq A \subseteq X$, define

$$m_l(A) = f(\emptyset) + \sum_{x_i \in A} (f(S_i) - f(S_{i-1})).$$

Clearly m_l is modular and

$$m_l(Y) = f(\emptyset) + \sum_{x_i \in Y} (f(S_i) - f(S_{i-1})) = f(Y).$$

Moreover, for any set $A \subseteq X$ with $A \neq \emptyset$, suppose $A = \{x_{i_1}, x_{i_2}, \ldots, x_{i_k}\}$ and then we have

$$
\begin{aligned}
m_l(A) &= f(\emptyset) + (f(S_{i_1}) - f(S_{i_1-1})) + (f(S_{i_2}) - f(S_{i_2-1})) \\
&\quad + \cdots + (f(S_{i_k}) - f(S_{i_k-1})) \\
&\leq f(\emptyset) + (f(\{x_{i_1}\}) - f(\emptyset)) + (f(\{x_{i_1}, x_{i_2}\}) - f(\{x_1\})) \\
&\quad + \cdots + (f(A) - f(\{x_{i_1}, \ldots x_{i_{k-1}}\})) \\
&= f(A).
\end{aligned}
$$

From the above construction of modular lower bound, we have found the following theorem.

Theorem 6 (Iyer and Bilmes [27]) *In the submodular–supermodular algorithm, compute m_{hl} by using the permutation of elements of X. At each iteration, try at most n permutations $\sigma_1, \ldots, \sigma_k$ such that $A = \{\sigma_1(|A| - 1), \ldots, \sigma_k(|A| - 1)\}$ and $X \setminus A = \{\sigma_1(|A| + 1), \ldots, \sigma_k(|A|+)\}$. Then the submodular–supermodular algorithm would stop at a local minimum satisfying condition 1.*

5 Data-Dependent Approximation

A typical data-dependent approximation algorithm is the sandwich method, which has been used frequently for solving nonsubmodular optimization problems [4, 25, 36, 39, 45]. It runs as follows for the minimization problem.
Sandwich Method:

- **Input** a set function $f : 2^X \to R$ and a collection Ω of subsets of X. The goal is to solve $\min_{A \in \Omega} f(A)$.
- Initially, find two submodular functions u and l such that $u(A) \geq f(A) \geq l(A)$ for $A \in \Omega$. Then carry out the following:

  - Compute a α-approximation solution S_u for $\min_{A \in Omega} u(A)$ and a β-approximation solution S_l for $\min_{A \in \Omega} l(A)$.
  - Compute a greedy solution S_o for $\min_{A \in \Omega} f(A)$.
  - Set $S = \mathrm{argmin}(f(S_u), f(S_o), f(S_l))$.

- **Output** S.

  This method has the following guaranteed performance.

Theorem 7 (Lu et al. [25]) *The solution S produced by the sandwich method satisfies the following:*

$$f(S) \leq \min \left\{ \frac{f(S_1)}{l(S_l)} \cdot \beta, \frac{opt_u}{opt_f} \cdot \alpha \right\} \cdot opt_f,$$

where opt_f (opt_u) is the objective function value of the minimum solution for $\min_{A \in \Omega} f(A)$ ($\min_{A \in \Omega} u(A)$).

Proof Since S_l is a β-approximation solution for $\min_{A \in \Omega} l(A)$, we have

$$f(S_l) = \frac{f(S_l)}{l(S_l)} \cdot l(S_l) \leq \frac{f(S_l)}{l(S_l)} \cdot \beta \cdot opt_l \leq \frac{f(S_l)}{l(S_l)} \cdot \beta \cdot l(OPT_f) \leq \frac{f(S_l)}{l(S_l)} \cdot \beta \cdot opt_f,$$

where OPT_f is an optimal solution for $\min_{A \in \Omega} f(A)$. Since S_u is an α-approximation solution for $\min_{A \in \Omega} u(A)$, we have

$$f(S_u) \leq u(S_u) \leq \alpha \cdot opt_u = \alpha \cdot \frac{opt_u}{opt_f} \cdot opt_f.$$

Therefore, the theorem holds.

From theoretical point of view, the sandwich method is always applicable for the set function optimization since we have the following.

Theorem 8 *For any set function f on 2^X, there exist two monotone nondecreasing submodular functions u and l such that $u(A) \geq f(A) \geq l(A)$ for every $A \in 2^X$.*

Proof By the DS decomposition theorem, there exist two monotone nondecreasing submodular functions g and h such that $f = g - h$. Note that for every $A \in 2^X$, $h(\emptyset) \leq h(A) \leq h(X)$. Set $u(A) = g(A) - h(\emptyset)$ and $l(A) = g(A) - h(X)$ for any $A \in 2^X$. Then u and l meet our requirement.

However, in the real world, it is often not easy to find easily computable upper bound u and lower bound l, especially in case that the DS decomposition is not found. Therefore, many efforts on specific design of sandwich methods are still significant.

References

1. Bai, W., Bilmes, J.A.: Greed is still good: maximizing monotone submodular+ supermodular functions (2018). Preprint. arXiv:1801.07413
2. Barhan, A., Shakhomirov, A.: Methods for sentiment analysis of twitter messages. In: 12th Conference of FRUCT Association (2012)
3. Calinescu, G., Chekuri, C., Pál, M., Vondrák, J.: Maximizing a monotone submodular function subject to a matroid constraint. SIAM J. Comput. **40**(6), 1740–1766 (2011)
4. Chen, W., Lin, T., Tan, Z., Zhao, M., Zhou, X.: Robust influence maximization. In: Proceedings of the 22nd ACM SIGKDD International Conference on Knowledge Discovery and Data Mining, pp. 795–804. ACM, New York (2016)
5. Cheng, X., Huang, X., Li, D., Wu, W., Du, D.-Z.: A polynomial-time approximation scheme for the minimum-connected dominating set in ad hoc wireless networks. Netw. Int. J. **42**(4), 202–208 (2003)
6. Ding, L., Gao, X., Wu, W., Lee, W., Zhu, X., Du, D.-Z.: Distributed construction of connected dominating sets with minimum routing cost in wireless networks. In: 2010 IEEE 30th International Conference on Distributed Computing Systems (ICDCS), pp. 448–457. IEEE, Piscataway (2010)
7. Ding, L., Wu, W., Willson, J.K., Du, H., Lee, W.: Construction of directional virtual backbones with minimum routing cost in wireless networks. In: INFOCOM, 2011 Proceedings IEEE, pp. 1557–1565. IEEE, Piscataway (2011)
8. Du, H., Wu, W., Lee, W., Liu, Q., Zhang, Z., Du, D.-Z.: On minimum submodular cover with submodular cost. J. Glob. Optim. **50**(2), 229–234 (2011)

9. Du, H., Ye, Q., Wu, W., Lee, W., Li, D., Du, D., Howard, S.: Constant approximation for virtual backbone construction with guaranteed routing cost in wireless sensor networks. In: INFOCOM, 2011 Proceedings IEEE, pp. 1737–1744. IEEE, Piscataway (2011)

10. Du, H., Ye, Q., Zhong, J., Wang, Y., Lee, W., Park, H.: Polynomial-time approximation scheme for minimum connected dominating set under routing cost constraint in wireless sensor networks. Theor. Comput. Sci. **447**, 38–43 (2012)

11. Du, H., Wu, W., Ye, Q., Li, D., Lee, W., Xu, X.: CDS-based virtual backbone construction with guaranteed routing cost in wireless sensor networks. IEEE Trans. Parallel Distrib. Syst. **24**(4), 652–661 (2013)

12. Edmonds, J., Pruhs, K.: Scalably scheduling processes with arbitrary speedup curves. In: Proceedings of the twentieth Annual ACM-SIAM Symposium on Discrete Algorithms, pp. 685–692. SIAM, Philadelphia (2009)

13. Feige, U., Izsak, R.: Welfare maximization and the supermodular degree. In: Proceedings of the 4th conference on Innovations in Theoretical Computer Science, pp. 247–256. ACM, New York (2013)

14. Feige, U., Mirrokni, V.S., Vondrak, J.: Maximizing non-monotone submodular functions. SIAM J. Comput. **40**(4), 1133–1153 (2011)

15. Feldman, M., Izsak, R.: Constrained monotone function maximization and the supermodular degree. (2014). Preprint. arXiv:1407.6328

16. Feldman, M., Izsak, R.: Building a good team: secretary problems and the supermodular degree. In: Proceedings of the Twenty-Eighth Annual ACM-SIAM Symposium on Discrete Algorithms, pp. 1651–1670. SIAM, Philadelphia (2017)

17. Feldman, M., Naor, J., Schwartz, R.: A unified continuous greedy algorithm for submodular maximization. In: 2011 IEEE 52nd Annual Symposium on Foundations of Computer Science (FOCS), pp. 570–579. IEEE, Piscataway (2011)

18. Fisher, M.L., Nemhauser, G.L., Wolsey, L.A.: An analysis of approximations for maximizing submodular set functions—II. In: Polyhedral Combinatorics, pp. 73–87. Springer, Berlin (1978)

19. Guha, S., Khuller, S.: Approximation algorithms for connected dominating sets. Algorithmica **20**(4), 374–387 (1998)

20. Guo, J., Wu, W. Viral marketing with complementary products. In: Du, D.-Z., Pardalos, P.M., Zhang, Z. (eds.) Nonlinear Combinatorial Optimization. Springer International Publishing, Cham (2019)

21. Iyer, R., Bilmes, J.: Algorithms for approximate minimization of the difference between submodular functions, with applications. (2012). Preprint. arXiv:1207.0560

22. Iyer, R.K., Bilmes, J.A.: Submodular optimization with submodular cover and submodular knapsack constraints. In: Advances in Neural Information Processing Systems, pp. 2436–2444 (2013)

23. Lee, J., Mirrokni, V.S., Nagarajan, V., Sviridenko, M.: Non-monotone submodular maximization under matroid and knapsack constraints. In: Proceedings of the Forty-First Annual ACM Symposium on Theory of Computing, pp. 323–332. ACM, New York (2009)

24. Leskovec, J., Krause, A., Guestrin, C., Faloutsos, C., VanBriesen, J., Glance, N.: Cost-effective outbreak detection in networks. In: Proceedings of the 13th ACM SIGKDD International Conference on Knowledge Discovery and Data Mining, pp. 420–429. ACM, New York (2007)

25. Lu, W., Chen, W., Lakshmanan, L.V.S.: From competition to complementarity: comparative influence diffusion and maximization. Proc. VLDB Endowment **9**(2), 60–71 (2015)

26. Maehara, T., Murota, K.: A framework of discrete dc programming by discrete convex analysis. Math. Program. **152**(1–2), 435–466 (2015)

27. Narasimhan, M., Bilmes, J.A.: A submodular-supermodular procedure with applications to discriminative structure learning (2012). Preprint. arXiv:1207.1404

28. Nemhauser, G.L., Wolsey, L.A., Fisher, M.L.: An analysis of approximations for maximizing submodular set functions I. Math. Program. **14**(1), 265–294 (1978)

29. Orlin, J.B.: A faster strongly polynomial time algorithm for submodular function minimization. Math. Program. **118**(2), 237–251 (2009)

30. Ruan, L., Du, H., Jia, X., Wu, W., Li, Y., Ko, K.-I.: A greedy approximation for minimum connected dominating sets. Theor. Comput. Sci. **329**(1–3), 325–330 (2004)
31. Schrijver, A.: A combinatorial algorithm minimizing submodular functions in strongly polynomial time. J. Comb. Theory Ser. B **80**(2), 346–355 (2000)
32. Sivakumar, R., Das, B., Bharghavan, V.: An improved spine-based infrastructure for routing in ad hoc networks. In: IEEE Symposium on Computers and Communications, vol., 98 (1998)
33. Stojmenovic, I., Seddigh, M., Zunic, J.: Dominating sets and neighbor elimination-based broadcasting algorithms in wireless networks. IEEE Trans. Parallel Distrib. Syst. **13**(1), 14–25 (2002)
34. Sviridenko, M.: A note on maximizing a submodular set function subject to a knapsack constraint. Oper. Res. Lett. **32**(1), 41–43 (2004)
35. Svitkina, Z., Fleischer, L.: Submodular approximation: sampling-based algorithms and lower bounds. SIAM J. Comput. **40**(6), 1715–1737 (2011)
36. Tong, A., Du, D.-Z., Wu, W.: On misinformation containment in online social networks. In: Advances in Neural Information Processing Systems, pp. 339–349 (2018)
37. Wan, P.-J., Alzoubi, K.M., Frieder, O.: Distributed construction of connected dominating set in wireless ad hoc networks. In: INFOCOM 2002. Twenty-First Annual Joint Conference of the IEEE Computer and Communications Societies. Proceedings. IEEE, vol. 3, pp. 1597–1604. IEEE, Piscataway (2002)
38. Wan, P.-J., Du, D.-Z., Pardalos, P., Wu, W.: Greedy approximations for minimum submodular cover with submodular cost. Comput. Optim. Appl. **45**(2), 463–474 (2010)
39. Wang, Z., Yang, Y., Pei, J., Chu, L., Chen, E.: Activity maximization by effective information diffusion in social networks. IEEE Trans. Knowl. Data Eng. **29**(11), 2374–2387 (2017)
40. Wolsey, L.A.: An analysis of the greedy algorithm for the submodular set covering problem. Combinatorica **2**(4), 385–393 (1982)
41. Wu, J., Li, H.: On calculating connected dominating set for efficient routing in ad hoc wireless networks. In: Proceedings of the 3rd International Workshop on Discrete Algorithms and Methods for Mobile Computing and Communications, pp. 7–14. ACM, New York (1999)
42. Wu, B., Lyu, S., Ghanem, B.: Constrained submodular minimization for missing labels and class imbalance in multi-label learning. In: AAAI, pp. 2229–2236 (2016)
43. Wu, C., Wang, Y., Lu, Z., Pardalos, P.M., Xu, D., Zhang, Z., Du, D.-Z.: Solving the degree-concentrated fault-tolerant spanning subgraph problem by DC programming. Math. Program. **169**(1), 255–275 (2018)
44. Wu, W., Zhang, Z., Du, D.Z.: Set function optimization. J. Oper. Res. China (to appear)
45. Zhu, J., Zhu, J., Ghosh, S., Wu, W., Yuan, J.: Social influence maximization in hypergraph in social networks. IEEE Trans. Netw. Sci. Eng. (2018)

On Block-Structured Integer Programming and Its Applications

Lin Chen

Abstract Integer programming in a variable dimension is a crucial research topic that has received a considerable attention in recent years. A series of fixed parameter tractable (FPT) algorithms have been developed for a variety of integer programming that has a special block structure, and such results were later applied successfully in many classical combinatorial optimization problems to derive FPT or approximation algorithms. From a theoretical point of view, it is important to understand the overall landscape, and distinguish the structures of integer programming that are tractable vs. intractable or unknown so far. From the application point of view, it is important to understand how the structure of such integer programming is related to the structure of concrete combinatorial optimization problems. The goal of this survey is to summarize recent progress in theory and application of integer programming that has a block structure and point to important open problems in this research direction.

1 Introduction

The NP-hardness of integer programming (IP), in general, is well-known decades ago [19]. Consequently, much effort has been devoted to the search for tractable special cases. Famous polynomially solvable cases are IPs with few rows and small coefficients as shown by Papadimitriou in 1981 [26], and IPs with few variables as shown by Lenstra in 1983 [23]. In recent years, many researches target at identifying tractable special cases of IP in variable dimensions, and arguably the most significant development towards this direction is the introduction of *iterative augmentation* methods which has led to the development of fast algorithms for wide classes of IPs whose constraint matrix has a special block structure, and to subsequent breakthrough applications in parameterized and approximation

L. Chen (✉)
Department of Computer Science, University of Houston, Houston, TX, USA

© Springer Nature Switzerland AG 2019

D.-Z. Du et al. (eds.), *Nonlinear Combinatorial Optimization*,
Springer Optimization and Its Applications 147,
https://doi.org/10.1007/978-3-030-16194-1_7

153

algorithms for various combinatorial optimization problems (see, e.g., [3, 18, 21]), which will be discussed in detail in this survey.

On a very high level, it has been shown, in a series of fundamental researches, that an IP can be solved in a polynomial oracle time given an oracle, which, roughly speaking, augments an existing solution by an amount no less than that of any augmentation in the direction of the so-called *Graver basis*, which is one basis of the integer kernel of the IP (see, e.g., the book [25], we will also provide details in the later part of this survey). Unfortunately, finding out the augmentation required by the oracle can be difficult, particularly as there can be an exponential number of different Graver basis elements, and each Graver basis element may be exponentially large. Recent researches (see, e.g., [3, 13–15, 21]) showed that, if the given IP has a special structure, particularly if it has some specific block structure, then it may admit a "nice" Graver basis in the sense that the Graver basis element may have a small ℓ_1- or ℓ_∞-norm. By utilizing such a fact, it may be possible to find out the augmentation required by the oracle in polynomial time, and consequently solve the given IP in polynomial time. In fact, essentially all the current known polynomial time algorithms along this research line are based on such a high-level idea. It is thus a fundamental question how far we can carry over such an idea. In this survey, we will review recent progress in this research direction, and point out several open problems where such an idea encounters difficulty.

This survey will also cover a wide range of applications of these block-structured IPs in concrete combinatorial optimization problems. In particular, we will review recent progress in applying block-structured IP in the research area of scheduling, computational social choice, string matching, etc. These applications, on the one hand, give a good motivation for us to further investigate into block-structured IP, and highlight the importance of those fundamental results. On the other hand, the research in these specific combinatorial optimization problems may also give us inspirations on further generalizations of block-structured IP.

The paper is organized as follows: In Section 2, we will provide preliminaries. In Sections 3 and 4, we review the algorithmic results for block-structured IPs with linear and non-linear objective functions, respectively. We also remark on relevant open problems at the end of each corresponding subsection. In Section 5, we review the applications of block-structured IP in combinatorial optimization problems. We conclude the survey with open problems in Section 6.

2 Preliminaries

2.1 Block-Structured Integer Programs

We consider the following general form of an integer program (IP):

$$(\text{IP})_{n,\mathbf{b},\mathbf{l},\mathbf{u},\mathbf{w}}: \quad \min(\text{or max}) \quad \{\phi(\mathbf{x}) : H\mathbf{x} = \mathbf{b}, \ \mathbf{l} \le \mathbf{x} \le \mathbf{u}, \ \mathbf{x} \in \mathbb{Z}^N\}. \tag{1}$$

Here the objective ϕ is some computable function. We will distinguish into linear functions and non-linear functions, and review relevant research works separately.

Throughout this paper, all the matrices and vectors in the IP, including $H, \mathbf{b}, \mathbf{l}, \mathbf{u}$, are integral. For two vectors $\mathbf{y}, \mathbf{z}$ of the same dimension, we write $\mathbf{y} \cdot \mathbf{z}$, or sometimes $\mathbf{yz}$ for their inner product, if it is clear from the context.

In this survey, we mainly focus on block-structured IP, where the constraint matrix H has a special structure as we elaborate below. In the following A, B, C, D, or A_i that form the constraint matrix H are small submatrices and 0 is a zero matrix.

4-Block n-Fold Matrix

$$H = \begin{pmatrix} C & D \\ B & A \end{pmatrix}^{(n)} := \begin{pmatrix} C & D & D & \cdots & D \\ B & A & 0 & & 0 \\ B & 0 & A & & 0 \\ \vdots & & & \ddots & \\ B & 0 & 0 & & A \end{pmatrix},$$

that is, H consists of one copy of c and n copies of A, B, D where all B's are on the first column, all D's are on the first row, and all A's are on the main diagonal. In particular, if C is a zero matrix, then H is a 3-block n-fold matrix.

n-Fold Matrix An n-fold matrix is a special case of 4-block n-fold matrix in which $B = \cdot$ and $C = \cdot$, i.e.,

$$H - \begin{pmatrix} D & D & D & \cdots & D \\ A & 0 & 0 & & 0 \\ 0 & A & 0 & & 0 \\ 0 & 0 & A & & 0 \\ \vdots & & & \ddots & \\ 0 & 0 & 0 & & A \end{pmatrix}.$$

Two-Stage Stochastic Matrix A two-stage stochastic matrix is a special case of 4-block n-fold matrix in which $C = \cdot$ and $D = \cdot$, i.e.,

$$H = \begin{pmatrix} B & A & 0 & & 0 \\ B & 0 & A & & 0 \\ \vdots & & & \ddots & \\ B & 0 & 0 & & A \end{pmatrix}.$$

Tree-Fold Matrix A tree-fold matrix is a generalization of n-fold matrix. The structure of an n-fold matrix could be viewed as a star with the root representing the row of $(D, D, \cdots, D)$ and each leaf representing one of the rows $(0, \cdots, 0, A, 0, \cdots, 0)$. More precisely, we can view each row i as a vertex i such that vertex i is a parent of vertex j if row i dominates row j, where by saying row i dominates row j, we mean row j is more "sparse" than row i as a vector, i.e., if the k-th coordinate of row j is non-zero, then the k-th coordinate of row i is also non-zero. Using this interpretation, we can generalize an n-fold matrix to a tree-fold matrix. The following is an example:

$$H = \begin{bmatrix}
A_1 & A_1 & A_1 & A_1 & A_1 & A_1 & A_1 & A_1 & A_1 & A_1 & A_1 & A_1 \\
A_2 & A_2 & A_2 & A_2 & A_2 & A_2 & A_2 & A_2 & 0 & 0 & 0 & 0 \\
0 & 0 & 0 & 0 & 0 & 0 & 0 & 0 & A_2 & A_2 & A_2 & A_2 \\
A_3 & A_3 & A_3 & 0 & 0 & 0 & 0 & 0 & 0 & 0 & 0 & 0 \\
0 & 0 & 0 & A_3 & A_3 & 0 & 0 & 0 & 0 & 0 & 0 & 0 \\
0 & 0 & 0 & 0 & 0 & A_3 & A_3 & A_3 & 0 & 0 & 0 & 0 \\
0 & 0 & 0 & 0 & 0 & 0 & 0 & 0 & A_3 & A_3 & A_3 & A_3 \\
A_4 & 0 & 0 & 0 & 0 & 0 & 0 & 0 & 0 & 0 & 0 & 0 \\
0 & A_4 & 0 & 0 & 0 & 0 & 0 & 0 & 0 & 0 & 0 & 0 \\
0 & 0 & A_4 & 0 & 0 & 0 & 0 & 0 & 0 & 0 & 0 & 0 \\
0 & 0 & 0 & A_4 & 0 & 0 & 0 & 0 & 0 & 0 & 0 & 0 \\
0 & 0 & 0 & 0 & A_4 & 0 & 0 & 0 & 0 & 0 & 0 & 0 \\
0 & 0 & 0 & 0 & 0 & A_4 & 0 & 0 & 0 & 0 & 0 & 0 \\
0 & 0 & 0 & 0 & 0 & 0 & A_4 & 0 & 0 & 0 & 0 & 0 \\
0 & 0 & 0 & 0 & 0 & 0 & 0 & A_4 & 0 & 0 & 0 & 0 \\
0 & 0 & 0 & 0 & 0 & 0 & 0 & 0 & A_4 & 0 & 0 & 0 \\
0 & 0 & 0 & 0 & 0 & 0 & 0 & 0 & 0 & A_4 & 0 & 0 \\
0 & 0 & 0 & 0 & 0 & 0 & 0 & 0 & 0 & 0 & A_4 & 0 \\
0 & 0 & 0 & 0 & 0 & 0 & 0 & 0 & 0 & 0 & 0 & A_4
\end{bmatrix}.$$

A tree-representation of the matrix above is:

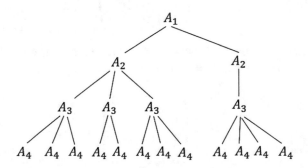

In general, a tree-fold matrix H consists of n copies of small matrices $A_1, A_2, \cdots, A_\tau$ with A_i being a $s_i \times t$-matrix. Every row consists of 0's and some A_i's in the form of $(0, \cdots, 0, A_i, A_i, \cdots, A_i, 0, \cdots, 0)$ (i.e., A_i appears consecutively). Every column consists of 0's and exactly one copy of each A_i. Furthermore, if we call a row containing A_i as an A_i-row, then any A_i-row is dominated by some A_{i-1}-row, that is, if at a certain row A_i appears consecutively from column ℓ to column k, then there exists some A_{i-1}-row such that A_{i-1} appears consecutively from ℓ' to k' such that $\ell' \leq \ell < k \leq k'$. Representing the matrix as a tree, every row is represented as a vertex and the vertex corresponding to each A_{i-1}-row will be the parent of the vertex corresponding to A_i-row it dominates.

Multistage Stochastic Matrix If the transpose H^T is a tree-fold matrix, then H is a multistage stochastic matrix. A multistage stochastic matrix is a generalization of a two-stage stochastic matrix.

An IP where the constraint matrix H belongs to one of the above types is named accordingly after the name of the matrix, e.g., an IP with H being a 4-block n-fold matrix is called a 4-block n-fold IP.

2.2 Graver Basis

Consider the general integer linear programming in the standard form (1). Let $\sqsubseteq$ be the *conformal order* in $\mathbb{R}^m$ defined such that $\mathbf{x} \sqsubseteq \mathbf{y}$ if $\mathbf{x}$ and $\mathbf{y}$ lie in the same orthant, i.e., $x_i \cdot y_i \geq 0$ for each $i = 1, \ldots, m$, and $|x_i| \leq |y_i|$ for each $i = 1, \ldots, m$. Given any subset $X \subseteq \mathbb{R}^n$, we say $\mathbf{x}$ is an $\sqsubseteq$-*minimal* element of X if $\mathbf{x} \in X$ and there does not exist $\mathbf{y} \in X$, $\mathbf{y} \neq \mathbf{x}$ such that $\mathbf{y} \sqsubseteq \mathbf{x}$. It is known that every subset of $\mathbb{Z}^m$ has finitely many $\sqsubseteq$-minimal elements.

Given an $n \times m$ integer matrix E, we denote by $\ker_{\mathbb{Z}}(E) = \{\mathbf{x} \in \mathbb{Z}^m \mid E\mathbf{x} = \mathbf{0}\}$ the *integer kernel of H*. We study the Graver basis:

Definition 1 (Graver Basis [12]) The *Graver basis* of an integer matrix E is the finite set $\mathcal{G}(E) \subseteq \ker_{\mathbb{Z}}(E)$ of all $\sqsubseteq$-minimal elements of $\ker_{\mathbb{Z}}(E) \setminus \{\mathbf{0}\}$.

We use the fact that any $\mathbf{x} \in \ker_{\mathbb{Z}}(E)$, $\mathbf{x} \neq 0$ can be written as $\mathbf{x} = \sum_i \alpha_i \mathbf{g}_i$, where $\alpha_i \in \mathbb{Z}_+$, $\mathbf{g}_i \in \mathcal{G}(E)$, and $\mathbf{g}_i \sqsubseteq \mathbf{x}$ [25, Lemma 3.4].

2.3 An Iterative Augmentation Framework

As we mentioned before, the tractability of a broad class of IP in variable dimension is built on an iterative augmentation framework. Briefly speaking, an algorithm that follows the iterative augmentation framework starts with an initial feasible solution $\mathbf{x}$ and iteratively finds *augmenting steps*, where a solution $\mathbf{x}$ for (1) is called feasible if $H\mathbf{x} = \mathbf{b}$ and $\mathbf{l} \leq \mathbf{x} \leq \mathbf{u}$, and an augmenting step $\mathbf{g} \in \mathbb{Z}^N$ is a vector such that $\mathbf{x} + \mathbf{g}$ is feasible and $\phi(\mathbf{x} + \mathbf{g}) < \phi(\mathbf{wx})$. A major question is *where* to obtain "good" augmenting steps. The *Graver basis of H*, $\mathcal{G}(H)$, has emerged as an excellent choice, with good guarantees on convergence to optimal solutions while still being algorithmically "tame." Specifically, at the heart of iterative augmentation techniques are bounds on the ℓ_1- and ℓ_∞-norm of elements of the Graver basis, which enable dynamic programming to enumerate Graver elements in an efficient way.

The general framework for solving (1) by utilizing Graver basis was developed in a series of papers, see, e.g., [3, 13, 14, 18, 21]. A very recent paper by Koutecký et al. [22] gives a nice summary and formalizes this framework for (1) where the objective function ϕ is linear. For non-linear objective functions, things are more complicated. We will discuss linear and non-linear objectives separately in the following sections.

2.4 Fixed Parameter Tractability

We give a very brief introduction to fixed parameter tractability here and the reader is referred to [8] for details.

A parameterized problem is a language $L \in \Sigma^* \times \mathbb{Z}_{\geq 0}$, where Σ is a finite alphabet. The second component is called the parameter of the problem. A parameterized problem L is fixed parameter tractable (FPT) if the question whether (x, k) is in L can be decided in running time $f(k) \cdot |x|^{O(1)}$ for some computable function f.

For this survey, we will revisit the existing algorithmic results for block-structured IP from the perspective of parameterized algorithms. As we will provide details later, the constraint matrix of a block-structured IP is composed of many copies of some small blocks (submatrices). We will take these small blocks as "parameters", more precisely, we take the number of rows and columns, and the largest absolute value over the entries of the small blocks as parameters. The number of variables or constraints, on the other hand, are *not* part of the parameters. We are interested in algorithms that run in time polynomially in the number of variables and constraints, while an exponential dependency on the parameters (small blocks) is acceptable.

3 Optimizing over a Linear Objective

Throughout this section, we consider linear objectives where $\phi(\mathbf{x}) = \mathbf{w} \cdot \mathbf{x}$. It should be clear that for linear objectives, the maximization problem can be solved by simply minimizing the additive inverse of the objective function. Without loss of generality, we will focus on minimization problem throughout this section, more precisely, we consider the following:

$$\min \quad \{\mathbf{wx} : H\mathbf{x} = \mathbf{b}, \mathbf{l} \leq \mathbf{x} \leq \mathbf{u}, \mathbf{x} \in \mathbb{Z}^N\}, \tag{2}$$

It is worth mentioning that all the results mentioned in this section also hold if we replace the equality $H\mathbf{x} = \mathbf{b}$ in IP (2) with inequalities $H\mathbf{x} \leq \mathbf{b}$ or $H\mathbf{x} \geq \mathbf{b}$ (indeed, inequalities can be handled using the standard technique of introducing slack variables).

3.1 Iterative Augmentation

We define an augmenting step $\mathbf{g}$ and a *step length* $\rho \in \mathbb{Z}$ as an $\mathbf{x}$-*feasible step pair* with respect to a feasible solution $\mathbf{x}$ if $\mathbf{l} \leq \mathbf{x} + \rho\mathbf{g} \leq \mathbf{u}$. An augmenting step $\mathbf{h}$ is a *Graver-best step* for $\mathbf{x}$ if $\mathbf{w}(\mathbf{x} + \mathbf{h}) \leq \mathbf{w}(\mathbf{x} + \rho\mathbf{g})$ for all $\mathbf{x}$-feasible step pairs $(\mathbf{g}, \rho) \in \mathcal{G}(H) \times \mathbb{Z}$. The next definition and theorem show that it is sufficient to focus all our attention on finding Graver-best steps.

Definition 2 (Graver-Best Oracle for IP (2)) A *Graver-best oracle* for IP (2) is one that, queried on $\mathbf{w}, \mathbf{b}, \mathbf{l}, \mathbf{u}$, and $\mathbf{x}$ feasible to (2), returns a Graver-best step $\mathbf{h}$ for $\mathbf{x}$.

Theorem 1 ([13, 14]) *Given a Graver-best oracle for IP* (2), *(1) can be solved with a number of augmentation steps that are polynomially bounded in the encoding lengths of* $H, \mathbf{b}, \mathbf{w}, \mathbf{l}, \mathbf{u}$.

Proof In each iteration step, consider the current solution $\mathbf{x}$ and the optimal solution $\mathbf{x}^*$. Using the property of Graver basis, we know that $\mathbf{x}^* - \mathbf{x} = \sum_{i=1}^{k} \gamma_i \mathbf{g}_i$, where $\gamma_i > 0$ and $\mathbf{g}_i$ are Graver basis elements that lie in the same orthant. As $\mathbf{l} \leq \mathbf{x}^* \leq \mathbf{u}$ and $\mathbf{l} \leq \mathbf{x} \leq \mathbf{u}$, we know $\mathbf{l} \leq \mathbf{x} + \gamma_i \mathbf{g}_i \leq \mathbf{u}$ and hence $\mathbf{x} + \gamma_i \mathbf{g}_i$ is feasible for any i. Moreover, by the integer Carathéodory theorem of [6, 27], we can assume that $k \leq 2N$. Let i be the index of a summand attaining the maximal absolute value, i.e., $i = argmax_i\{|\mathbf{w}\gamma_i\mathbf{g}_i|, 1 \leq i \leq k\}$, and let $\mathbf{h}$ be a step returned by the Graver-best oracle for $\mathbf{x}$, then it follows that

$$|\mathbf{w}(\mathbf{x} + \mathbf{h}) - \mathbf{w}\mathbf{x}| \geq |\mathbf{w}(\mathbf{x} + \gamma_i \mathbf{g}_i) - \mathbf{w}\mathbf{x}| \geq \frac{1}{2N}|\mathbf{w}\mathbf{x}^* - \mathbf{w}\mathbf{x}|,$$

that is, the gap between $\mathbf{w}\mathbf{x}$ and $\mathbf{w}\mathbf{x}^*$ is reduced by at least a fraction of $1/(2N)$ via each call of the Graver-best oracle. Consequently, after $O(N \log |\mathbf{w}^*\mathbf{x} - \mathbf{w}\mathbf{x}|)$ calls of the oracle, we achieve an optimal solution and the theorem is proved.

It is worth mentioning that the above theorem also takes care of finding an initial feasible solution, if it is not known in advance.

The number of calls of Graver-best oracle is polynomially dependent on $\log \|\mathbf{w}\|_\infty$ and $\log \|\mathbf{l}\|_\infty, \log \|\mathbf{u}\|_\infty$. Applying the techniques used by Tardos in [28], it is further possible to reduce the running time to strongly polynomial, as is implied by the following.

Theorem 2 ([22]) *Given a Graver-best oracle for H, (1) can be solved with a number of augmentation steps that are polynomially bounded in the encoding lengths of H.*

It is recently observed by Altmanová et al. [1] and Eisenbrand et al. [9] that Theorem 2 also holds for an approximate version of Graver-best oracle, as we describe below.

Definition 3 (c-Approximate Graver-Best Oracle [1]) Let $c \in \mathbb{R}$, $c \geq 1$. A c-*approximate Graver-best oracle for an integer matrix H is one that, queried on* $\mathbf{w}, \mathbf{b}, \mathbf{l}, \mathbf{u}$, and $\mathbf{x}$ feasible to (1), returns a feasible augmentation step $\mathbf{h}$ such that $\mathbf{wh} \leq 1/c \cdot \mathbf{w}\rho\mathbf{g}$ for all $\mathbf{x}$-feasible step pairs $(\mathbf{g}, \rho) \in \mathcal{G}(H) \times \mathbb{Z}$.

The following lemma states that an approximate Graver-best oracle suffices for an augmentation algorithm.

Lemma 1 ([1]) *Given a c-approximate Graver-best oracle for H, (1) can be solved with a number of augmentation steps that are polynomially bounded in c and the encoding lengths of H, $\mathbf{b}, \mathbf{w}, \mathbf{l}, \mathbf{u}$.*

It is worth mentioning that the proof of Theorem 2 in [22] can be used to establish a similar strong polynomial bound on the number of augmentation steps.

The advantage of a c-approximate Graver-best oracle is that we can restrict our attention to the following subproblem:

Given IP (1) and $\omega \leq O(\log \|\mathbf{u}\|_\infty + \log \|\mathbf{l}\|_\infty)$,[1] find $\mathbf{h}$ such that $\mathbf{wh} \leq 1/c \cdot 2^\omega \mathbf{wg}$ for all $\mathbf{x}$-feasible step pairs $(\mathbf{g}, 2^\omega) \in \mathcal{G}(H) \times \mathbb{Z}$.

If we solve the subproblem for each individual $\omega \in \{0, 1, 2, \cdots, O(\log \|\mathbf{u}\|_\infty + \log \|\mathbf{l}\|_\infty)\}$, then the best $\mathbf{h}$ among all the subproblems serves as a feasible output of a $2c$-approximate Graver-best oracle.

3.2 Designing the Graver-Best Oracle

We have shown that solving an IP essentially reduces to finding a (approximate) Graver-best oracle. In general, finding a (approximate) Graver-best step is essentially as hard as solving an arbitrary IP. However, for IP that has a special structure, an efficient algorithm may exist, and this is particularly the case for IP with a block structure as we discussed in the preliminary. The crucial observation here is that for certain types of block-structured IP, its Graver basis element has a nice property that it has a small ℓ_1- or ℓ_∞-norm. Based on such a bound, an efficient algorithm for finding a (approximate) Graver-best step may be designed. Indeed, all existing algorithms for block-structured IP are of this kind, as we provide details in the following subsections.

---

[1]Here we remark that by using proximity results or the standard bound from linear programming, we can always restrict that $\log \|\mathbf{u}\|_\infty + \log \|\mathbf{l}\|_\infty$ is bounded by a polynomial.

3.2.1 *n*-Fold IP

The following lemma is a restatement of the finiteness theorem in [17].

Lemma 2 ([17]) *Let H be an n-fold matrix. Then $\|\mathbf{g}\|_1 \le \lambda$ for some λ that only depends on the small matrices A and D, i.e., on s_i, t_i which are the number of rows and columns of i where $i = A, D$, and Δ which is the maximal absolute value of entries of A and D.*

Note that the structure of an *n*-fold matrix requires that $t_A = t_D$. The value of λ is discussed and improved in a series of follow-up papers [9, 14, 22], with the current best upper of $\lambda \le O(s_A s_D \Delta)^{s_A s_D}$ [9].

Based on Lemma 2, a dynamic program can be designed to solve the following subproblem: Given IP (1) and $\omega \le O(\log \|\mathbf{u}\|_\infty + \log \|\mathbf{l}\|_\infty)$, find $\mathbf{h}$ such that $\|\mathbf{h}\| \le O(\lambda)$, $\mathbf{wh} \le 2^\omega \mathbf{wg}$ for all $\mathbf{x}$-feasible step pairs $(\mathbf{g}, 2^\omega) \in \mathcal{G}(H) \times \mathbb{Z}$. The basic idea for the dynamic program is similar to that of the classical knapsack problem. We skip the technical details here. Interested readers may refer to [14].

Combining the dynamic program and Lemma 2, we have the following theorem.

Theorem 3 ([9, 14, 22]) *There exists an algorithm that solves an n-fold integer program in $f_{np}(s_A, s_D, t_A, t_D, \Delta)|I|^{O(1)}$ time, where f_{np} is some computable function and $|I|$ is the encoding length of the input.*

The exact running time was shown by Hemmecke et al. [14] to be $\Delta^{O(s_A s_D t_A + s_A t_A^2)} n^3 |I|$, and was improved independently in recent papers [9, 22]. In particular, an algorithm of running time $\Delta^{O(s_A^2 s_D + s_A s_D^2)} (n t_A)^3 + \mathbf{LP}$ was provided by Koutecký et al. [22], and an algorithm of running time $n^2 t_A^2 \beta \log^2(n t_A) \cdot (s_A s_D \Delta)^{O(s_A^2 s_D + s_A s_D^2)} + \mathbf{LP}$ was provided by Eisenbrand et al. [9], where β is the logarithm of the largest number occurring in the input, and $\mathbf{LP}$ is the time for solving the LP relaxation of (1).

Remark We do not know yet whether the running time, particularly its (exponential) dependency on s_A and s_D is necessary and/or optimal. The exponential dependency on s_A is unavoidable assuming exponential time hypothesis [21], yet it is not clear whether we can have an algorithm of running time like, e.g., $(s_A s_D \Delta)^{O(s_A + s_D)} |I|^{O(1)}$.

3.2.2 Tree-Fold IP

Tree-fold IP is a generalization of *n*-fold IP. A recent paper by Chen and Marx [3] generalizes Lemma 2 as follows.

Lemma 3 ([3]) *Let H be a tree-fold matrix consisting of $A_1, A_2, \cdots, A_\tau$. Then $\|\mathbf{g}\|_1 \le \lambda$ for some λ that only depends on the small matrices $A_1, A_2, \cdots, A_\tau$, i.e., on s_i, t_i which are the number of rows and columns of A_i where $i = 1, 2, \cdots, \tau$, and Δ which is the maximal absolute value of entries of all A_i's.*

Further generalizing the dynamic programming for n-fold IP, Chen and Marx were able to present the following algorithm.

Theorem 4 ([3]) *There exists an algorithm that solves a tree-fold integer program in $f_{tf}(s_1, t_1, s_2, t_2, \cdots, s_\tau, t_\tau, \Delta)|I|^{O(1)}$ time, where f_{tf} is some computable function and $|I|$ is the encoding length of the input.*

The running time was further improved very recently in [9, 22]. In particular, Koutecký et al. [22] showed that the running time can be further made into strongly polynomial, and Eisenbrand et al. [9] showed an algorithm of an explicit running time $n^2 \beta \log^2(nt_A) \cdot (s\Delta)^{O(\sigma s)} + \mathbf{LP}$, where β is the logarithm of the largest number occurring in the input, $s = \prod_{i=1}^{\tau}(s_i + 1)$, $\sigma = \sum_{i=1}^{\tau} s_i$, and $\mathbf{LP}$ is the time for solving the LP relaxation of (2).

Remark The current best algorithm for tree-fold IP has a running time doubly exponential in τ. It is an important open problem whether there exists an algorithm of running time singly exponential in τ.

3.2.3 Multistage Stochastic IP

The following lemma is a restatement of the finiteness theorem (Proposition 8.11) of [2].

Lemma 4 ([2]) *Let H be a multistage stochastic matrix consisting of $A_1, A_2, \cdots, A_\tau$. Then $\|\mathbf{g}\|_\infty \leq \lambda$ for some λ that only depends on the small matrices $A_1, A_2, \cdots, A_\tau$, i.e., on s_i, t_i which are the number of rows and columns of A_i where $i = 1, 2, \cdots, \tau$, and Δ which is the maximal absolute value of entries of all A_i's.*

It is worth mentioning that unlike n-fold IP, the general concrete upper bound on λ is unknown even for the special case of two-stage stochastic IP (i.e., even if $\tau = 2$). Nevertheless, for any fixed H we can still compute λ [2]. Based on Lemma 4, Aschenbrenner and Hemmecke were able to show the following.

Theorem 5 ([2]) *There exists an algorithm that solves multistage stochastic integer program in $f_{ms}(s_1, t_1, s_2, t_2, \cdots, s_\tau, t_\tau, \Delta)|I|^{O(1)}$ time, where f_{ms} is some computable function and $|I|$ is the encoding length of the input.*

Recently Koutecký et al. [22] showed that the running time can be further made into strongly polynomial.

Remark Even for the special case of two-stage stochastic IP, we do not yet know an algorithm with an explicit running time. It is not even clear whether the running time is doubly or triply exponential in parameters like τ. This is an important open problem. To derive such an algorithm, if we still use the iterative augmentation framework, then we may need an explicit bound for λ in Lemma 4.

3.2.4 4-Block n-Fold IP

Recall the structure of a 4-block n-fold matrix. Suppose H consists of n copies of A, B, D and one copy of C, then $N = t_C + nt_B$. We write the $(t_C + nt_B)$-dimensional vector $\mathbf{x}$ into $n + 1$ "bricks" $\mathbf{x} = (\mathbf{x}^0, \mathbf{x}^1, \cdots, \mathbf{x}^n)$, where $\mathbf{x}^0$ is t_C-dimensional and $\mathbf{x}^i$ is t_B-dimensional for $1 \leq i \leq n$. According to the augmentation framework, if $\|\mathbf{g}^0\|_\infty \leq \lambda$ for any $\mathbf{g} \in \mathcal{G}(H)$, then it suffices to solve the following subproblem:

Given IP (2) and $\omega \leq O(\log \|\mathbf{u}\|_\infty + \log \|\mathbf{l}\|_\infty)$, find $\mathbf{h}$ such that $\|\mathbf{h}^0\|_\infty \leq \lambda$ and $\mathbf{wh} \leq 1/c \cdot 2^\omega \mathbf{wg}$ for all $\mathbf{x}$-feasible step pairs $(\mathbf{g}, 2^\omega) \in \mathcal{G}(H) \times \mathbb{Z}$.

Let $\mathbf{h}_*$ be the optimal solution for the subproblem, as $\|\mathbf{h}_*^0\|_\infty \leq \lambda$, we can guess, via $O(\lambda)^{t_B}$ enumerations, the t_B-dimensional vector $\mathbf{h}_*^0$. Once we guess the correct $\mathbf{h}_*^0$, the subproblem reduces to the following:

Given IP (2) and $\omega \leq O(\log \|\mathbf{u}\|_\infty + \log \|\mathbf{l}\|_\infty)$, find $\mathbf{h}$ such that $\mathbf{h}^0 = \mathbf{h}_*^0$, and $\mathbf{wh} \leq 1/c \cdot 2^\omega \mathbf{wg}$ for all $\mathbf{x}$-feasible step pairs $(\mathbf{g}, 2^\omega) \in \mathcal{G}(H) \times \mathbb{Z}$.

It is easy to see that the subproblem reduces to exactly a subproblem for an n-fold integer program, which, according to Theorem 3, can be solved in $f_{np}(s_A, s_D, t_A, t_D, \Delta)|I|^{O(1)}$ time. Hence, we have the following observation, which is also implicitly used in [5, 15]:

Observation 1 *If $\|\mathbf{g}^0\|_\infty \leq \lambda$ for any $\mathbf{g} \in \mathcal{G}(H)$, then 4-block n-fold integer programming can be solved in $O(\lambda)^{t_B} f_{np}(s_A, s_D, t_A, t_D, \Delta)|I|^{O(1)}$ time, where $|I|$ is the length of the input.*

The observation allows us to focus on the ℓ_∞-norm of the first brick $\mathbf{g}^0$ of a Graver basis element. Is the upper bound λ only dependent on the small matrices A, B, C, D? Unfortunately, this is not true. In 2014, Hemmecke et al. showed an upper bound of $n^{2^{O(s_C)}} f_b(s_A, s_B, s_C, s_D, t_A, t_B, t_C, t_D, \Delta)$ for some computable function f_b, which is doubly exponential in the parameter s_C. Very recently, Chen et al. [5] is able to show the following:

Lemma 5 ([5]) *Let H be a 4-block n-fold matrix consisting of A, B, C, D. Then $\|\mathbf{g}\|_\infty \leq n^{s_C} f_b(s_A, s_B, s_C, s_D, t_A, t_B, t_C, t_D, \Delta)$ for some computable function f_b, where s_i, t_i are the number of rows and columns of matrix i for $i = A, B, C, D$, and Δ which is the maximal absolute value of entries of A, B, C, D. Furthermore, there exists some 4-block n-fold matrix H such that $\|\mathbf{g}^0\|_\infty \geq \Omega(n^{s_C})$ for any Graver basis element $\mathbf{g}$.*

The above lemma indicates that, unlike its two special cases, n-fold IP and two-stage stochastic IP, we do not have a similar finiteness theorem for 4-block n-fold IP. Nevertheless, we have the following theorem by combining Lemma 5 and Observation 1.

Theorem 6 ([5]) *There exists an algorithm that solves 4-block n-fold integer program in $n^{O(t_B s_C)} f_b(s_A, s_B, s_C, s_D, t_A, t_B, t_C, t_D, \Delta)|I|^{O(1)}$ time, where f_b is some computable function and $|I|$ is the encoding length of the input.*

Remark The reader can see that the unboundness of the Graver basis elements of 4-block n-fold IP, as implied by Lemma 5, causes the algorithm to have a much higher dependency of n in its running time, in particular, it is $n^{O(t_B s_C)}$ instead of $n^{O(1)}$ in the running times of algorithms for n-fold and two-stage stochastic IPs. Nevertheless, this fact only indicates that the traditional approach may not be so efficient for 4-block n-fold IP. It is still possible that a completely new approach can yield a running time of $f(s_A, s_B, s_C, s_D, t_A, t_B, t_C, t_D, \Delta)n^{O(1)}$ for some computable function f, which is one of the most important open problems in this research direction.

4   Optimizing over a Non-linear Objective

When ϕ is a non-linear function, the problem is less understood. In particular, polynomial time algorithms are only known for a restricted class of non-linear functions. Moreover, even for those non-linear functions that admit a polynomial time algorithm, it is not yet clear how these running times depend on the parameters. These known algorithms still rely on the iterative augmentation framework, but we will need a modified version of the Graver-best oracle.

4.1   Minimizing an Appropriate Convex Function

We define a function $\psi : \mathbb{Z} \to \mathbb{Z}$ to be $\mathbb{Z}$-convex, if for all $x, y \in \mathbb{Z}$ and for all $0 \leq \lambda \leq 1$ with $\lambda x + (1 - \lambda)y \in \mathbb{Z}$, the inequality $\psi(\lambda x + (1 - \lambda)y) \geq \lambda \psi(x) + (1 - \lambda)\psi(y)$ holds. We define a function $\bar{\psi} : \mathbb{Z}^{k+1} \to \mathbb{Z}$ to be a separable $\mathbb{Z}$-convex function if

$$\bar{\psi}(y_0, y_1, y_2, \cdots, y_k) = y_0 + \sum_{i=1}^{k} \psi_i(y_i),$$

where each ψ_i is a $\mathbb{Z}$-convex function.

Let $\mathbf{c}_i \in \mathbb{Z}^N$ for $0 \leq i \leq k$. We consider the objective of $\min \phi(\mathbf{x})$ where $\phi(\mathbf{x}) = \bar{\psi}(\mathbf{c}_0\mathbf{x}, \mathbf{c}_1\mathbf{x}, \cdots, \mathbf{c}_k\mathbf{x})$, or more precisely, we consider the following:

$$\min \quad \{\mathbf{c}_0\mathbf{x} + \sum_{i=1}^{k} \psi_i(\mathbf{c}_i\mathbf{x}) : H\mathbf{x} = \mathbf{b}, \mathbf{l} \leq \mathbf{x} \leq \mathbf{u}, \mathbf{x} \in \mathbb{Z}^N\}, \tag{3}$$

Hemmecke et al. [13] showed a similar result to Theorem 1, which roughly states that an integer program that minimizes a separable $\mathbb{Z}$-convex function can be solved by calling a polynomial number of a slightly modified Graver-best oracle. More precisely, We define $\hat{H}$ as the augmented matrix such that

$$\hat{H} := \begin{pmatrix} H & 0 \\ C & I_k \end{pmatrix} ,$$

where I_k is a $k \times k$ identity matrix and $C = (\mathbf{c}_1, \mathbf{c}_2, \cdots, \mathbf{c}_k)^T$, which consists of every $\mathbf{c}_i$ as rows.

Similar to minimizing linear objectives, we define a Graver-best oracle for separable $\mathbb{Z}$-convex functions as follows. Towards that, recall that any Graver basis element of $\mathcal{G}(\hat{H})$ is an $(N + k)$-dimensional vector. We define $\mathcal{G}(H, C)$ to be the projection of vectors in $\mathcal{G}(\hat{H})$ onto its first N coordinates, whereas vectors in $\mathcal{G}(H, C)$ are N-dimensional. Furthermore, by the definition of $\hat{H}$, it is easy to see that for any $\mathbf{g} \in \mathcal{G}(H, C)$, we have $H\mathbf{g} = 0$.

Now we are able to define the Graver-best oracle for IP (3). Let the objective function ϕ and $\mathcal{G}(H, C)$ be defined as above. We define an augmenting step $\mathbf{g} \in \mathcal{G}(H, C)$ and a *step length* $\rho \in \mathbb{Z}$ as an $\mathbf{x}$-*feasible step pair* with respect to a feasible solution $\mathbf{x}$ if $\mathbf{l} \leq \mathbf{x} + \rho\mathbf{g} \leq \mathbf{u}$. An augmenting step $\mathbf{h}$ is a *Graver-best step* for $\mathbf{x}$ if $\phi(\mathbf{x} + \mathbf{h}) \leq \phi(\mathbf{x} + \rho\mathbf{g})$ for all $\mathbf{x}$-feasible step pairs $(\mathbf{g}, \rho) \in \mathcal{G}(H, C) \times \mathbb{Z}$. The next definition and theorem show that it is sufficient to focus all our attention on finding Graver-best steps.

Definition 4 (Graver-Best Oracle for IP (3)) A *Graver-best oracle* for IP (3) is one that, queried on $\mathbf{w}, \mathbf{b}, \mathbf{l}, \mathbf{u}$, and $\mathbf{x}$ feasible to (3), returns a Graver-best step $\mathbf{h}$ for $\mathbf{x}$.

Then the following is true:

Theorem 7 ([13]) *Let* $\phi(\mathbf{x}) = \bar{\psi}(\mathbf{c}_0\mathbf{x}, \mathbf{c}_1\mathbf{x}, \cdots, \mathbf{c}_k\mathbf{x}) = \mathbf{c}_0\mathbf{x} + \sum_{i=1}^{k} \psi_i(\mathbf{c}_i\mathbf{x})$ *be a separable $\mathbb{Z}$-convex function given by a polynomial time comparison oracle which, when queried on* $\mathbf{y}, \mathbf{z} \in \mathbb{Z}^{k+1}$, *decides whether* $\bar{\psi}(\mathbf{y}) < \bar{\psi}(\mathbf{z})$, $\bar{\psi}(\mathbf{y}) = \bar{\psi}(\mathbf{z})$, *or* $\bar{\psi}(\mathbf{y}) > \bar{\psi}(\mathbf{z})$ *holds in time polynomial in the encoding length of $\mathbf{y}$ and $\mathbf{z}$. Let $\bar{\psi}_{max}$ be an upper bound for the difference of maximum and minimum value of $\bar{\psi}$ over the feasible set* $\{\mathbf{x} : H\mathbf{x} = \mathbf{b}, \mathbf{l} \leq \mathbf{z} \leq \mathbf{u}, \mathbf{x} \in \mathbb{Z}^N\}$ *and assume that the encoding length of H is of polynomial size in the encoding lengths of* $H, \mathbf{b}, \mathbf{l}, \mathbf{u}, \mathbf{c}_0, \mathbf{c}_1, \cdots, \mathbf{c}_k$. *Given a Graver-best oracle for IP (3), (3) can be solved with a number of augmentation steps that are polynomially bounded in the encoding lengths of* $H, \mathbf{b}, \mathbf{l}, \mathbf{u}, \mathbf{c}_0, \mathbf{c}_1, \cdots, \mathbf{c}_k$.

Combining Theorem 7 and the finiteness results shown by Lemmas 2 and 4, n-fold and two-stage stochastic IP with certain type of non-linear objective functions can be solved in polynomial time, as we specify the details in the following.

4.1.1 Convex n-Fold IP

Consider n-fold IP. Recall that in an n-fold IP H consists of n copies of $D \in \mathbb{Z}^{s_D \times t_D}$ and $A \in \mathbb{Z}^{s_A \times t_A}$, where $t_A = t_D := t$, and any feasible solution $\mathbf{x}$ is an nt-dimensional vector. We write $\mathbf{x} = (\mathbf{x}^1, \mathbf{x}^2, \cdots, \mathbf{x}^n)$ where each $\mathbf{x}^i \in \mathbb{Z}^t$. We consider the objective function ϕ such that

$$\phi(\mathbf{x}) = \sum_{i=1}^{n} \phi^i(\mathbf{x}^i),$$

where each ϕ^i satisfies that

$$\phi^i(\mathbf{z}) = \sum_{j=1}^{k} \phi_j^i(\mathbf{c}_j \mathbf{z}) + \mathbf{c}_0^i \mathbf{z}$$

for convex functions ϕ_j^i, $j = 1, 2, \cdots, k$ where k is a *fixed constant*. Note that here we have the same vectors $\mathbf{c}_j$'s in each ϕ^i, while the linear coefficient $\mathbf{c}_0^i$ is different for each ϕ^i. We call an IP with H being an n-fold matrix and the objective function ϕ being defined as above the problem of *minimization convex n-fold IP*. Then we have the following.

Theorem 8 ([13]) *Given a polynomial time comparison oracle for each $\phi^i : \mathbb{Z}^t \to \mathbb{Z}$ which, when queried on $\mathbf{y}, \mathbf{z} \in \mathbb{Z}^t$, decides whether $\phi^i(\mathbf{y}) < \phi^i(\mathbf{z})$, $\phi^i(\mathbf{y}) = \phi^i(\mathbf{z})$, or $\phi^i(\mathbf{y}) > \phi^i(\mathbf{z})$ holds in time polynomial in the encoding length of $\mathbf{y}$ and $\mathbf{z}$, there exists an algorithm that solves the problem of minimization convex n-fold IP in time polynomial in the input length.*

We remark that the explicit running time of the algorithm for Theorem 8 is not known yet, but the readers may expect a similar running time to that of Theorem 3. That is, the running time is polynomial if $s_A, s_D, t = t_A = t_D$ and Δ are all constant. It is not clear, though, whether the running time can be made into polynomial in t.

4.1.2 Convex Splittable Two-Stage Stochastic IP

Consider two-stage stochastic IP. Recall that in a two-stage stochastic IP H consists of n copies of $A \in \mathbb{Z}^{s_A \times t_A}$ and $B \in \mathbb{Z}^{s_B \times t_B}$, and any feasible solution $\mathbf{x}$ is an $(t_B + n t_A)$-dimensional vector. We write $\mathbf{x} = (\mathbf{x}^0, \mathbf{x}^1, \cdots, \mathbf{x}^n)$, where $\mathbf{x}^0 \in \mathbb{Z}^{t_B}$ and $\mathbf{x}^i \in \mathbb{Z}^{t_A}$ for $1 \leq i \leq n$. We consider the objective function ϕ such that

$$\phi(\mathbf{x}) = \sum_{i=1}^{n} \phi^i(\mathbf{x}^0, \mathbf{x}^i),$$

where each $\phi^i : \mathbb{R}^{t_A + t_B} \to \mathbb{R}$ satisfies that

$$\phi^i(\mathbf{x}^0, \mathbf{x}^i) = \sum_{j=1}^{k} \phi_j^i(\mathbf{c}_j \mathbf{x}^0 + \mathbf{d}_j \mathbf{x}^i)$$

for convex functions ϕ_j^i, $j = 1, 2, \cdots, k$, where k is a *fixed constant*. Note again that we have the same vectors $\mathbf{c}_j$ and $\mathbf{d}_j$.

We further define a convex function $\phi^i : \mathbb{R}^{t_A + t_B} \to \mathbb{R}$ that maps $\mathbb{Z}^{t_A + t_B}$ to $\mathbb{Z}$ *splittable*, if for all fixed vectors $\mathbf{y} \in \mathbb{Z}^{t_B}$, $\mathbf{z}, \mathbf{h}_1, \mathbf{h}_2 \in \mathbb{Z}^{t_A}$, and for all finite intervals $[a, b] \subseteq \mathbb{R}$, there exists polynomially many (in the encoding length of the problem data) intervals $I_1, I_2, \cdots, I_r$ such that:

- $[a, b] = \cup_{j=1}^{r} I_j$,
- $I_j \cap I_{j'} \cap \mathbb{Z} = \emptyset$ for all $1 \le j < j' \le r$,
- for each $j = 1, 2, \cdots, r$, either $\phi^i(\mathbf{y}, \mathbf{z} + \alpha \mathbf{h}_1) \le \phi^i(\mathbf{y}, \mathbf{z} + \alpha \mathbf{h}_2)$ or $\phi^i(\mathbf{y}, \mathbf{z} + \alpha \mathbf{h}_1) \ge \phi^i(\mathbf{y}, \mathbf{z} + \alpha \mathbf{h}_2)$ holds for all $\alpha \in I_j$.

It is worth mentioning that convex polynomials of a constant maximal degree are splittable.

We call an IP with H being a two-stage stochastic matrix and the objective function $\phi(\mathbf{x}) = \sum_{i=1}^{n} \phi^i(\mathbf{x}^0, \mathbf{x}^i)$ being such that each ϕ^i is a convex splittable function for the problem of minimization convex splittable two-stage stochastic IP, then we have the following theorem:

Theorem 9 ([13]) *There exists an algorithm that solves the problem of minimization convex splittable two-stage stochastic IP in time polynomial in the input length.*

Note that the running time is polynomial when $s_A, s_B, t_A, t_B,$ and Δ are all constant. The explicit running time is unknown yet.

4.2 Maximizing an Appropriate Convex Function for n-Fold IP

The problem of maximizing a convex function is much less understood. It is not clear, e.g., for two-stage stochastic IP, whether maximizing a convex function, under appropriate restrictions, can be solved in polynomial time. Nevertheless, there exists some result for the n-fold IP due to De Leora et al. [7], as we elaborate below.

Consider n-fold IP. Recall that in an n-fold IP H consists of n copies of $D \in \mathbb{Z}^{s_D \times t_D}$ and $A \in \mathbb{Z}^{s_A \times t_A}$, where $t_A = t_D := t$, and any feasible solution $\mathbf{x}$ is an nt-dimensional vector. We write $\mathbf{x} = (\mathbf{x}^1, \mathbf{x}^2, \cdots, \mathbf{x}^n)$, where each $\mathbf{x}^i \in \mathbb{Z}^t$.

We consider the objective function ϕ such that:

$$\phi(\mathbf{x}) = \psi(\mathbf{c}_1 \mathbf{x}, \mathbf{c}_2 \mathbf{x}, \cdots, \mathbf{c}_k \mathbf{x}),$$

where k is a *fixed constant*, $\mathbf{c}_i \in \mathbb{Z}^{nt}$ for $1 \le i \le k$, and $\psi : \mathbb{R}^k \to \mathbb{R}$ is a convex function. We call an IP with H being an n-fold matrix and the objective function ϕ being defined as above the problem of maximization convex n-fold IP. We have the following.

Theorem 10 ([7]) *Given a polynomial time comparison oracle for $\psi : \mathbb{R}^k \to \mathbb{R}$ which, when queried on $\mathbf{y}, \mathbf{z} \in \mathbb{Z}^{nt}$, decides whether $\psi(\mathbf{y}) < \psi(\mathbf{z})$, $\psi(\mathbf{y}) = \psi(\mathbf{z})$ and $\psi(\mathbf{y}) > \psi(\mathbf{z})$ holds in time polynomial in the encoding length of $\mathbf{y}$ and $\mathbf{z}$, there exists an algorithm that solves the problem of maximization convex n-fold IP in time polynomial in the input length.*

Note that the running time is polynomial if $k, s_A, s_D, t = t_A = t_D$, and Δ are all constant.

5 Application to Combinatorial Optimization Problems

Integer programming is a powerful tool for solving many combinatorial optimization problems. However, solving a general IP is NP-hard [19], as we mentioned before in the introduction. To derive polynomial time (approximation or FPT) algorithms, a common approach is to model the problem with IPs that belong to the tractable sub-classes. In particular, most previous researches focus on modeling the problem with IPs that consist of only few variables and utilize Lenstra's elegant algorithm [23]. While this approach has proved to be successful in many problems, sometimes we do encounter difficulty in modeling problems with only a constant number of integral variables. Block-structured IP, on the other hand, provides us with a new alternative solution that allows us to model problems by utilizing an arbitrary number of integral variables. In this section, we will review recent progress in the application of the block-structured IP in many classical combinatorial optimization problems. Some of the application involves a highly non-trivial modeling of the problem, and may inspire future research in this direction.

We remark that, this section of the survey is dedicated to theoretical algorithmic results. There do exist researches targeting at implementation. For example, n-fold IP applied for scheduling and other related problems has been implemented and evaluated through extensive experiments in [1], where a Github code is also provided.

5.1 Scheduling

Scheduling is a classical problem in combinatorial optimization whose research dates back to the 1960s [11]. Most of the existing researches in the past decades focus on approximation or exact algorithms for different variants of the scheduling problem. The study of FPT algorithms for scheduling, however, is relatively new and was initiated in a recent paper by Mnich and Wiese [24]. A very recent paper by Knop et al. [20] establishes an interesting connection between n-fold IP and the scheduling problems, which yields FPT algorithms for a broad class of scheduling problems. In the following we will first briefly introduce the classical scheduling problem, and then elaborate their results.

In a classical scheduling problem, there are n independent jobs (tasks) and m machines (processors). We focus on non-preemptive scheduling in this section, where each job has to be processed non-preemptively on one of the machines. The processing time of job j on machine i is $p_{ij} \in \mathbb{Z}_{\geq 0}$, giving rise to the processing time matrix $PM = (p_{ij})_{n \times m}$. A feasible solution (schedule) to the scheduling problem is an assignment $\mathcal{A}$ that assigns each job j to a specific machine $A(j) \in \{1, 2, \cdots, m\}$.

There are three common machine environments commonly studied in the literature, namely:

- Identical machines (P): $p_{ij} = p_j$. In this case, the processing time of job j is independent of machines.
- Related machines (Q): $p_{ij} = p_j/s_i$ for $s_i \in \mathbb{Z}_+$, $1 \leq i \leq m$.
- Unrelated machines (R): p_{ij}'s are arbitrary non-negative integers.

We consider two popular objective functions, minimizing the *makespan* and minimizing the *total weighted completion times*. Given a feasible schedule, we denote by C_j the completion time of job j, which is the time when the processing of job j finishes. The makespan, denoted as C_{max}, is the latest completion time among all jobs, i.e., $C_{max} = \max_j C_j$. The weighted total completion times, denoted as $\sum w_j C_j$, is the weighted sum over all the completion times of jobs, where w_j is the weight of job j, which is a given integer. Adopting the classical three field notation, the problem of minimizing the makespan on identical machines is denoted as $P||C_{max}$. Similarly, the problem of minimizing total weighted completion times on unrelated machines is denoted as $R||\sum w_j C_j$.

To give a simple illustration on how we can handle scheduling problems via n-fold IP, we will take $P||C_{max}$ as an example. We will show that, an FPT algorithm, parameterized by the largest job processing time, p_{max}, can be derived by utilizing Theorem 3. It is worth mentioning that such an FPT algorithm was first discovered by Mnich and Wiese [24] through a different approach. Nevertheless, n-fold IP provides an alternative approach to derive an FPT algorithm for $P||C_{max}$, and is interesting on its own due to its straightforwardness—the reader will see that, the natural IP formulation of $P||C_{max}$ already belongs to n-fold IP, and an FPT algorithm follows directly via Theorem 3.

Consider $P||C_{max}$ and recall that p_{max} is the largest job processing time. Let N_j be the number of jobs whose processing time is $j \in \{1, \cdots, p_{max}\}$. Let $T \in \mathbb{Z}_{\geq 0}$ be a target makespan, we establish the following integer program to determine whether there exists a feasible schedule of makespan bounded by T:

$$\sum_{i=1}^{m} x_{ij} = N_j, \quad \forall 1 \le j \le p_{max} \tag{4a}$$

$$\sum_{j=1}^{p_{max}} p_j x_{ij} \le T, \quad \forall 1 \le m, \tag{4b}$$

where $x_{ij} \in \mathbb{Z}_{\ge 0}$ denotes the number of jobs whose processing time is j and is scheduled on machine i.

To see the IP above falls into the class of n-fold IP, we let $\mathbf{x}^i = (x_{i1}, x_{i2}, \cdots, x_{i,p_{max}})$ and $\mathbf{x} = (\mathbf{x}^1, \mathbf{x}^2, \cdots, \mathbf{x}^m)$. Define $D = I_{p_{max}}$ and $A = (1, 2, \cdots, p_{max})$, it is easy to see that Equation (4a) is equivalent to $\sum_{i=1}^{m} D\mathbf{x}^i = \mathbf{b}^0$, where $\mathbf{b}^0 = (N_1, N_2, \cdots, N_{p_{max}})$, and Equation (4b) is equivalent to $A\mathbf{x}^i \le T$. That is, the IP above is an n-fold IP. Consequently, applying Theorem 3, we have the following theorem.

Theorem 11 ([24]) $P||C_{max}$ *admits an FPT algorithm parameterized by* p_{max}.

It is worth mentioning that a PTAS for $P||C_{max}$ can be derived by utilizing the theorem above and the classical rounding technique. More precisely, the classical rounding technique allows us to transform a given scheduling instance into a modified instance in which the largest job processing time is $O(1/\epsilon \log 1/\epsilon)$ at the cost of a multiplicative $1 + O(\epsilon)$ factor in the objective value (see, e.g., [16]). Applying the theorem above with $p_{max} = O(1/\epsilon \log 1/\epsilon)$, a PTAS follows directly.

Using a similar idea, we may leverage n-fold IP to derive FPT algorithms for different variants of scheduling problems, despite that the natural IP formulation of the problem no longer belongs to the n-fold IP, and consequently much more effort is needed to derive a clever IP formulation. Prior to presenting these more general results, we first introduce the parameters needed.

Recall the unrelated machine scheduling. We consider a restricted version of this problem in which the m machines can be divided into K different groups, where machines of the same group are identical. That is, we have K different *types* of machines. In addition to the parameter K, we also consider the following parameters.

- p_{max}: largest job processing time,
- w_{max}: largest job weights,
- m: number of machines, and
- θ: number of distinct job processing times for the objective of C_{max}, or number of distinct job processing times and job weights for the objective of $\sum w_j C_j$.

The following results are due to Knop and Koutecký [20].

Theorem 12 ([20]) *The following scheduling problems admit FPT algorithms parameterized by τ that is defined as follows for each specific problem:*

- $Q||C_{max}$: $\tau = p_{max}$,
- $R||C_{max}$: $\tau = p_{max}^K$,
- $R||\sum w_j C_j$: $\tau = \max\{p_{max}^K, w_{max}^K\}$, *and*
- $R||\sum w_j C_j$: $\tau = m\theta^m$.

It is remarkable that the above results are all based on the application of n fold IP. The parameterized algorithms for $Q||C_{max}$ and $R||C_{max}$ can be further generalized. Recently, Chen et al. [4] considered a new parameter rk, which is the rank of the processing time matrix PM. Note that for $Q||C_{max}$, the rank of PM is 1. For $R||C_{max}$ where machines have K different types, the rank of PM is at most K. Chen et al. [4] showed the following:

Theorem 13 ([4]) $R||C_{max}$ *admits an FPT algorithm parameterized by p_{max} and* rk.

Although the algorithm of Chen et al. does not use n-fold IP directly, it utilizes Theorem 12.

5.2 Computational Social Choice

Computational social choice is an interdisciplinary research field that combines social choice theory, theoretical computer science, and the analysis of multi-agent systems. A very recent paper by Knop et al. [21] established a very interesting connection between n-fold IP and a classical problem in computational social choice, namely the bribery problem. Prior to presenting the results, we first introduce some basic concepts in this field.

Election Model Consider a set of m candidates C and a set of n voters V. Each voter v has a preference list of candidates, which is essentially a total order $\succ_v$ over candidates. The rank of candidate c in the preference list of voter v is given by $r(c, v)$.

Voting Rules A voting rule is a function that maps an election, specified by C, V and the preference lists, to a subset of winners $W \subseteq C$. There are various different voting rules. For the purpose of the result to be presented in this section, we focus on two broad sub-classes of voting rules.

Scoring Rules A scoring rule $\mathcal{R}$ assigns a score or point for each candidate based on their position in the preference list of a voter. More precisely, if a voter v ranks a candidate c at its h-th position, then the scoring rule assigns a value of $s(h)$, implying that c_i receives $s(h)$ points from v_j. Summarizing the scores of c_i received from all the voters, we get the total score of c_i. A candidate with the maximal score becomes the winner. For example, one of the most popular voting rules, *Plurality*, is one kind of scoring rules where each candidate only receives one score from a voter that prefers him/her the most.

*C*1 *Rules* A candidate $c \in C$ is called a *Cordorcet winner* if any other $c' \in C \setminus \{c\}$ satisfies that $|\{v \in V | c \succ_v c'\}| > |\{v \in V | c' \succ_v c\}|$. A voting rule $\mathcal{R}$ is Condorcet-consistent if it selects the Condorcet winner in case there is one. Fishburn [10] gives a classification of Condorcet-consistent voting rules. For our sake, we only focus on one kind known as *C*1 rules defined as follows. For any $c, c' \in C$, we define $v(c, c') = |\{v \in V | c \succ_v c'\}|$. A voting rule is a *C*1 rule if the Condorcet winner can be determined as long as we know whether $v(c, c') > v(c', c)$, $v(c, c') = v(c', c)$, or $v(c, c') < v(c', c)$ for each pair c, c'.

Bribery via Swaps For any pair of candidates $c, c' \in C$, a swap $(c, c')_v$ is defined to be the exchange of positions c and c' in the preference list $\succ_v$. Suppose there is an attacker/briber who tries to manipulate the result of an election by a sequence of swaps. Each swap $(c, c')_v$ is associated with a cost $\sigma_v(c, c')$. The $\mathcal{R}$-swap bribery problem asks for a set of swaps with a minimum total cost that makes some designated candidate c^* a winner. More precisely, consider the following problem:

R-swap Bribery Given are an election C, V together with the preference lists of voters, a designated candidate c^*, a voting rule $\mathcal{R}$, and a cost function σ. The goal is to find out a set of swaps such that c^* becomes the winner and total cost is minimized.

By establishing a suitable IP formulation of the $\mathcal{R}$-swap bribery problem and applying the algorithmic results from n-fold IP, Knop et al. [21] were able to show the following:

Theorem 14 ([21]) *R-swap bribery can be solved in time*

- $|C|^{O(|C|^2)} O(|V|^3 (\log |V| + \log |\sigma_{max}|))$ *for any natural scoring rule $\mathcal{R}$, and*
- $|C|^{O(|C|^4)} O(|V|^3 (\log |V| + \log \sigma_{max}))$ *for any C1 rule $\mathcal{R}$,*

where σ_{max} is the largest swap cost.

We remark that Knop et al. [21] were able to model the $\mathcal{R}$-swap bribery problem into a special case of n-fold IP where $A = (1, 1, \cdots, 1)$, i.e., A is a $1 \times t$ matrix with each coordinate being 1. They call such a special n-fold IP as combinatorial n-fold IP. While we can directly utilize the current best-known algorithm for the n-fold IP in [9] to derive the results as is shown in the above theorem, it is still possible, though, that the combinatorial n-fold IP admits a better algorithm, yielding an improved FPT result for $\mathcal{R}$-swap bribery.

5.3 Multiple TSP on a Tree

TSP (traveling salesman problem) is a fundamental problem in combinatorial optimization. Very recently, Chen and Marx [3] were able to establish the relationship between a variant of TSP and the tree-fold IP, which is very interesting and also of great importance, as this is arguably the first application of block-structured IP to derive FPT algorithms for TSP-related routing problems, with the potential further extensions to more general graphs. We give a brief introduction to their results in the following.

We consider the multiple traveling salesman problem on a tree. In this problem, we are given a tree $T = (V, E)$ with V and E being the set of vertices and edges. There is a non-negative weight (length) $w(e_j) \in \mathbb{Z}_{\geq 0}$ associated with each edge e_j. Let rt be the root of the tree. Initially there are m salesmen located at the root. Each of the salesmen needs to travel along edges to visit a subset of the vertices and return to the root at last. The problem of multiple TSP on a tree asks for the tour of every salesman such that every vertex of the tree is visited by at least one salesman, and the length of the longest tour among all the salesmen is minimized.

It is easy to observe that if the given tree T is a star, then each salesman actually travels a distinct subset of edges, and the length of the tour is twice the total length of the edges. If we view the length of an edge as a processing time of a job, it is not difficult to see that the multiple TSP problem on a star is essentially the identical machine scheduling problem with the objective of minimizing the makespan.

Notice that for a salesman to visit a subset of vertices and return to the root, if he has to pass an edge, then he has to pass this edge at least twice, and it is possible for him to pass this edge exactly twice. Hence, the problem of multiple TSP on a tree is equivalent to the rooted subtree cover problem in which the goal is to find out m subtrees rooted at rt whose union covers T such that the largest total weight among all the subtrees is minimized.

Chen and Marx [3] showed the following in their recent paper.

Theorem 15 ([3]) *There exists an FPT algorithm parameterized by k that determines whether the problem of multiple TSP on a tree admits a feasible solution of objective value at most k.*

The above theorem is proved via formulating multiple TSP on a tree into a tree-fold IP. The connection between TSP and block-structured IP is quite hindsight and such a formulation is also non-trivial and requires a lot of technical details, which are skipped in this survey.

5.4 Other Applications

Besides the applications mentioned in the previous subsections, block-structured IP is also used to derive FPT algorithms for many other combinatorial optimization problems. We give a very brief introduction here.

Non-linear Multicommodity Transportation The problem was defined in [14]. We briefly recap their description. There are of ℓ commodities, m suppliers, and n consumers. Each supplier i has a supply vector $\mathbf{s}^i \in \mathbb{Z}_+^\ell$ with $\mathbf{s}_k^i$ its supply in commodity k. The amount of commodity k to be routed from supplier i to consumer j is an integer decision variable $x_{i,k}^j$. The total amount $\sum_{k=1}^\ell x_{i,k}^j$ of commodities routed on the channel from i to j should not exceed the channel capacity $u_{i,j}$ and has cost $f_{i,j}(\sum_{k=1}^\ell x_{i,k}^j)$ for suitable univariate functions $f_{i,j}$. By modeling the non-linear multicommodity transportation problem into an n-fold IP and using Theorem 7, Hemmecke et al. showed the following.

Theorem 16 ([14]) *For every fixed ℓ commodities, m suppliers, and p, there exists an algorithm that, given n consumers, supplies, and demands $\mathbf{s}^i, \mathbf{c}^j \in \mathbb{Z}_+^\ell$, capacities $u_{i,j} \in \mathbb{Z}_+$, and convex p-piecewise affine costs $f_{i,j} : \mathbb{Z} \to \mathbb{Z}$, solves in polynomial time $O(n^3 L)$, with L being the input length, the integer multicommodity transportation problem*

$$\min\{\sum_{i=1}^m \sum_{j=1}^n f_{i,j}(\sum_{k=1}^\ell x_{i,k}^j) : x_{i,k}^j \in \mathbb{Z}_+, \sum_j x_{i,k}^j = \mathbf{s}_k^i, \sum_i x_{i,k}^j = \mathbf{c}_k^j, \sum_{k=1}^\ell x_{i,k}^j \le u_{i,j}\}.$$

Weighted Set Multicover This is a generalization of the traditional set cover problem. In this problem, there are a universe U of size k, a set system represented by a multiset $\mathcal{F} = \{F_1, F_2, \cdots, F_n\} \in 2^U$, weights $w_1, w_2, \cdots, w_n \in \mathbb{Z}_{\ge 0}$, demands $d_1, d_2, \cdots, d_k \in \mathbb{Z}_{\ge 0}$. The goal is to find a multisubset $\mathcal{F}' \subseteq \mathcal{F}$ minimizing $\sum_{i:F_i \in \mathcal{F}'} w_i$ and satisfying $|\{i : F_i \in \mathcal{F}', j \in F_i\}| \ge d_j$ for all $j = 1, 2, \cdots, k$. By modeling the weighted set multicover problem into an n-fold IP and using Theorem 3, Knop et al. [21] showed the following.

Theorem 17 *There is an algorithm that solves weighted set multicover in time $k^{O(k^2)} O(\log n + \log w_{max})$, where w_{max} is the maximal set weight.*

String Matching Many variants of string matching problems are shown to admit an FPT algorithm via n-fold IP. In the following we briefly describe the problem of δ-multi-strings, which is shown to admit an FPT algorithm by Knop et al. [21]. It is worth mentioning that although δ-multi-strings are somehow an artificial problem, its generality allows us to derive FPT algorithms for other closely related string matching variants, including closest string, d-dismatches, etc. The reader may refer to [21] for these extensions. In this subsection, however, we will focus on the problem of δ-multi-strings.

In this problem, given are a set of strings $S = \{s_1, s_2, \cdots, s_k\}$, each of length L over alphabet $\Sigma \cup \{\star\}$, distance lower and upper bounds $d_1, d_2, \cdots, d_k \in \mathbb{Z}_{\geq 0}$ and $D_1, D_2, \cdots, D_k \in \mathbb{Z}_{\geq 0}$, distance function $\delta : \Sigma^* \times \Sigma^* \to \mathbb{Z}_{\geq 0}$, and a binary parameter $b \in \{0, 1\}$. The goal is to find a string $y \in \Sigma^L$ with $d_i \leq \delta(y, s_i) \leq D_i$ for each $s_i \in S$ such that $b \cdot (\sum_{i=1}^{k} \delta(y, s_i))$ is minimized.

We define a distance function $\delta : \Sigma^* \times \Sigma^* \to \mathbb{Z}_{\geq 0}$ to be *character-wise wildcard-compatible* if $\delta(x, y) = \sum_{i=1}^{L} \delta(x[i], y[i])$ for any two strings $x, y \in \Sigma^L$, and $\delta(e, \star) = 0$ for all $e \in \Sigma$.

Knop et al. [21] showed the following.

Theorem 18 ([21]) *There is an algorithm that solves δ-multi-strings in $K^{O(k^2)} O(\log L)$ time, where $K = \max\{|\Sigma|, k, \max_{a,b \in \Sigma} \delta(a, b)\}$ and δ is any character-wise wildcard-compatible function.*

6 Conclusion

In this paper, we have reviewed recent progress in the algorithmic research of block-structured IP and its application in various combinatorial optimization problems. There are several important open problems, which have been mentioned in the remarks at the end of corresponding subsections, and we give a short summary here. From a theoretical point of view, the concrete running time of the algorithm for n-fold IP with a linear objective has been shown and improved over a series of papers; however, for other block-structured IP, e.g., the two-stage or multistage stochastic IP, we do not know a concrete running time yet. Furthermore, for the more general 4-block n-fold IP, the current best algorithm has a running time whose dependency on n is n^{sc}, it is not clear whether an algorithm of a running time $f(s_A, s_B, s_C, s_D, t_A, t_B, t_C, t_D, \Delta)n^{O(1)}$ exists for some computable function f. For non-linear objectives, we know even less, e.g., 4-block n-fold IP with a non-linear objective is much less understood. From the application point of view, we have applied block-structured IP successfully in many combinatorial optimization problems and derived interesting results, including scheduling, computational social choice, string matching, multiple TSP, etc. The investigation of these problems may also inspire our research in block-structured IP. In particular, the development of the tree-fold IP is motivated by the problem of multiple TSP on a tree, while the research on tree-fold IP is also important towards a better understanding of block-structured IP. With so many different variants of combinatorial optimization problems, it is always desirable that a unified approach can be derived, and block-structured IP, particularly n-fold IP, does serve as a general tool for a broad class of problems. It will be interesting to further investigate and discover the relationship between block-structured IP and combinatorial optimization problems.

References

1. Altmanová, K., Knop, D., Koutecký, M.: Evaluating and tuning n-fold integer programming. arXiv preprint arXiv:1802.09007 (2018)
2. Aschenbrenner, M., Hemmecke, R.: Finiteness theorems in stochastic integer programming. Found. Comput. Math. **7**(2), 183–227 (2007)
3. Chen, L., Marx, D.: Covering a tree with rooted subtrees–parameterized and approximation algorithms. In: Proceedings of the Twenty-Ninth Annual ACM-SIAM Symposium on Discrete Algorithms, pp. 2801–2820. SIAM, Philadelphia (2018)
4. Chen, L., Marx, D., Ye, D., Zhang, G.: Parameterized and approximation results for scheduling with a low rank processing time matrix. In: Proceedings of the Thirty-Fourth Symposium on Theoretical Aspects of Computer Science, STACS, pp. 22:1–22:14 (2017)
5. Chen, L., Xu, L.. Shi, W.: On the graver basis of block-structured integer programming. arXiv preprint arXiv:1805.03741 (2018)
6. Cook, W., Fonlupt, J., Schrijver, A.: An integer analogue of Caratheodory's theorem. J. Comb. Theory B **40**(1), 63–70 (1986)
7. De Loera, J.A., Hemmecke, R., Onn, S., Rothblum, U.G., Weismantel, R.: Convex integer maximization via graver bases. J. Pure Appl. Algebra **213**(8), 1569–1577 (2009)
8. Downey, R.G., Fellows, M.R.: Parameterized Complexity. Springer Science & Business Media, New York (2012)
9. Eisenbrand, F., Hunkenschröder, C., Klein, K.-M.: Faster algorithms for integer programs with block structure. arXiv preprint arXiv:1802.06289 (2018)
10. Fishburn, P.C.: Condorcet social choice functions. SIAM J. Appl. Math. **33**(3), 469–489 (1977)
11. Graham, R.L.: Bounds on multiprocessing timing anomalies. SIAM J. Appl. Math. **17**(2), 416–429 (1969)
12. Graver, J.E.: On the foundations of linear and integer linear programming i. Math. Program. **9**(1), 207–226 (1975)
13. Hemmecke, R., Onn, S., Weismantel, R.: A polynomial oracle-time algorithm for convex integer minimization. Math. Program. **126**(1), 97–117 (2011)
14. Hemmecke, R., Onn, S., Romanchuk, L.: N-fold integer programming in cubic time. Math. Program. **137**(1–2), 325–341 (2013)
15. Hemmecke, R., Köppe, M., Weismantel, R.: Graver basis and proximity techniques for block-structured separable convex integer minimization problems. Math. Program. **145**(1–2), 1–18 (2014)
16. Hochbaum, D.S., Shmoys, D.B.: Using dual approximation algorithms for scheduling problems theoretical and practical results. J. ACM (JACM) **34**(1), 144–162 (1987)
17. Hoşten, S., Sullivant, S.: A finiteness theorem for Markov bases of hierarchical models. J. Comb. Theory A **114**(2), 311–321 (2007)
18. Jansen, K., Klein, K.-M., Maack, M., Rau, M.: Empowering the configuration-IP-new PTAS results for scheduling with setups times. arXiv preprint arXiv:1801.06460 (2018)
19. Karp, R.M.: Reducibility among combinatorial problems. In: Complexity of Computer Computations, pp. 85–103. Springer, Boston (1972)
20. Knop, D., Koutecký, M.: Scheduling meets n-fold integer programming. J. Sched. **21**(5), 493–503 (2018)
21. Knop, D., Koutecký, M., Mnich, M.: Combinatorial n-fold integer programming and applications. arXiv preprint arXiv:1705.08657 (2017)
22. Koutecký, M., Levin, A., Onn, S.: A parameterized strongly polynomial algorithm for block structured integer programs. arXiv preprint arXiv:1802.05859 (2018)
23. Lenstra Jr., H.W.: Integer programming with a fixed number of variables. Math. Oper. Res. **8**(4), 538–548 (1983)
24. Mnich, M., Wiese, A.: Scheduling and fixed-parameter tractability. Math. Program. **154**(1–2), 533–562 (2015)

25. Onn, S.: Nonlinear Discrete Optimization. Zurich Lectures in Advanced Mathematics. European Mathematical Society, Zurich (2010)
26. Papadimitriou, C.H.: On the complexity of integer programming. J. ACM (JACM) **28**(4), 765–768 (1981)
27. Sebö, A.: Hilbert bases, Caratheodory's theorem and combinatorial optimization. In: Proceedings of the 1st Integer Programming and Combinatorial Optimization Conference, pp. 431–455. University of Waterloo Press, Ontario (1990)
28. Tardos, E.: A strongly polynomial algorithm to solve combinatorial linear programs. Oper. Res. **34**(2), 250–256 (1986)

Online Combinatorial Optimization Problems with Non-linear Objectives

Zhiyi Huang

Abstract We survey some recent progress on the design and the analysis of online algorithms for optimization problems with non-linear, usually convex, objectives. We focus on an extension of the online primal dual technique, and highlight its application in a number of applications, including an online matching problem with concave returns, an online scheduling problem with speed-scalable machines subjective to convex power functions, and a family of online covering and packing problems with convex objectives.

1 Introduction

Online combinatorial optimization problems are ubiquitous. In these problems, partial decisions must be made irrevocably based on the information revealed so far. For example, in the online bipartite matching problem, only one side of the vertices is given at the beginning. Then, the vertices on other side arrive one by one. On the arrival of an online vertex, its incident edges are revealed and the algorithm must irrevocably decide how to match it without any knowledge of the vertices that will arrive later. Due to the uncertainty of future, it is impossible in general to guarantee a maximum cardinality matching in the online setting. The performance of the algorithm is measured by the ratio of the size of the obtained matching to that of the maximum matching in hindsight. The competitive ratio of the algorithm is defined to be the above ratio in the worst case.

The online bipartite matching problem and its many variants are extensively studied in the literature. So are other classic online combinatorial optimization problems, including online covering and packing problems, online caching and

Z. Huang (✉)
The University of Hong Kong, Pokfulam, Hong Kong
e-mail: zhiyi@cs.hku.hk

© Springer Nature Switzerland AG 2019
D.-Z. Du et al. (eds.), *Nonlinear Combinatorial Optimization*,
Springer Optimization and Its Applications 147,
https://doi.org/10.1007/978-3-030-16194-1_8

paging, and online scheduling. Most of the previous work have focused on problems with linear objectives in the sense that the objective function can be written as a linear function of the decision variables. For example, the standard formulation of online bipartite matching uses an indicator variable $x_e \in \{0, 1\}$ to denote whether an edge e is chosen in the matching. The objective, i.e., the cardinality of the matching, is simply the sum of all such indicator variables. It remains linear even if we consider the generalization which allows edge weights and seeks to maximize the total weight instead of the cardinality of the matching.

However, there are also a wide range of problems whose natural formulations involve non-linear (often convex or concave) objectives. For instance, a variant of the online bipartite matching problem originated from the Adwords problem allows an offline vertex to match multiple online vertices, but imposes a cap (e.g., an advertiser's budget in the Adwords problem) on the total gain of an offline vertex. In other words, the contribution of an offline vertex to the objective function is the smaller of its cap and the sum of weights of matched edges incident to it. The objective is then the sum of such cap-additive functions of the decision variables, which are concave instead of linear. Other examples of online combinatorial optimization problems with non-linear objectives include online scheduling problems with speed-scalable machines in which the energy consumption of a machine is a convex function of its speed, and a generalized online resource allocation problems in which each resource has a "soft capacity constraint" specified by a convex production cost function.

In this chapter, we will survey some recent progress in the design and analysis of online algorithms for online combinatorial optimization problems with non-linear objectives. We will talk about a line of research on generalizing the online primal dual technique by Buchbinder and Naor [6], which was originally designed for linear objectives, to handle convex and concave objectives. The generalization allows us to solve the above-mentioned problems that do not admit natural linear program relaxations.

In particular, we will focus on convex programs and Fenchel's duality in the non-stochastic setting. Readers are also referred to other interesting work along this line, including using the online primal dual technique (or dual fitting) with Lagrangian duality by Anand et al. [2], Gupta et al. [12], Nguyen [17], and online stochastic convex optimization by Agrawal and Devanur [1].

1.1 Organization

We will first recap the online primal dual technique (Section 2), and explain how to extend it to handle convex programs via Fenchel's duality (Section 3). Then, we will talk about three applications of the extension: online matching with concave returns [11] (Section 4), online scheduling with speed scaling [10] (Section 5), and online covering and packing problems with convex objectives [3] (Section 6).

2 Online Primal Dual for Linear Objectives

We first give a brief introduction to the original online primal dual technique for problems with linear objectives by demonstrating its application in the ski-rental problem (e.g., Buchbinder and Naor [6]). Readers who are familiar with the original online primal dual technique may skip this section.

In the ski-rental problem, a skier arrives at a ski resort but does not know when the ski season will end. Every day, if the skier has not bought skis yet, she needs to decide whether to rent skis at a cost of $1 or to buy skis at a cost of B. We will assume for simplicity of our discussion that B is a positive integer. The goal is to minimize the total cost.

We follow the standard framework of competitive analysis of online algorithms. That is, we use the optimal cost in hindsight, which equals T if $T \leq B$ and B otherwise, as the benchmark. The performance of an algorithm for the ski-rental problem and, in general, for any cost minimization problem, is measured by the ratio of the expected cost of the algorithm to the optimal cost. We say that an algorithm is F-competitive or it has competitive ratio F if the aforementioned ratio is at most F for any instance of the problem. Obviously, the competitive ratio of an algorithm is always greater than or equal to 1, and the smaller the better.

Consider a natural linear program formulation of the ski-rental problem and the corresponding dual program below.

$$
\begin{aligned}
& \text{minimize} && B \cdot x + \sum_{t=1}^{T} y_t \\
& \text{subject to} && x + y_t \geq 1 && t = 1, \ldots, T \\
& && x, y_t \geq 0 && t = 1, \ldots, T
\end{aligned}
$$

$$
\begin{aligned}
& \text{maximize} && \sum_{t=1}^{T} \alpha_t \\
& \text{subject to} && \sum_{t=1}^{T} \alpha_t \leq B \\
& && 0 \leq \alpha_t \leq 1 && t = 1, \ldots, T
\end{aligned}
$$

Here, x is the indicator of whether the algorithm buys skis, and y_t is the indicator of whether the algorithm rents skis on day t. For simplicity, we will only discuss solving the linear programs online to minimize the expected primal objective value. Readers are referred to the survey by Buchbinder and Naor [6, Section 3] for an online rounding algorithm which convert the online fractional solution into a randomized integral algorithm for the ski-rental problem with the same competitive ratio. Roughly speaking, the fractional value of x denotes the probability that the algorithm buys skis and the fractional value of y_t denotes the probability that the algorithm rents skis on day t.

2.1 High-Level Plan

As the input information being revealed over time piece by piece, more variables and constraints of the primal and dual programs are presented to the algorithm. On each day t, a new primal variable $y_t \geq 0$, a new primal constraint $x + y_t \geq 1$, and a new dual variable $1 \geq \alpha_t \geq 0$ arrive. Online primal dual algorithms maintain at all time a feasible primal assignment and a feasible dual assignment simultaneously. Let $\Delta_t P$ and $\Delta_t D$ be the changes of the primal and dual objectives, respectively, on day t. The goal is to update the primal and dual variables to satisfy:

$$\Delta_t P \leq F \cdot \Delta_t D \tag{1}$$

for some fixed parameter $F \geq 1$. If Equation (1) holds and the initial values of the primal and dual objectives are zero, the algorithm is F-competitive because the final primal objective P is at most F times the final dual objective D, which by weak duality is less than or equal to the optimal primal objective.[1]

2.2 Relaxed Complementary Slackness

In order to understand how an online primal dual algorithms can be derived from the structure of the linear programs, we need to look into their optimality conditions. It is known that the offline optimal primal and dual solutions satisfy the complementary slackness conditions, which state that a primal (resp., dual) variable must be zero unless the corresponding dual (resp., primal) constraint is tight (e.g., [9]). Specifically, for the ski-rental problem, we have the followings:

(a) x must be zero unless $\sum_{t=1}^{T} \alpha_t = B$;
(b) y_t must be zero unless $\alpha_t = 1$;
(c) α_t must be zero unless $x + y_t = 1$.

However, it is generally impossible to satisfy all complementary slackness conditions exactly in an online problem. In particular, it is not possible to guarantee satisfying conditions (a) and (c) exactly in the ski-rental problem. The best we could hope for is to satisfy the complementary slackness conditions approximately. Online primal dual algorithms are therefore driven by satisfying these conditions approximately, where the value of a primal (resp., dual) variable depends on the tightness of the corresponding dual (resp., primal) constraint. Concretely, consider

---

[1]Some applications of the online primal dual technique maintain an alternative set of invariants, e.g., one may consider keeping primal and dual objectives equal and guaranteeing primal feasibility, while showing approximate dual feasibility. However, such variants can be easily rewritten to fit into the framework in this chapter.

the following relaxed conditions for the ski-rental problem. Note that condition (b) will remain the same as it can be satisfied exactly even in the online setting.

(a') x depends on the tightness of the corresponding dual constraint $\sum_{t=1}^{T} \alpha_t \leq B$, i.e., it is an increasing function of $\sum_{t=1}^{T} \alpha_t$;

(c') α_t must be zero unless $x + y_t = 1$, at the end of day t (the constraint may have slack in the future because the algorithm may increase x).

2.3   Online Primal Dual Algorithms

Fix any day $1 \leq i \leq T$, first consider the new dual variable α_i. To maximize the dual objective, letting $\alpha_i = \min\{1, B - \sum_{t=1}^{i-1} \alpha_t\}$ is the most natural choice in light of the dual constraints. Recall that B is an integer, this is equivalent to letting $\alpha_i = 1$ if $i \leq B$ and $\alpha_i = 0$ otherwise. As a result, $\sum_t \alpha_t$ increases by 1 on each day $i \leq B$ and the algorithm increases x according to condition (a'). Let x_i denote the value of x after day i for $i = 1, \ldots, B$, and $x_0 = 0$. After day B, $\sum_t \alpha_t$ and, thus, the value of x must remain constant because of condition (a'). Further, let $y_i = 1 - x_i$ to satisfy condition (c'). Note that this is also the most natural choice to minimize the primal objective. Finally, since we let $\alpha_i = 0$ for $i > B$, we must have $y_i = 0$ and $x = 1$ on any day $i > B$ according to condition (b) and, thus, $x_B = 1$ most hold.

   In sum, for every monotone sequence x_t, $t = 0, 1, \ldots, B$, such that $x_0 = 0$ and $x_B = 1$, there is an online primal dual algorithm as follows:

1. On day $i = 1, \ldots, B$, let $\alpha_i = 1$, $x = x_i$, and $y_i = 1 - x_i$.
2. On day $i > B$, let $\alpha_i = 0$, x remains the same (i.e., equals 1), and $y_i = 0$.

2.4   Online Primal Dual Analysis

It remains to find the best monotone sequence $\{x_t\}_{t=1,\ldots,B}$ such that Equation (1) holds with the smallest possible $F \geq 1$. Note that both the primal and dual objectives remain the same after day B, so it suffices to analyze Equation (1) on the first B days. On each day $t = 1, \ldots, B$, x changes from x_{t-1} to x_t, and $y_t = 1 - x_t$. So the change of primal objective is equal to:

$$\Delta_t P = B \cdot (x_t - x_{t-1}) + (1 - x_t) = (B - 1) \cdot x_t - B \cdot x_{t-1} + 1$$

On the other hand, the algorithm sets $\alpha_t = 1$. So the change of dual objective is

$$\Delta_t D = 1$$

So Equation (1) becomes $(B - 1) \cdot x_t - B \cdot x_{t-1} \leq F - 1$. Reorganizing terms, it is equivalent to the followings:

$$x_t + F - 1 \leq \tfrac{B}{B-1}(x_{t-1} + F - 1),$$

Using the above inequality for $1, 2, \ldots, t$, we get that:

$$x_t \leq \left(\tfrac{B}{B-1}\right)^t (x_0 + F - 1) - (F - 1) = \left(\left(\tfrac{B}{B-1}\right)^t - 1\right)(F - 1) \tag{2}$$

Let $e(B) = \left(\tfrac{B}{B-1}\right)^B$. We have $e(B) \geq e$ and $\lim_{B \to +\infty} e(B) = e \approx 2.718$. By $x_B = 1$ and the above inequality, we have $F \geq \tfrac{e(B)}{e(B)-1}$. Let $F = \tfrac{e(B)}{e(B)-1}$ and let x_t be such that Equation (2) holds with equality. Then, we have an online primal dual algorithm with competitive ratio $\tfrac{e(B)}{e(B)-1} \leq \tfrac{e}{e-1} \approx 1.582$. This competitive ratio is in fact the best possible (e.g., Buchbinder and Naor [6]).

As a concluding remark of the section, we highlight that the derivation of the online primal dual algorithm for the ski-rental problem and its analysis follow mechanically from the primal and dual linear programs and the corresponding relaxed complementary slackness conditions. No cleverness is needed to derive the optimal competitive ratio. This is, in my opinion, the main strength of the online primal dual framework. In more complicated problems, it is non-trivial to obtain a good enough understanding of the mathematical programs and to find the right relaxation of their optimality conditions. Once we figure them out, however, the design of the algorithm and the analysis will again become mechanical.

3 Online Primal Dual for Convex and Concave Objectives

Let us first introduce some necessary background on the conjugates of convex and concave functions and a duality theory for convex programs.

3.1 Conjugates

Let $f : \mathbb{R}^n_+ \mapsto \mathbb{R}_+$ be a convex function. Its *convex conjugate* is defined as:

$$f^*(x^*) = \max_{x \geq 0} \left\{ \langle x, x^* \rangle - f(x) \right\}.$$

Here, x^* is also an n-dimensional vector. $\langle x, x^* \rangle$ denotes the inner product of vectors x and x^*. For example, suppose $f(x) = \tfrac{1}{\alpha}x^\alpha$ is a polynomial. Then, $f^*(x^*) = (1 - \tfrac{1}{\alpha})x^{*\frac{\alpha}{\alpha-1}}$ is also a polynomial. Here, we add the coefficient $\tfrac{1}{\alpha}$ to f so that the coefficient of the conjugate f^* is simple, without changing the nature of the functions. We will have similar treatments throughout the chapter.

For simplicity, we will further assume that f is non-negative, non-decreasing, strictly convex, differentiable, and normalized such that $f(0) = 0$ in this chapter. In this case, the conjugate satisfies the following properties:

- f^* is non-negative, non-decreasing, strictly convex, differentiable, and normalized such that $f^*(0) = 0$;
- $f^{**} = f$;
- ∇f and ∇f^* are inverse of each other, and we say that x and x^* form a complementary pair if $x = \nabla f^*(x^*)$ and $x^* = \nabla f(x)$.

Next, consider a concave function $g : \mathbb{R}^n_+ \mapsto \mathbb{R}$. Its *concave conjugate* is defined similarly as follows:

$$g_*(x^*) = \min_{x \geq 0} \left\{ \langle x, x^* \rangle - g(x) \right\} .$$

Similar to the convex case, we will further assume g to be non-negative, non-decreasing, strictly concave, differentiable, and normalized such that $g(0) = 0$ in this chapter. In this case, the concave conjugate satisfies the following properties:

- g_* is non-positive, non-decreasing, strictly concave, and differentiable;
- $g_{**} = g$;
- ∇g and ∇g_* are inverse of each other, and we say that x and x^* form a complementary pair if $x = \nabla g_*(x^*)$ and $x^* = \nabla g(x)$.

Assuming some mild conditions which hold for all problems in this chapter and, hence, are omitted, the following strong duality holds. It is known as Fenchel's duality theorem.

$$\underset{x \geq 0}{\text{minimize}} \left\{ f(x) - g(x) \right\} = \underset{x^* \geq 0}{\text{maximize}} \left\{ g_*(x^*) - f^*(x^*) \right\} . \tag{3}$$

3.2 An Example: Online Auction of an Item with Production Cost

In the rest of the chapter, we will restrict our attentions to convex programs with linear constraints. In this section, we will consider an online auction of one item with production cost as a simple running example to demonstrate the application of Fenchel's duality as well as how the online primal dual technique works.

Let there be a seller with one item for sale. Let there be n buyers who arrive online. Each buyer i has a value $v_i \in \mathbb{R}_+$ that specifies the maximum price i is willing to pay for a copy of the item. The technique can actually handle much more general settings with multiple heterogeneous items and combinatorial valuations of agents. We consider this simple case an illustrative example in this section, and refer readers to Huang and Kim [13] for further discussions on the general case. For simplicity, let us omit the strategic behaviors of buyers and assume that the valuation of each buyer is revealed to the seller at the buyer's arrival. On the arrival of a buyer i, the seller decides whether to allocate a copy of the item to buyer i or not. The seller may produce an arbitrary number of copies of the item subject to a production cost function f, i.e., producing y copies of the item leads to a production cost of $f(y)$. The goal is to maximize the social welfare, i.e., the sum of values of the buyers for the allocated bundle of items less the production cost.

Below is a natural convex program relaxation of the online combinatorial auction with production costs. Here, we assume for simplicity that f is defined for all non-negative real numbers. Readers may think of it as, e.g., $f(y) = \frac{1}{2}y^2$, for concreteness.

$$
\begin{aligned}
\text{maximize} \quad & \textstyle\sum_{i=1}^{n} v_i x_i - f(y) \\
\text{subject to} \quad & \textstyle\sum_{i=1}^{n} x_i = y \\
& 0 \le x_i \le 1 && i \in [n] \\
& y \ge 0
\end{aligned}
$$

Here, x_i is the indicator of whether buyer i gets a copy of the item, and y is the total number of allocated copies.

The Fenchel's dual convex program can be derived from the Lagrangian dual and the definition of convex conjugates. Taking the Lagrangian dual, we have

$$
\underset{u \ge 0, p}{\text{minimize}} \ \underset{x, y \ge 0}{\text{maximize}} \quad \textstyle\sum_{i=1}^{n} v_i x_i - f(y) + \sum_{i=1}^{n} u_i (1 - x_i) + p\left(y - \sum_{i=1}^{n} x_i\right)
$$

First, consider the maximization problem w.r.t. x_i, namely

$$
\underset{x_i \ge 0}{\text{maximize}} \ (v_i - u_i - p) \cdot x_i =
\begin{cases}
0 & \text{if } u_i + p \ge v_i, \\
+\infty & \text{otherwise.}
\end{cases}
$$

Thus, it imposes a linear constraint $u_i + p \ge v_i$ in the dual problem. Next, consider the maximization problem w.r.t. y, namely

$$
\underset{y \ge 0}{\text{maximize}} \ py - f(y)
$$

By the definition of convex conjugates, the optimal value of the above maximization problem is $f^*(p)$.

In sum, the Lagrangian dual can be simplified as follows:

$$
\begin{aligned}
\text{minimize} \quad & \textstyle\sum_{i=1}^{n} u_i + f^*(p) && \quad (4) \\
\text{subject to} \quad & u_i + p \ge v_i && i \in [n] \\
& u_i \ge 0 && i \in [n]
\end{aligned}
$$

We leave it to interested readers to verify the above program is equivalent to the Fenchel's dual as defined in Equation (3) of the online auction of an item with production cost. In this dual program, we can interpret p as the price for a copy of the item and u_i as the utility of buyer i, i.e., his value for the allocated bundle less the total price.

3.3 Optimality Conditions

Similar to their counterparts for linear programs, the online primal dual algorithms for convex programs are also driven by the optimality conditions of the programs and their dual programs. There are different ways to formulate such conditions. We will use the most familiar one known as the Karush–Kuhn–Tucker (KKT) conditions [14, 15]. We refer readers to Boyd and Vandenberghe [5] for an extensive discussion on the optimality conditions of convex programs. In this chapter, we will explain the conditions only on a problem-by-problem basis. For the running example of online auction of an item with production cost, the conditions are

(a) x_i must be zero unless $u_i + p = v_i$;
(b) u_i must be zero unless $x_i = 1$;
(c) y and p form a complementary pair.

Here, the first two conditions concern primal/dual linear constraints and the corresponding dual/primal variables. They are complementary slackness conditions just like in the case of linear programs. The third condition is about variables involved in the non-linear parts of the primal and dual objectives. It states that they must form complementary pairs in the sense that we defined at the beginning of the section. Next, we will show how one can derive an online algorithm from the principle of satisfying these conditions approximately.

3.4 High-Level Plan

A meta online primal dual algorithm, much like their counterparts for linear programs, proceeds as follows. It maintains a feasible dual at all time. At the beginning, it is just the value of p since none of the u_i's has arrived yet. On the arrival of a buyer i, it decides whether to allocate a copy of the item to i, i.e., the value of x_i, based on the current dual, sets values to the new dual variable u_i, and updates dual variable p. The high-level principle guiding these decisions is to satisfy the aforementioned optimality conditions as much as possible. We shall elaborate how shortly. Finally, the competitive ratio follows by comparing the increments in the primal and dual objectives in each step.

3.5   (Approximate) Complementary Slackness

Recall that the KKT conditions for linear constraints are the same as complementary slackness. An online primal dual algorithm handles these conditions the same way as in the original approach for linear programs.

In particular, let us consider conditions (a) and (b) in our running example. On buyer i's arrival, a new dual variable u_i shows up and it is subject to two constraints $u_i + p \geq v_i$ and $u_i \geq 0$. Therefore, if $p > v_i$, the first constraint cannot hold with equality and x_i must be 0, i.e., we must not allocate a copy of the item to i, according to condition (a). On the other hand, if $p < v_i$, the value of u_i must be positive, and x_i must be 1, i.e., we must allocate a copy of the item to i, according to condition (b). The decision in the tie-breaking case when $v_i = p$ does not matter; we will assume that the algorithm does allocate a copy in the tie-breaking case. To this end, the algorithm shall interpret the current value of the dual variable p as a take-it-or-leave-it price for a copy of the item. It allocates a copy to buyer i if and only if its value v_i is at least the price p. Further, it shall let $u_i = \max\{v_i - p, 0\}$. Doing so will satisfy the complementary slackness conditions (a) and (b) at the end of i's arrival. However, the algorithm may increase p in the future at which point condition (a) will be violated. Condition (b), on the other hand, will be satisfied exactly since x_i and u_i will not change in the future.

3.6   (Approximate) Complementary Pairs

So far, we have pinned down how to decide the allocation, i.e., the value of x_i, at each buyer i's arrival based on the current dual, in particular the current price p. We have also explained how to set u_i accordingly. There is still a missing piece, namely how to update the price p, after i's arrival.

This last piece of the algorithm is driving by the last condition. Recall it says that in the offline primal and dual solutions, y and p form a complementary pair, i.e., $p = f'(y)$. This coincides with the economic intuition, namely the unit price of a copy of the item shall equal to its marginal production cost. In the online problem, however, the algorithm knows only the current demand y of the item, but not its final value at the end of the day. As a result, the algorithm needs to in some sense predict the final demand according to the current demand, and set the price according to the predicted final demand. For example, we may simply predict the final demand to be twice the current demand. It turns out this simple heuristic already gives reasonably good competitive ratio for nice production cost functions such as polynomials. For simplicity of exposition, we will use a slight variant of this simple heuristic and assume a specific production function $f(y) = \frac{1}{2}y^2$ to explain the primal dual analysis in the rest of the section.

3.7 Online Primal Dual Algorithms

Putting together, consider the following online primal dual algorithm:

1. Initially, $y = 0$, $p = f'(2(y + 1)) = 2$.
2. On the arrival of each buyer i:

    2a. Let $x_i = 1$ if $v_i \geq p$, and let $x_i = 0$ otherwise.
    2b. Let $u_i = \max\{v_i - p, 0\}$.
    2c. Update $y = y + x_i$ and, subsequently, $p = f'(2(y + 1)) = 2(y + 1)$.

We remark that a more principled approach is to leave the prediction mapping as an unknown function g, i.e., let $g(y)$ be the predicted final demand if the current demand is y and set $p = f'(g(y))$, and to derive the optimal prediction mapping from the analysis. Interested readers are referred to Huang and Kim [13] for the details.

3.8 Primal Dual Analysis

Recall that $f(y) = \frac{1}{2}y^2$. Hence, we have $f^*(p) = \frac{1}{2}p^2$. Clearly, the algorithm maintains feasible primal and dual at all time by design. It remains to compare the increments of the objectives due to the arrival of each buyer i.

If i does not get a copy of the item, both the primal and the dual objectives remain the same. So the increments are both zero and, thus, equal.

If i does get a copy of the item, the primal objective changes by v_i, the gain from allocating a copy to i, less $f(y + 1) - f(y) = y + \frac{1}{2}$, the marginal cost of producing an extra copy of the item, where y denotes the demand for the item before i's arrival. The changes in the dual, on the other hand, equals the utility of i, $u_i = v_i - p = v_i - 2(y + 1)$, plus the change due to the update of price p, $f^*(2(y + 2)) - f^*(2(y + 1)) = 4y + 6$. We claim that the increment in the dual objective is at most 4 times that in the primal objective. That is,

$$4\left(v_i - (y + \tfrac{1}{2})\right) \geq v_i - 2(y + 1) + (4y + 6)$$

By the coefficient of v_i and that $v_i \geq p = 2(y + 1)$ (otherwise, i would not have gotten a copy of the item), it suffices to show the inequality for $v_i = 2(y + 1)$. In that case, the inequality becomes

$$4y + 6 \geq 4y + 6$$

and holds with equality.

Note that the initial value of the primal objective is 0 while that of the dual objective is 2 because the initial value of $p = 2$. We have that:

$$4 \cdot \text{Primal} \geq \text{Dual} - 2$$

So the online primal dual algorithm is 4-competitive.

4 Application: Online Matching with Concave Returns

In this section, we will talk about the problem of online matching with concave returns. The problem and its analysis were originally from Devanur and Jain [11]. To be consistent with the other parts of the chapter, our exposition of the results and techniques will be somewhat different from the original version. Nonetheless, the core maths and the underlying ideas are the same.

4.1 Problem Definition

Let there be m agents (left-hand side offline vertices) that are known upfront. Let there be n items (right-hand side online vertices) that arrive online. On the arrival of an item, the online algorithm must immediately decide which agent shall get the item. Let v_{ij} denote agent j's value for item i. Let there be a concave, non-decreasing function $g : \mathbb{R}_+ \mapsto \mathbb{R}_+$ such that an agent j's value for getting a bundle S of items is $g(\sum_{i \in S} v_{ij})$. We shall interpret g as a discount function of an agent's value for bundles of items. Further, we do not discount the value if no items have been allocated to the agent so far, i.e., $g'(0) = 1$. This is without loss of generality up to scaling of the g function. The goal is to allocate the items to the agents to maximize the sum of values of all agents.

We remark that the online primal dual technique can in fact handle a more general version where different agents have different discount functions. We will omit the general version in this chapter for simplicity of discussion. Interested readers are referred to Devanur and Jain [11].

Next, recall that the concave conjugate of g is denoted as g_*. We give a convex program relaxation of the problem and its dual program below:

$$
\begin{array}{lll}
\text{maximize} & \sum_{j=1}^m g(y_j) & \\
\text{subject to} & \sum_{i=1}^n x_{ij} = 1 & i \in [n] \\
& y_j = \sum_{i=1}^n v_{ij} x_{ij} & j \in [m] \\
& x_{ij}, y_j \geq 0 & i \in [n], j \in [m]
\end{array}
$$

$$
\begin{array}{lll}
\text{minimize} & \sum_{i=1}^n \alpha_i - \sum_{j=1}^m g_*(\beta_j) & \\
\text{subject to} & \alpha_i \geq v_{ij} \beta_j & i \in [n], j \in [m]
\end{array}
$$

Here, x_{ij} is the indicator of whether item i is allocated to agent j. y_j denote the sum of values of the items allocated to agent j. We will focus on solving the fractional problem. That is, x_{ij}'s may take fractional values between 0 and 1.

4.2 Relaxed KKT Conditions

The online primal dual algorithms are driven by the optimality conditions of the primal and dual programs. We list below the relaxed conditions.

(a) x_{ij} must be zero unless $\alpha_i = v_{ij}\beta_j$ when item i comes;
(b) $\beta_j = \beta(y_j)$ is a function of the current total value y_j.

Before moving on to the design of online algorithms based on these conditions, let us briefly discuss how they differ from the exact versions and why relaxations are necessary in the online setting. The exact version of (a) requires the condition to hold at the end of the algorithm. In the online problem, however, the constraint may gain slack in the future because the algorithm may decrease β_j's in the future. The exact version of (b) requires that y_j and β_j form a complementary pair. Similar to the previous example of online auction with production cost, instead of setting β_j based on the current value of y_j, a smarter algorithm shall anticipate y_j to further increase in the future and predict its final value. Hence, we set $\beta_j = \beta(y_j)$ to be a function of y_j where the function will be chosen based on the analysis.

4.3 Online Primal Dual Algorithms

The algorithm maintains a pair of feasible primal and dual at all time. Initially, the primal has only the y_j's which will be set to 0. The dual has only the β_j's which will be set to $\beta(0)$ according to condition (b).

Then, on the arrival of an item i, the relaxed optimality conditions (a) and (b) determine the allocation of the item. Concretely, conditions (a) and (b), together with the need of maintaining a feasible dual, indicate that i shall be allocated to the agent j such that $v_{ij}\beta_j$ is maximized. Here, β_j serves as a discount factor of the value of allocating an item to agent j. Therefore, it makes sense to have the discount function start from $\beta(0) = g'(0) = 1$ and be non-increasing in the agent's total value. Once the item is allocated to an agent, the algorithm updates y_j accordingly. Finally, we need to update the dual variables β_j as a result of the change in y_j.

The actual algorithm is slightly different from the above in that it allocates each item fractionally to multiple agents since we consider the fractional problem. We present the algorithm below.

1. Initially, let $y_j = 0$ and $\beta_j = \beta(y_j) = 1$ for $j = 1, \ldots, m$.
2. Maintain $\beta_j = \beta(y_j)$ and $y_j = \sum_{i=1}^{n} v_{ij} x_{ij}$ throughout.
3. On the arrival of each item i, initialize all x_{ij}'s to be zero, and continuously increase x_{ij}'s as follows until $\sum_{j=1}^{m} x_{ij} = 1$:

   3a. Let $j^* = \arg\max_{j \in [m]} v_{ij}\beta_j$, breaking ties arbitrarily.
   3b. Increase x_{ij^*} by dx and α_i by $v_{ij^*}\beta_{j^*} \cdot dx$.
   3c. Update y_j's and, thus, β_j's accordingly.

4.4   Online Primal Dual Analysis

Again, since the algorithm mains a pair of feasible primal and dual assignments at all time, it remains to compare the changes in the primal and dual objectives due to the arrival of each item. In particular, we would like to show that as the algorithm continuously increases x_{ij}'s in step 3 of the algorithm, it always holds that $F \cdot dP \geq dD$ for some fixed parameter $F \geq 1$, where dP and dD are the changes of the primal and dual objectives, respectively.

When the algorithm increases x_{ij*} by some infinitesimal amount dx in step 3b (recall that j^* is the agent with the largest discounted value $v_{ij}\beta_j$), by simple calculus, the primal objective increases by:

$$dP = g'(y_{j*})dy_{j*}$$

The dual objective, on the other hand, increases by:

$$dD = \underbrace{v_{ij*}\beta_{j*}dx}_{\text{due to }\alpha_i} - \underbrace{g'_*(\beta_{j*})d\beta_{j*}}_{\text{due to }\beta_{j*}} = \beta_{j*}dy_{j*} - g'_*(\beta_{j*})d\beta_{j*}$$

Therefore, to ensure $F \cdot dP \geq dD$, it boils down to choosing a non-increasing function β such that the following differential equation is feasible for the smallest possible $F \geq 1$:

$$\forall y \geq 0: \quad F \cdot g'(y) \cdot dy \geq \beta(y) \cdot dy - g'_*(\beta(y)) \cdot d\beta(y) \tag{5}$$

subject to the boundary conditions that $\beta(0) = g'(0) = 1$ and $\beta(y) \geq 0$ for all $y \geq 0$.

Then, we have recovered the main algorithmic result in Devanur and Jain [11].

Theorem 1 *Suppose there is a non-increasing function β that satisfies the differential equation (5) for some $F \geq 1$. Then, there is an F-competitive online primal dual algorithm for the online matching problem with concave return function g.*

4.5   An Example: Additive Agents with Budgets

Differential equation (5) may look mysterious to the readers. For concreteness, we next present the optimal solution to the differential equation for a concrete example, namely $g(y) = \min\{y, B\}$ or some $B > 0$. This corresponds to the Adwords problem, where the agents are the advertisers and the items are ad slots. An advertiser has a budget B that caps its contribution to the seller's revenue. Again, one may consider a more general case when different advertisers have different budgets, which can be solved under the same framework. We will assume equal budget for the simplicity of our discussions.

In this case, the conjugate function g_* is

$$g_*(y^*) = \begin{cases} B(y^* - 1) & \text{if } y^* \leq 1 \\ 0 & \text{if } y^* > 1 \end{cases}$$

Therefore, in the special case, the differential equation becomes the following (recall that $\beta(0) = 1$ and therefore $\beta(y) \leq 1$ for all y):

$$\forall 0 \leq y \leq B: \quad F \cdot dy \geq \beta(y) \cdot dy - B \cdot d\beta(y)$$

and

$$\forall y > B: \quad 0 \geq \beta(y) \cdot dy - B \cdot d\beta(y)$$

with boundary conditions that $\beta(0) = 1$ and $\beta(y) \geq 0$ for all $y \geq 0$.

Note that β is non-increasing. It implies that the second part concerning $y > B$ is feasible only when $\beta(y) = 0$ for all $y = B$.

Next, we solve for the best β to satisfy the first part for $0 \leq y \leq B$ for the smallest possible F, subject to the boundary condition that $\beta(0) = 1$ and $\beta(B) = 0$. It is easy to check that it suffices to consider the differential equation with equality. Rearrange terms, it becomes

$$\forall 0 \leq y \leq B: \quad \big(F - \beta(y)\big) \cdot dy = B \cdot d\big(F - \beta(y)\big)$$

Together with $\beta(0) = 1$, we get that:

$$F - \beta(y) = (F - 1) \cdot e^{y/B}$$

Plugging in $\beta(B) = 0$, we get that

$$F = F - \beta(B) = (F - 1) \cdot e$$

which implies $F = \frac{e}{e-1}$. This is the best possible competitive ratio as it matches known hardness results from previous work (e.g., [8, 16]).

The corresponding discount function β that achieves this optimal ratio is

$$\beta(y) = \frac{1}{e-1}\big(e - e^{y/B}\big)$$

As a concluding remark, the primal dual technique gives optimal competitive ratio not only for the above example of additive agents with budgets, but in fact for arbitrary concave functions g. The cool thing about it is that Devanur and Jain [11] established the optimality of the algorithm without explicitly giving the optimal competitive ratio, other than the characterization using the differential equation (5). Instead, they constructed a family of hard instances that are parameterized by a function and showed that solving for the worst function is equivalent to solving the differential equation (5). This chapter focuses on positive results and will not get into further details of the hardness. Readers are referred to Devanur and Jain [11].

5 Application: Online Scheduling with Speed Scaling

This section considers another application of the online primal dual framework in online optimization problems with non-linear objectives. We will consider an online scheduling problem with speed-scalable machines, where the energy consumption is naturally a convex function of the speed. The results were originally from Devanur and Huang [10], which built on an earlier dual fitting approach by Anand et al. [2].

5.1 Problem Definition

Let there be m machines that are known upfront. Let there be n jobs that arrive online. Each machine j can run at different speeds subject to different energy consumptions. Let f be the power function such that running a machine at speed y consumes $f(y)$ energy per unit of time. For concreteness, readers may think of $f(y) = \frac{1}{3}y^3$ as a typical cubic energy function. Each job i is defined by its arrival time r_i, and its volumes v_{ij} when allocated to each machine j. That is, we consider the unrelated machine setting in which different machines may take different amount of computational resources to complete the same job. A feasible schedule allocates each job to a machine and more specifically a subset of time slots on the machine, and specifies how fast the machine runs in each time slot, such that the total computational resources assigned to the job, i.e., the sum of speeds of the allocated time slots, are equal to its volume on the machine.

Again, the primal dual framework can in fact handle a more general setting where (1) jobs have weights and (2) machines may have different energy functions. We will omit such generalizations and refer interested readers to Devanur and Huang [10].

There are two natural objectives for this problem. The first one is to minimize total energy consumption, and the second is to minimize the total delay experienced by the jobs. Here, there are different ways to define delay. We will consider a popular one known as flow time. The flow time of a job is the difference between its completion time and arrival time. The flow time of a feasible schedule is the sum of the flow time of all jobs. Minimizing either objective without considering the other is trivial. One could get arbitrarily small flow time by running the machines at extremely high speed, but paying an arbitrarily high energy consumption. One could also have arbitrarily small energy consumption by running the machines at extremely low speed, but suffering an arbitrarily large flow time. We will use the standard combination of minimizing the sum of flow time and energy.

This problem is more complicated than the previous examples in the sense that the online algorithm needs to make multiple types of online decisions. At each time slot, the online algorithm must decide which job shall be run on each machine (*job selection*) and at what speed (*speed scaling*). Further, on the arrival of a job, the online algorithm must immediately dispatch the job to one of the machines (*job assignment*). We consider the preemptive model in which the algorithm may preempt the current job processing on a machine with the new job and resume after the new job is finished.

The convex program relaxation and its dual are given below.

$$\text{minimize} \quad \sum_{i=1}^{n} \sum_{j=1}^{m} \int_{r_i}^{+\infty} (t - r_i)(x_{ijt}/v_{ij})dt$$

$$+ \sum_{j=1}^{m} \int_{0}^{+\infty} f(y_{jt})dt$$

$$+ \sum_{i=1}^{n} \sum_{j=1}^{m} \int_{r_i}^{+\infty} (f^*)^{-1}(1)x_{ijt}dt$$

$$\text{subject to} \quad \sum_{j=1}^{m} \int_{r_i}^{+\infty} (x_{ijt}/v_{ij})dt = 1 \qquad\qquad i \in [n]$$

$$y_{jt} = \sum_{i=1}^{n} x_{ijt} \qquad\qquad j \in [m], t \geq 0$$

$$x_{ijt}, y_{jt} \geq 0 \qquad\qquad i \in [n], j \in [m], t \geq 0$$

$$\text{maximize} \quad \sum_{i=1}^{n} \alpha_i - \sum_{j=1}^{m} \int_{0}^{+\infty} f^*(\beta_{jt})dt$$

$$\text{subject to} \quad \alpha_i \leq v_{ij}\beta_{jt} + t - r_i + v_{ij}(f^*)^{-1}(1) \qquad i \in [n], j \in [m], t \geq r_i$$

Here, x_{ijt} is the speed of machine j at time t that is devoted to processing job i. y_{jt} is the total speed of machine j at time t. The first primal constraint is a relaxation of the feasibility constraint. Recall the original constraint says that each job must be completed on a machine j such that the sum of speed devoted to the job is equal to v_{ij}, i.e., $\int_{r_i}^{+\infty} x_{ijt}dt = v_{ij}$. The relaxed constraint, on the other hand, allows the job to be fractionally processed on different machines so long as the overall effort processes the entire job. The second primal constraint says that the speed of a machine at any time must be high enough to execute the schedule given by x_{ijt}'s.

The first term in the primal objective is a relaxation of flow time in the sense that the fraction of a job i that is completed at time t experiences a flow time defined by the current time, i.e., $t - r_i$, instead of the flow time given by the completion of the job. This is usually known as the fractional flow time of the schedule, which is widely used as an intermediate objective in the study of flow time minimization.

The second term is exactly the energy consumption.

It turns out that these two terms on their own suffer from a large integrability gap. Consider the case of having only one job arriving at time 0 with volume v on every machine. An integral schedule must process the job on the same machine and hence pays a non-zero flow time plus energy. Interested readers may verify that the minimum cost equals $v \cdot (f^*)^{-1}(1)$. A feasible schedule of the relaxed linear program, however, can split the workload and process a $1/m$ fraction of the job on each machine. Such a fractional schedule effectively pays 0 in the objective when m goes to infinity. Introducing the third term fixes this problem. Since the minimum cost of assigning a job of volume v is $v \cdot (f^*)^{-1}(1)$, the extra unit cost of $(f^*)^{-1}(1)$ for the x_{ijt} volume of a job i that is processed on a machine j at time t sums to at most the actual flow time plus energy for any feasible integral schedule. Hence, adding the third term increases the optimal by at most a factor of 2.

5.2 A Simple Online Primal Dual Algorithm

As usual, we start with a set of relaxed optimality conditions that drive the design of the online primal dual algorithms.

(a) If a job i is dispatched to a machine j and is tentatively scheduled at time t when it comes, it must be that $\alpha_i = v_{ij}\beta_{jt} + t - r_i + v_{ij}(f^*)^{-1}(1)$ (there may be some slack in the future because the algorithm may increase β_{jt}'s);
(b) $\beta_{jt} = \beta(y_{jt})$ is a function of the current total speed y_t.

Condition (a) is a relaxation in the sense that the actual complementary slackness condition holds for the optimal primal and dual assignments at the end of the instance, while here it holds only at the moment when i comes. Again, this is unavoidable and shall look standard to the readers after seeing similar conditions in the previous two examples. Condition (b) is a relaxation of the complementary pair condition saying that y_{jt} and β_{jt} will form a complementary pair. Intuitively, the online algorithm will predict the final value of y_{jt}, i.e., the speed of machine j at time t, based on its current value, and set the dual variable β_{jt} to form a complementary pair with the predicted final value.

We present below the meta online primal dual algorithm driven by the above relaxed optimality conditions, given any function $\beta : \mathbb{R}_+ \mapsto \mathbb{R}_+$.

1. Initially, let $y_{jt} = 0$ and $\beta_{jt} = \beta(0)$ for $j = 1, \dots, m, t \geq 0$.
2. Maintain $\beta_{jt} = \beta(y_{jt})$ and $y_{jt} = \sum_{i=1}^{n} x_{ijt}$ throughout.
3. On the arrival of each job i:

   3a. Let j^* and t^* be such that $v_{ij^*}\beta_{j^*t^*} + t^* - r_i$ is minimized. Dispatch the job i to machine j^* and tentatively schedule it at time t^*.
   3b. Update y_{j^*t}'s and, thus, β_{j^*t}'s accordingly.

5.3 A Simple Greedy Algorithm

In this chapter, we will consider a particularly simple greedy algorithm and analyze it using the online primal dual framework.

The greedy algorithm processes the current jobs using the optimal schedule assuming that no other jobs will arrive in the future. In particular, given how the current jobs are assigned to the machines, the greedy algorithm processes the jobs on each machine from the shortest to the longest. The shortest job first principle is the best strategy assuming no future jobs will arrive. To see the intuition behind its optimality, note that the processing time of the first job will contribute to the flow time of all remaining jobs, that of the second job will contribute to the flow time of all but one remaining jobs, and so on. Hence, it makes sense to prioritize on shorter jobs.

In terms of the choice of speed, it will run the machine at a speed that depends on the number of remaining jobs on the machines. Concretely, suppose there are k remaining jobs. Then, the speed y is set such that $f^*(f'(y)) = k$. To see why this is the right choice, suppose we process a job, say, with volume v, at speed y. Then, the amount of time needed to process the job is v/y. All k jobs will suffer from this delay in their flow time and, thus, the total contribution to the objective in terms of flow time is $(v/y) \cdot k$. On the other hand, the energy consumption during this period of time is $(v/y) \cdot f(y)$. Therefore, the optimal speed is the one that minimizes $(v/y) \cdot (k + f(y))$. Then, it is easy to verify that our choice of speed minimizes this quantity, which equals $v \cdot (f^*)^{-1}(k)$.

Finally, we will use a simple β mapping such that $\beta_{jt} = f'(y_{jt})$ forms a complementary pair with y_{jt}. When a new job arrives, the algorithm dispatches the job according to step 3a in the meta algorithm. By our choice of speed, for any machine j and any time t, we have $\beta_{jt} = (f^*)^{-1}(k)$ where k is the number of remaining jobs on machine j at time t. Hence, the right-hand side of the dual constraint becomes

$$v_{ij}\beta_{jt} + t - r_i + v_{ij}(f^*)^{-1}(1) = t - r_i + v_{ij}\big((f^*)^{-1}(k) + (f^*)^{-1}(1)\big)$$

$$\geq t - r_i + v_{ij}(f^*)^{-1}(k+1)$$

Here, the inequality follows by the concavity of $(f^*)^{-1}$, which is due to the convexity of f^*. Having a closer look at the right-hand side of the above inequality, the $t - r_i$ term corresponds to the flow time for having job i wait until time t before we process it. The last term, on the other hand, is the energy consumption of processing job i on machine j plus the delay it causes to the flow time of the existing k jobs and itself, provided that we run the machine with the aforementioned greedy speed. Hence, we shall interpret the right-hand side as the total increase in the objective if we decide to dispatch the job to machine j and tentatively schedule it at time t. Therefore, our job dispatch rule (step 3a) simply greedily minimizes this increment.

5.4 Online Primal Dual Analysis

Following the above discussion, we shall let α_i be the increment in the objective due to the arrival of job i. Recall that the dual is equal to

$$\sum_{i=1}^{n} \alpha_i - \sum_{j=1}^{m} \int_0^{+\infty} f^*(\beta_{jt})dt$$

The first term, by definition, sums to the objective of flow time plus energy consumption since each α_i accounts for the increment in the objective due to the corresponding job i. The second term, on the other hand, is equal to the flow time because $f^*(\beta_{jt}) = f^*(f'(y_{jt}))$ equals the number of remaining jobs on machine j at time t by our choice of y_{jt} and β_{jt}. Hence, the dual objective is in fact extremely simple for the greedy algorithm: It equals the energy consumption.

So far, we did not use any special property of the power function f. For arbitrary power functions, no competitive algorithms exist for this problem. Next, we focus on polynomial power functions $f(y) = \frac{1}{\alpha} y^\alpha$, where α is between 2 and 3 for typical power functions in practice. In this case, we have $f^*(\beta) = (1 - \frac{1}{\alpha})\beta^{\alpha/(\alpha-1)}$ and, thus, our choice of speed implies that $k = f^*(f'(y)) = (1 - \frac{1}{\alpha})y^\alpha$. Therefore, at any time, assuming there are k jobs remaining, the contribution (per unit of time) to flow time, i.e., $k = (1 - \frac{1}{\alpha})y^\alpha$ is exactly $\alpha - 1$ times the contribution to energy consumption, i.e., $f(y) = \frac{1}{\alpha} y^\alpha$. Therefore, the fact that the dual objective equals the energy consumption means that it is exactly a $1/\alpha$ fraction of the flow time plus energy of the algorithm. This implies the following theorem.

Theorem 2 *The greedy algorithm is $O(\alpha)$-competitive for minimizing flow time plus energy when the power function is $f(y) = \frac{1}{\alpha} y^\alpha$.*

5.5 Better Online Primal Dual Algorithms

The main drawback of the simple algorithm lies in that it fails to anticipate future jobs and is conservative in its choice of speed. We can improve the greedy algorithm with a speed-up. In particular, the greedy algorithm achieves the optimal flow time plus energy if there are indeed no future jobs, but is far from optimal when there are a lot of future jobs. A smarter algorithm shall run faster than the optimal schedule of the remaining instance to hedge between the two cases: If there are no future jobs, this approach pays more in energy comparing to the simple greedy algorithm; if there are a lot of future jobs, however, it is better than the simple greedy algorithm because it has predicted the arrival of future jobs and has run faster in the past.

Specifically, Devanur and Huang [10] considered a family of such algorithms parameterized by a parameter $C \geq 1$ such that the algorithm runs C times faster than the optimal schedule of the remaining instance. Such an algorithm is called the C-aggressive greedy algorithm, as given below.

Speed Scaling:

– Choose speed s.t. $f^*(f'(\frac{y_{jt}}{C}))$ equals the number of remaining jobs.
– Let $\beta_{jt} = \frac{1}{C} f'(\frac{y_{jt}}{C})$ s.t. $f^*(C\beta_{jt})$ equals the number of remaining jobs.

Job Selection:

– Schedule jobs from the shortest to the longest.

Job Assignment:

– Dispatch the job to a machine that minimizes the increment in the objective.

Moreover, Devanur and Huang [10] showed an online primal dual analysis of the C-aggressive greedy algorithms for all $C \geq 1$ and chose C to balance the case with no future jobs and the case with a lot of future jobs. By doing so, they showed the following result. We refer readers to Devanur and Huang [10] for the proofs.

Theorem 3 *For polynomial power function $f(y) = \frac{1}{\alpha}y^\alpha$, the C-aggressive greedy online primal dual algorithm is $O\left(\frac{\alpha}{\log \alpha}\right)$-competitive for minimizing flow time plus energy on unrelated machines with $C \approx 1 + \frac{\log \alpha}{\alpha}$.*

If there are at least two machines, then Devanur and Huang [10] further showed that the above competitive ratio is asymptotically tight.

Theorem 4 *For polynomial power function $f(y) = \frac{1}{\alpha}y^\alpha$, there are no $o\left(\frac{\alpha}{\log \alpha}\right)$-competitive online algorithm for minimizing fractional flow time plus energy with at least two machines.*

Recall that in the previous example of online matching, the upper and lower bounds match exactly as they both reduce to the same differential equation. In this example of online scheduling for minimizing flow time plus energy, however, there is a constant gap between the upper and lower bounds. One of the reasons is that the algorithms in this example assign jobs to machines integrally, i.e., a job cannot be processed on multiple machines in parallel as in a fractional solution to the convex program relaxation. We do not know whether we can derive the same form of tight upper and lower bounds as in the previous two examples if we allow parallel processing and truly focus on solving the primal convex program online.

Further, if there is only one machine, Bansal et al. [4] gave a 2-competitive online algorithm for minimizing fractional flow time plus energy with arbitrary power functions using a potential function argument. It is an interesting open question whether there is a 2-competitive online primal dual algorithm for single machine.

Finally, essentially the same algorithm and analysis can be further used to derive better competitive ratios in the resource augmentation setting, where the online algorithm can run the machines $1 + \epsilon$ times faster than the offline benchmark using the same amount of energy.

Theorem 5 *There is a $(1 + \epsilon)$-speed and $O\left(\frac{1}{\epsilon}\right)$-competitive online primal dual algorithm for minimizing flow time plus energy with arbitrary power functions.*

6 Application: Online Covering and Packing Problems with Convex Objectives

As the last example of the chapter, we present a result by Azar et al. [3] that gave online algorithms for a large family of covering and packing problems with convex objectives. It generalizes the original online covering and packing problems that considers linear objectives (e.g., Buchbinder and Naor [7]).

6.1   Problem Definition

We study the following online covering and packing problems with convex objectives. The (offline) covering problem is modeled by the following convex program:

$$\min_{x \in \mathbb{R}^n_+} f(x) \text{ s.t. } Ax \geq 1$$

where $f : \mathbb{R}^n_+ \mapsto \mathbb{R}_+$ is a monotone and convex cost function, and A is an $m \times n$ matrix with non-negative entries. Each row of the constraint matrix A corresponds to a covering constraint. In the online problem, the rows of A come online and the algorithm must maintain a feasible assignment x that is non-decreasing over time.

The (offline) packing problem, on the other hand, is modeled by the following convex program:

$$\max_{y \in \mathbb{R}^m_+} \sum_{j=1}^{m} y_j - f^*(A^T y)$$

where $f^* : \mathbb{R}^n_+ \mapsto \mathbb{R}_+$ is the convex conjugate of f. It is the Fenchel's dual program of covering problem. In the online problem, each variable y_j arrives online and the algorithm must decide the value of y_j on its arrival.

When f is a linear function in the covering problem, e.g., $f(x) = \langle c, x \rangle$ for some positive vector c, f^* becomes a 0-∞ step function that imposes supply constraints $A^T y \leq c$ in the packing problem. Then, the problems become the original online covering and packing problems with linear objectives.

In the rest of the section, we will focus on the packing problem, and refer readers to Azar et al. [3] for the algorithms and the corresponding analysis for the covering problem. In other words, we will take the packing program as the primal program and the covering one as the dual.

6.2   Relaxed Optimality Conditions

As usual, we will start with a set of relaxed optimality conditions for the (primal) packing program and the (dual) covering program. Let $A_j = (a_{j1}, a_{j2}, \ldots, a_{jn})$ denote the j-th row of matrix A.

(a) y_j must be zero unless $A_j x = 1$ at the end of the round when covering constraint j and, thus, packing variable j, arrive.
(b) $x = \nabla g(A^T y)$ for some convex function g.

Condition (a) is a relaxation of the actual complementary condition, which holds for the final value of x and y. Condition (b) is a relaxation of the condition that x and $A^T y$ form a complementary pair with respect to f and f^*, i.e., $x = \nabla f^*(A^T y)$. Just like in the previous examples, a smart algorithm shall anticipate that the variables y and, thus, $A^T y$ will further increase in the future and set x accordingly.

6.3   Online Primal Dual Packing Algorithm

Driven by the above relaxed optimality conditions, we can define the following meta algorithm that depends on the choice of a convex function g.

1. Initialize $x = z = 0$.
2. Maintain $x = \nabla g(z)$ and $z = A^T y$ throughout.
3. When y_k and $A_k = (a_{k1}, \ldots, a_{kn})$ arrive in round k, do the followings:

   2a. Initialize $y_k := 0$;
   2b. While $\sum_{i=1}^{n} a_{ki} x_i < 1$, continuously increase y_k and do the followings:

   (2b.1) Simultaneously for each $i \in [n]$, increase z_i at rate $\frac{dz}{dy_k} = a_{ki}$.
   (2b.2) Increase x according to $x = \nabla g(z)$.

Here, the vector x plays an auxiliary role and is initialized to 0. We can interpret x as a price vector such that x_i is the unit price of the i-th resource in the packing problem. Then, we keep including more of the k-th item into the packing solution as long as its unit gain of 1 can pay for the total price of the resources that it demands.

We will consider a particularly simple mapping function in this chapter:

$$g(z) = \tfrac{1}{\rho} f^*(\rho z) = \tfrac{1}{\rho} f^*(\rho A^T y)$$

and, thus, $x = \nabla g(z) = \nabla f^*(\rho z)$, for some parameter $\rho > 1$ to be determined in the analysis. Intuitively, this means that the algorithm anticipates $A^T y$ will further increase by a factor of ρ in the future, and pick x accordingly.

In round $k \in [m]$, the vector $a_k = (a_{k1}, a_{k2}, \ldots, a_{kn})$ is revealed. The variable y_k is initialized to 0, and is continuously increased while $\sum_{i \in [n]} a_{ki} x_i < 1$, i.e., the total price for the amount of resources needed to produce a unit of resource k is less than 1. To maintain $z = A^T y$, for each $i \in [n]$, z_i is increased at rate $\frac{dz}{dy_k} = a_{ki}$. As the coordinates of z are increased, the vector x is increased according to the invariant $x = \nabla f^*(\rho z)$. We shall assume that ∇f^* is monotone. Hence, both x and z increase monotonically as y_k increases.

6.4   Online Primal Dual Analysis

For simplicity of our discussion, we will make an assumption that f^* is a homogeneous polynomial with non-negative coefficients (so that the gradient is monotone) and degree $\lambda > 1$. The competitive ratio will depend on λ.

We first show that the algorithm will not keep increasing y_k forever in any round k and it maintains a feasible primal solution throughout. Concretely, the following lemma indicates that unless the offline packing problem is unbounded, eventually $\sum_{i \in [n]} a_{ki} x_i$ will reach 1. At this moment, y_k will stop increasing and we complete round k. Recall that the coordinates of x increase monotonically as the algorithm proceeds. It implies that the covering constraints $\sum_{i \in [n]} a_{ji} x_i \geq 1$ are satisfied for all $j \in [m]$ at the end of the algorithm since each constraint is satisfied at the end of the corresponding round. Hence, the vector x is feasible for the dual covering problem.

Lemma 1 *Suppose that the offline optimal packing objective is bounded. Then, in each round $k \in [m]$, eventually we have $\sum_{i \in [n]} a_{ki} x_i \geq 1$, and y_k will stop increasing.*

Proof We have

$$\frac{\partial P(y)}{\partial y_k} = 1 - \langle a_k, \nabla f^*(z) \rangle$$

$$= 1 - \frac{1}{\rho^{\lambda - 1}} \cdot \langle a_k, \nabla f^*(\rho z) \rangle$$

$$= 1 - \frac{1}{\rho^{\lambda - 1}} \cdot \langle a_k, x \rangle$$

$$\geq 1 - \frac{1}{\rho^{\lambda - 1}}$$

Here, the first equality is due to $z = A^T y$, which indicates that when the algorithm increases y_k, it also increases each z_i at rate $\frac{dz_i}{dy_k} = a_{ki}$. The second equality follows by that f^* is a homogeneous polynomial of degree λ. The third equality is because of our choice of x. Finally, the last inequality follows because $\langle a_k, x \rangle < 1$ when y_k is increased by the definition of the algorithm.

Suppose for contrary that $\langle a_k, x \rangle$ never reaches 1. Then, the objective function $P(y)$ increases at least at some positive rate $1 - \frac{1}{\rho^{\lambda - 1}}$ (recalling $\rho > 1$ and $\lambda > 1$) as y_k increases, which means the offline packing problem is unbounded, contradicting our assumption.

To complete the competitive analysis, it remains to compare the primal and dual objectives. In the rest of the section, for $k \in [m]$, we let $z^{(k)}$ denote the vector z at the end of round k, where $z^{(0)} := 0$. Let $P(y)$ denote the packing objective for any vector y and the induced vector $z = A^T y$. Similarly, let $C(x)$ denote the covering objective for any feasible covering solution x.

First, let us look into the contribution to the packing objective from y.

Lemma 2 *At the end of round k when y_k stops increasing (by Lemma 1)*

$$y_k \geq \frac{1}{\rho} \cdot \left(f^*(\rho z^{(k)}) - f^*(\rho z^{(k-1)}) \right) .$$

In particular, since $f^*(0) = 0$, this implies that at the end of the algorithm,

$$\sum_{k \in [m]} y_k \geq \frac{1}{\rho} \cdot f^*(\rho z^{(m)}) \ .$$

Proof Recall again that y_k increases only when $\langle a_k, x \rangle < 1$, we have

$$1 \geq \sum_{i \in [n]} a_{ki} x_i = \sum_{i \in [n]} x_i \cdot \frac{dz_i}{dy_k} \ .$$

Hence, integrating this with respect to y_k throughout round k, and observing that $x = \nabla f^*(\rho z)$, we have

$$y_k \geq \int_{z=z^{(k-1)}}^{z^{(k)}} \langle \nabla f^*(\rho z), dz \rangle = \frac{1}{\rho} \cdot (f^*(\rho z^{(k)}) - f^*(\rho z^{(k-1)})) \ ,$$

where the last equality comes from the fundamental theorem of calculus for path integrals of vector fields.

Therefore, the dual objective is lower bounded by:

$$\frac{1}{\rho} f^*(\rho z^{(m)}) - f^*(z^{(m)}) = (\rho^{\lambda-1} - 1) f^*(z^{(m)})$$

Next, we consider the dual (covering) objective. Suppose $z^{(m)} = A^T y$ is the vector at the end of the algorithm, and $x^{(m)} = \nabla f^*(\rho z^{(m)})$. We have the following.

Lemma 3 *The covering objective satisfies that:*

$$C(x^{(m)}) = (\lambda - 1) \rho^{\lambda} f(z^{(m)})$$

Proof By how our algorithm maintains the covering solution, we have

$$\begin{aligned}
C(x^{(m)}) &= f(\nabla f^*(\rho z^{(m)})) \\
&= \max_{z \geq 0} \{ \langle z, \nabla f^*(\rho z^{(m)}) \rangle - f^*(z) \} \\
&= \rho z^{(m)} \nabla f^*(\rho z^{(m)}) - f^*(\rho z^{(m)}) \\
&= (\lambda - 1) f^*(\rho z^{(m)}) \\
&= (\lambda - 1) \rho^{\lambda} f^*(z^{(m)}) \ ,
\end{aligned}$$

where the second equality follows by the definition of convex conjugate functions, the third equality is because the maximum is achieved at $z = \rho z^{(m)}$ by first order condition, the last two equalities are due to our assumption that f^* is a homogeneous polynomial of degree λ.

Putting together our bounds on the packing and covering objectives, we conclude that they are with the following bounded factor from each other:

$$\frac{C(x)}{P(y)} \leq \frac{(\lambda - 1)\rho^\lambda}{\rho^{\lambda-1} - 1} \;, \tag{6}$$

Choosing $\rho := \lambda^{\frac{1}{\lambda-1}}$, the above ratio becomes

$$\lambda^{\frac{\lambda}{\lambda-1}} = O(\lambda)$$

Hence, we conclude that:

Theorem 6 *Suppose f^* is a convex, homogeneous polynomial with non-negative coefficients and degree λ. Then, there is an $O(\lambda)$-competitive online algorithm for the online packing problem with a convex objective defined with f^*.*

References

1. Agrawal, S., Devanur, N.R.: Fast algorithms for online stochastic convex programming. In: Proceedings of the 26th Annual Symposium on Discrete Algorithms. SIAM, Philadelphia (2015)
2. Anand, S., Garg, N., Kumar, A.: Resource augmentation for weighted flow-time explained by dual fitting. In: Proceedings of the 23rd Annual Symposium on Discrete Algorithms, pp. 1228–1241. SIAM, Philadelphia (2012)
3. Azar, Y., Buchbinder, N., Chan, T.H., Chen, S., Cohen, I.R., Gupta, A., Huang, Z., Kang, N., Nagarajan, V., Naor, J., Panigrahi, D.: Online algorithms for covering and packing problems with convex objectives. In: 2016 IEEE 57th Annual Symposium on Foundations of Computer Science (FOCS), pp. 148–157. IEEE, Piscataway (2016)
4. Bansal, N., Chan, H.L., Pruhs, K.: Speed scaling with an arbitrary power function. In: Proceedings of the 20th Annual symposium on discrete algorithms, pp. 693–701. SIAM, Philadelphia (2009)
5. Boyd, S., Vandenberghe, L.: Convex Optimization. Cambridge University Press, Cambridge (2004)
6. Buchbinder, N., Naor, J.: The design of competitive online algorithms via a primal: dual approach. Found. Trends Theor. Comput. Sci. 3(2–3), 93–263 (2009)
7. Buchbinder, N., Naor, J.: Online primal-dual algorithms for covering and packing. Math. Oper. Res. 34(2), 270–286 (2009)
8. Buchbinder, N., Jain, K., Naor, J.S.: Online primal-dual algorithms for maximizing ad-auctions revenue. In: Algorithms–ESA, pp. 253–264. Springer, Berlin (2007)
9. Dantzig, G.B.: Linear Programming and Extensions. Princeton University Press, Princeton (1998)
10. Devanur, N.R., Huang, Z.: Primal dual gives almost optimal energy efficient online algorithms. In: Proceedings of the 25th Annual Symposium on Discrete Algorithms, pp. 1123–1140. SIAM, Philadelphia (2014)
11. Devanur, N.R., Jain, K.: Online matching with concave returns. In: Proceedings of the 44th Annual Symposium on Theory of Computing, pp. 137–144. ACM, New York (2012)

12. Gupta, A., Krishnaswamy, R., Pruhs, K.: Online primal-dual for non-linear optimization with applications to speed scaling. In: Approximation and Online Algorithms, pp. 173–186. Springer, Berlin (2013)
13. Huang, Z., Kim, A.: Welfare maximization with production costs: a primal dual approach. In: Proceedings of the 26th Annual Symposium on Discrete Algorithms. SIAM, Philadelphia (2015)
14. Karush, W.: Minima of functions of several variables with inequalities as side constraints. PhD thesis, Master's Thesis, Department of Mathematics, University of Chicago (1939)
15. Kuhn, H.W., Tucker, A.W.: Nonlinear programming. In: Proceedings of 2nd Berkeley Symposium (1951)
16. Mehta, A., Saberi, A., Vazirani, U., Vazirani, V.: Adwords and generalized online matching. J. ACM **54**(5), 22 (2007)
17. Nguyen, K.T.: Lagrangian duality in online scheduling with resource augmentation and speed scaling. In: Algorithms–ESA, pp. 755–766. Springer, Berlin (2013)

Solving Combinatorial Problems with Machine Learning Methods

Tiande Guo, Congying Han, Siqi Tang, and Man Ding

Abstract With the development of machine learning in various fields, it can also be applied to combinatorial optimization problems, automatically discovering generic and fast heuristic algorithms based on training data, and requires fewer theoretical and empirical knowledge. Pointer network improves the attention mechanism, instead of allocating different attention to hidden states of encoder to generate context vectors, using attention as a pointer to select an element of the input sequence at every step of decoding, which solves the problem of variable dictionary size of the output sequence. Pointer net (Ptr-Net) applied to three combinatorial optimization problems, convex hull, Delaunay triangulation, and traveling salesman problem (TSP), obtains good approximate solutions. Point matching is also a special kind of combinatorial optimization problems that is to obtain the optimal corresponding references, which can be modeled by Ptr-Net. However, Ptr-Net can't be used to solve point matching problem because it doesn't take full advantage of the correspondences between the two point sets. We propose multi-pointer network, which draws the idea from multi-label classification, to address this limitation by pointing out a set of input elements. These applications are all based on supervised learning to approximate expected known solutions. However, high-quality labeled data is often expensive, unreliable, or simply unavailable and may be infeasible for new problem statements, making supervised learning being unpractical. Reinforcement learning, as another research hotspot in the field of machine learning, does not require labeled sample data. It interacts with the environment through trial-and-error mechanism and focuses more on learning problem-solving strategies. We introduce a framework to tackle combinatorial optimization problems using neural networks and reinforcement learning, focusing on the traveling salesman problem. We also introduce a framework, a unique combination of reinforcement learning and graph embedding network, to solve graph optimization problems, focusing on maximum cut (MAXCUT) and minimum vertex cover (MVC) problems.

T. Guo · C. Han (✉) · S. Tang · M. Ding
School of Mathematical Sciences, University of Chinese Academy of Sciences, Beijing, China
e-mail: hancy@ucas.ac.cn

© Springer Nature Switzerland AG 2019
D.-Z. Du et al. (eds.), *Nonlinear Combinatorial Optimization*,
Springer Optimization and Its Applications 147,
https://doi.org/10.1007/978-3-030-16194-1_9

1 Introduction

In the early stage, simple combinatorial optimization problems, such as minimum spanning tree problem [13] and shortest path problem [7], can design convenient and fast algorithm to obtain its optimal solution. With the development of practice, most problems about combination and sequential optimization are in fact NP-hard, so, it's impossible to find a general polynomial time algorithm to solve it. However, NP-hard problems have important applications in many fields, such as social networks, transportation, communications, and scheduling. Therefore, many specialists and scholars apply themselves to the research of the theory and achieve many perfect productions.

Traditional methods to tackle NP-hard optimization problem have two main flavors: exact algorithm and approximation algorithm. Exact algorithm can get optimal solution of the primal problem, frequently used methods are dynamic programming, branch and bound, enumeration, and so on. However, these methods are ideal solutions to smaller scale data, which are difficult to solve the optimizations because computing time rises exponentially as the problem scale increases. Approximation algorithms can always provide an optimal approximation solution for NP-hard problems within a reasonable time limit. Approximation algorithms have three different types: mathematical programming, heuristics algorithm, and intelligent optimization algorithm. Algorithm based on mathematical programming establishes mathematical programming model and obtains approximate solutions by using Lagrange relaxation, column generation, and other algorithm. According to the characteristic of problem, heuristics algorithm is designed with some scheduled rules and experience. This kind of algorithm is more intuitive and fast, but due to the lack of theoretical basis, the quality of solution is not necessarily good. Intelligent optimization algorithm is a new global search strategy based on certain optimization search mechanism. The algorithm includes genetic algorithm [12], chaos optimization algorithm [3], ant algorithm [9], particle swarm algorithm [4], simulated annealing [17], and so on. This kind of algorithm has advantages of high efficiency performance, no special information of the problem, ease of implementation, and fast speed, but it can also not guarantee the global optimal solution.

Actually, in many applications, values of coefficients in the objective function or constraints can be thought of being sampled from the same underlying distribution. For instance, a courier needs to select an optimal path covering all customers and return to the starting point in a given area, thousands of similar optimizations need to be solved, since the daily delivery address is changing. The match problem between two images maximizes the similarity degree under the constraint of one-to-one correspondence, such matching problem requires to be solved repeatedly, since the pair images may be different each time. The objective function of these problems is fixed, which are maintaining the same combinatorial structure, but differing mainly in their data. We need to solve the same type of problem over and over again using traditional methods, which is a waste of time and resources. We need to find a general way, which can learn to mine essential information offline, update solution strategy online automatically, and raise the efficiency of the learning and the adaptability to real problem resolving.

Machine learning may be applicable to many optimization tasks by automatically discovering generic and fast heuristic algorithms based on training data. For a long time before machine learning, feature engineering is achieved manually, which is time-consuming, specific, and incomplete. Now the representation learning can realize end-to-end learning of automatic feature engineering, greatly reducing or even eliminating dependence on domain knowledge. The versatility, expressiveness, and flexibility of deep neural networks make some tasks easier or possible [29]. There have been great improvements with deep learning in many fields, such as image recognition [18] and object detection [11].

Recurrent neural networks (RNNs), which are used for sequential data such as text and voice, have made breakthroughs in speech recognition [14] and natural language processing [31]. RNN processes an input sequence at every step of the sequence, and maintains a state vector in the hidden layers, containing all elements of the past historical information. The sequence-to-sequence model [31] maps the input sequence to a fixed-sized vector with one RNN as encoder and then maps the vector to the target sequence with another RNN as decoder. Bahdanau et al. [1] enhanced the decoder by using content-based attention mechanism to focus on relevant contextual information of input as context vector instead of fixed-sized vector.

Many advances in artificial intelligence have been achieved with supervised learning which is trained using a large amount of real label data. Recently, there has been some seminal work on using deep neural networks with supervised learning to learn heuristics for combinatorial problems, including the traveling salesman problem (TSP). Vinyals et al. [34] introduced pointer network which solves the problem of variable size output dictionary using the recently proposed neural attention mechanism. However, label data is usually expensive, unreliable, or unusable. Even with reliable data, the performance of the system trained in this way may be limited. Moreover, the architecture used in combinatorial problems usually requires a large number of instances in order to learn how to extend it to new instances.

Hence, following the reinforcement learning (RL) to tackle combinatorial optimization. In reinforcement learning, agent is trained from its own experience, making it possible to surpass human capabilities and realize in fields lacking human expertise. Recently, deep neural networks using reinforcement learning have made rapid progress, such as computer games Atari [23] and 3D virtual environments, performing better than humans. The program AlphaGo Zero is based entirely on reinforcement learning, without human data, guidance, or domain knowledge beyond the rules of the game, and achieves superhuman performance [30]. The success AlphaGo Zero achieved by using reinforcement learning in the game of Go inspires us to use reinforcement learning to solve the NP-hard problem of combinatorial optimization, because Go can be seen as a combinatorial optimization problem requiring decision-making at every step.

2   Background

For human, we can not only be effective at identifying individual examples, like handwritten numeral recognition and target classification, but also be good at analyzing horizontal and vertical logical relation between the input information sequences, like speech recognition and natural language processing (NLP). As these problems can't be solved by traditional multi-layer perception (MLP), recurrent neural network (RNN) comes into being.

RNN is a neural network structure model put forward by Jordan, Pineda, Williams, and Elman [25, 37] in the 1980s. But it was a pity that the existence of vanishing gradient and exploding gradient problem and computing resources prevented the sound growth of RNN. Only in the last three decades, long short term memory [15], bidirectional RNN [27], gated recurrent unit [5], attention mechanism [1], and computation efficiency and stability are introduced to make a quick breakthrough. Now RNNs have recently shown impressive performance in several sequence prediction problems including machine translation [31], contextual parsing [35], image captioning [5], and even video description [8].

2.1   Recurrent Neural Network

RNN is a class of artificial neural network which can use their internal memory to process arbitrary sequences of inputs. In general, we capture hidden states through the following equation:

$$h_t = f(h_{t-1}, x_t | \theta) \tag{1}$$

where h is the hidden state to store the contextual information, f is the activate function, and θ represents the network parameters. Figure 1 shows the general form of unfolding RNN according to the time series.

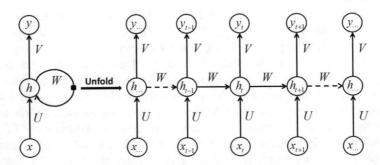

Fig. 1 The figure shows the general unfolding form of RNN, where W, U, V are the shared parameters. Given the input sequence $\{x_1, x_2, \cdots, x_{T_x}\}$, output the target sequence $\{y_1, y_2, \cdots, y_{T_y}\}$

The network of discrete time could be described as, given a set of N pairs $(X^i, Y^i)_{i=1}^N$, where (X^i, Y^i) is the ith pair of input and its corresponding target. The goal of the model is to estimate the conditional probability $P(Y^i|X^i)$, where $X^i = \{x_1^i, x_2^i, \cdots, x_{T_x}^i\}$ and $Y^i = \{y_1^i, y_2^i, \cdots, y_{T_y}^i\}$. It is noteworthy that the length of the input sequence T_x may differ from its corresponding output sequence T_y. The parameters θ of RNN can be estimated by maximizing the following cost function:

$$J(\theta) = \frac{1}{N} \sum_{n=1}^N \log P\left(Y^n|X^n; \theta\right) \tag{2}$$

In this case, it is reasonable to model each example using the chain rule to decompose it as follows (for the sake of brevity, we omit the index i):

$$P(Y|X; \theta) = \prod_{t=1}^{T_y} P(y_t|y_1, y_2, \cdots, y_{t-1}, X; \theta) \tag{3}$$

2.2 Encoder–Decoder Model

The strategy of encoder–decoder model is to map the input sequence to a fixed-sized vector c with one RNN, and then map the vector to the target sequence with another RNN. We call the former RNN the encoder and the latter the decoder. Figure 2 shows the general encoder–decoder model.

The most common encoder approach is to use an RNN such that

$$e_t = f_1(x_t, e_{t-1}) \tag{4}$$

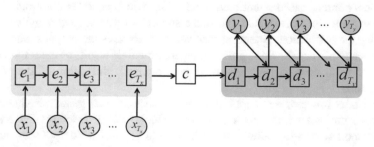

Fig. 2 The figure shows the general form of encoder–decoder model, where the blue part represents the encoder process and the violet represents the decoder process

where e_{t-1}, e_t is the hidden state of the encoder at time $t-1$, t, respectively, and f_1 is the activation function of encoder RNN. Generally, we will choose LSTM combined with Bi-RNN architecture as the activation function f_1, which is more suitable for long range temporal dependencies. Then we get the vector c,

$$c = q(e_1, e_2, \cdots e_{T_x}) \tag{5}$$

where q is an another nonlinear function. In paper [31], the fixed-dimensional representation c is the last hidden state of the LSTM. Then we use another standard LSTM formulation f_2, whose initial hidden state is c, to obtain the hidden state of the decoder d_t at time t,

$$d_t = f_2(d_{t-1}, y_{t-1}, c) \tag{6}$$

the conditional distribution of the next symbol is

$$P(y_t|y_1, \cdots y_{t-1}, X; \theta) = g(d_t, y_{t-1}, c), \tag{7}$$

where g is an activation function. Generally, the function is represented with a softmax over all the words in the dictionary.

2.3 Attention Mechanism

However, compressing all necessary information of the input sequence into a fixed-length vector is a bottleneck problem that limits the performance of the encoder–decoder model. There are two major concerns about the model. Firstly, the fixed-length vector could not represent complete information of the whole input sequence. And if the input sequence is too long, the later information will overwrite the previous contents. In order to address this issue, Bahdanau et al. [1] propose a content-based attentional mechanism which encodes the input sentence into a sequence of vectors and chooses a subset of these vectors adaptively while decoding the translation. Attention mechanism associates the output sequence with the sequence of vectors which preserve the intermediate results of input information produced from encoder. See Figure 3 for graphical illustration of the attention mechanism.

In encoder process, we adopt bidirectional RNN or other general RNN structure to encode the input into vector sequence. Unlike the decoder part in encoder–decoder model, we adopt attention mechanism to get the final output sequence,

$$P(y_t|y_1, \cdots y_{t-1}, X; \theta) = g(d_t, y_{t-1}, c_t) \tag{8}$$

Fig. 3 The figure illustrates
the general form of attention
mechanism to generate the
t-th target output given a
source sentence
$\{x_1, x_2, \cdots, x_{T_x}\}$

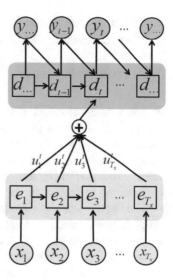

where d_t represents the decoder hidden state at time t. Attention mechanism
differs from traditional model in that the output y_t of each time t is related to the
corresponding content vector c_t. Actually, c_t is computed as a weighted sum of
encoder hidden state $(e_1, e_2, \cdots, e_{T_x})$,

$$c_t = \sum_{i=1}^{T_x} \alpha_i^t e_i \tag{9}$$

Taking a weighted sum of all the encoder hidden state indicates that attention
distribution is different at the output step t. The higher the α_i^t value describes the
output of time t, pay more attention to the ith input vector. α_i^t is the result of the
combined output hidden state d_t and all input hidden states,

$$\alpha_i^t = \frac{\exp(u_i^t)}{\sum_{k=1}^{T_x} u_k^t} \tag{10}$$

$$u_i^t = a(d_t, e_i) = v^T \tanh(W_1 d_t + W_2 e_i) \tag{11}$$

where v^T, W_1, W_2 are the weighting parameters.

Intuitively, attention mechanism relieves the encoder from the burden of having
to encode all information in the source sentence into a fixed-length vector. In the
meanwhile, the implements a mechanism of attention in the decoder help us to better
understand the internal operation mechanism of the model.

3 Pointer Network for Solving Some Combinatorial Optimization Problems

3.1 Pointer Network

Pointer network (Ptr-Net) [34] is an effective model repurposing a recently proposed mechanism of neural attention [1] to solve combinatorial optimization problem where the output dictionary size is equal to the length of the input sequence. It differs from the previous attention attempt in that, using attention as a pointer to select a member of the source sequence as the target, instead of using attention to compute the weighted sum of these encoder hidden units at each decoder step.

The encoder–decoder model normally uses a softmax distribution over a fixed-sized output dictionary to compute $P(y_t|c, y_1, \cdots y_{t-1}; \theta)$. This prevents us from learning solutions to problem that output dictionary size depends on the number of elements in the input sequence. To overcome this, Ptr-Net does not use attention to blend hidden units of an encoder e_i to a context vector, but instead uses u_i^t as pointer to select input elements at each decoder step. The modified attention mechanism is shown as follows:

$$u_i^t = v^T \tanh (W_1 e_i + W_2 d_t) \quad i \in (1, 2, \cdots, T_x) \tag{12}$$

$$P(y_t|y_1, y_2, \cdots, y_{t-1}, X; \theta) = soft \max(u^t) \tag{13}$$

where softmax normalizes the vector u^t (of length T_x) to be an output distribution over the dictionary of inputs, and v, W_1, and W_2 are learnable parameters of the output model. Ptr-Net provides a novel method for complex combinatory optimization, in which the output sequence corresponds to the positions in an input sequence. Figure 4 depicts the pointer network to solve simple scheduling problem.

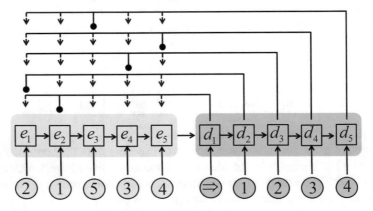

Fig. 4 The figure describes the pointer network to solve the simple scheduling problem, which accepts series of numbers and sorts them in ascending order

3.1.1 Experiments

In [34], they trained the Ptr-Net to output satisfactory solutions to three challenging geometric problems, computing planar convex hulls, Delaunay triangulations, and the symmetric planar traveling salesman problem (TSP). The experiments performed an evaluation of Ptr-Net algorithms on randomly synthesized graphs. All the training and testing data are sampled from a uniform distribution in $[0, 1] \times [0, 1]$ to generate Cartesian coordinates of the point $p = [x, y]$. The inputs of all three problems are planar point sets $P = [p_1, p_2, \cdots, p_n]$ with n elements each, where $p_i = [x_i, y_i]$, and the output are sequences representing the solution associated with the point set P. We briefly introduce the result of the network, specific data, and error analysis presented in the paper [34].

Convex hull is a concept in computational geometry (graphics), which is a task to find the minimal convex polygon of a finite number of points. In order to reduce the ambiguities during training, Vinyals ranks the output sequence uniformly which starts with the lowest index and arranges counterclockwise. They report two metrics: accuracy and area covered of the true convex hull, to carry out the verification on the effectiveness of the put forward model. The model has the key advantage of being inherently variable length. The result shows that model trained on a variety of lengths ranging from 5 to 50 can be generalized to solve even for 500 nodes. Although the accuracy is only 1.3%, the area coverage achieved with the model is close to 100%. In fact, this is a common source of errors in most algorithms to solve the convex hull.

Delaunay triangulation is another intensively studied problem in computer science and mathematics. Given a set P of points in a plane, a Delaunay triangulation is a triangulation such that there is no point from the set P in its interior. During the training phase, the labels of the output $S^P = \{S_1, S_2, \cdots, S_{m(P)}\}$ are the corresponding sequences representing the triangulation of the point set. Each S_i represents the vertices of the i triangle, the integers of the triple which is from 1 to n corresponding to the position of set P. Without taking into account the order of vertex, the accuracy and triangle coverage of $n = 5$, $n = 10$, $n = 50$ are $80.7\%/93.0\%$, $22.6\%/81.3\%$, $FAIL/52.8\%$, respectively.

Traveling salesman problem is a NP-hard, "A complete graph with n points, which each edge has a weight (tour length), one need to search an optimal sequence of points with the shortest total length." Experimental comparisons between the novel network and three traditional algorithm A1, A2, A3 illustrate that the novel algorithms are correct and feasible on a small scale ($n = 5$, $n = 10$, $n = 50$). The experiments also indicate that the quality of algorithm is closely related to the labels.

Pointer network is a novel network structure, which provides a new method for complicated combinatorial optimization problems. We can use this network structure to deal with a series of problems that depend on the input dictionary. In conclusion, using deep learning combined with suitable network structure can

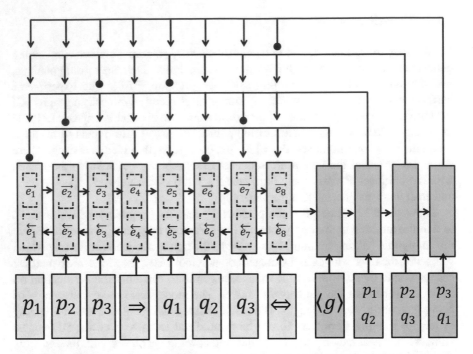

Fig. 5 Set $A = \{p_1, p_2, p_3\}$, set $B = \{q_1, q_2, q_3\}$, and the correspondence of the two sets is $(p_1, q_2), (p_2, q_3), (p_3, q_1)$. We use the symbol $\Rightarrow$ to represent the end of set A, the symbol $\Leftrightarrow$ to represent the end of encoder, and the symbol $\langle g \rangle$ to represent the input of the first decoder step. The two dashed rectangle boxes in the bigger boxes represent the process of the bidirectional RNN. At each step, the multi-pointer network produces an output that is a sigmoid distribution to point to two vertices of two sets, respectively

make good use of the computer's resources, improve efficiency, and conquer the complexities. At the same time, we should also pay attention to improve the accuracy and broaden the scope of algorithm (Figure 5).

3.2 Multi-Pointer Network

Point matching is also a special kind of combinatorial optimization problems that is to obtain the optimal corresponding references, which can be modeled by Ptr-Net. However, the Ptr-Net doesn't take full advantage of the correspondences between two point sets owing to the fact that the output is a member of the input sequence at each time series. We propose multi-pointer network, which draws the idea from multi-label classification [38], to address this limitation by pointing out a set of input elements.

For point sets matching, we define the matching point-pair as the output target of each time series in training process. After sorting the points of the two sets, we construct a series of objective vectors that the elements of each one are zero besides the position of the matching point-pair. Hence, we need to output a set of the input sequences simultaneously at the decoder part. Drawing the idea on multi-label classification, we present a simple modification of the Ptr-Net.

Firstly, we still use Equation (12) to calculate the vector u^i to be the "attention" mask over the inputs. Then a different approach is adapted to obtain the conditional distribution in every time step.

$$P(y_t|y_1, y_2, \cdots, y_{t-1}, X; \theta) = sigmoid(u^i) \qquad (14)$$

What makes our model different from Ptr-Net is that we use the sigmoid function instead of softmax function to fit the multi-label classification loss function. This allows us to maintain that the output of the every time series is the matching point-pair or the points of other geometry structure.

Our model is specifically aiming at problems whose outputs of every time series are high correlation. The nature of point matching intents to determine the correspondence between two sets. Ptr-Net tends to solve problems where outputs are discrete and depend on its location in the input sequences. Modification of the our model mines the underlying information between two point sets. By bringing about changes of this kind, the information can be spread throughout the two point sets, which can be greatly improved by the accuracy of matching problems. Besides, the multi-pointer network can also be applied to the problem of combined structural optimization, like Delaunay triangulation.

3.2.1 Experiments

The validity of the network is established to solve point matching and Delaunay triangulation by the synthetic datasets. The accuracy of network on the point sets matching problem is discussed at different sizes ($n = 6, n = 5$–$10, n = 10$) and different kinds of transformation (transformation, rotation, similarity). The results of experiments verify that our new method can effectively solve the matching problem with various transformations. The results are presented in Table 1. Compared to the Ptr-Net, multi-pointer network doesn't have to consider the problem as a consequence, any permutation of its elements will represent the same triangulation. The experiment is about 5 points of the set to achieve in average 0.45% improvement.

Multi-pointer network is a new end-to-end network, which is based on multi-label classification to improve the pointer network. The results show that the proposed method can effectively solve the translation and other rigid transformation with large-scale. In this way, we can rapidly gain the corresponding relations of the point sets in the inference. Moreover, the method can be extended to solve other problem of combined structural optimization.

Table 1 Comparison between Ptr-Net and our multi-pointer network on translation, and different translation

(a) Translation			
Number	N = 6	N = 5–10	N = 10
Ptr-Net model	98.59%	96.02%	95.79%
Our model	99.47%	98.18%	97.31%
(b) Rotation (different interval)			
Max angle θ	$\theta = 45°$	$\theta = 90°$	$\theta = 135°$
N = 5–10	98.87%	97.83%	97.13%
(c) Rotation (different database $\theta = 30°$)			
Number	N = 6	N = 5–10	N = 10
Accuracy	99.78%	99.48%	97.83%
(d) Translation + Rotation ($\theta = 30°$)			
Number	N = 6	N = 5–10	N = 10
Accuracy	98.45%	97.33%	96.78%
(e) Similarity transformation			
Number	N = 6	N = 5–10	N = 10
Accuracy	97.35%	96.77%	96.24%

3.3   Objective-Based Learning

For NP-hard problem, the time needed for the exact solution is increasing exponentially with the problems scale. How to take advantages of the resource of approximate algorithm to obtain highly accurate solution at minimal computational cost is our goal. In paper [21], Milan introduces a simple, yet effective technique for improving the initial training set by incorporating the objective cost into the training procedure.

The training objective of classical supervised learning procedure is the proximity of the network's prediction to the label, like the mean squared error (MSE) or the cross-entropy loss for regression and classification problems, respectively. This type of loss function is differentiable and thus convenient to be optimized with back propagation (BP). However, it is not an effective training method to simply use the supervised method for NP-hard problem. Considering the well-known traveling salesman problem, we will get the similar node order, as measured by the log-likelihood classification error [34], but results in a highly non-optimal path measured by the tour length. Therefore, the objective-based learning scheme is introduced to calculate the task's objective function in each iteration. It is only when the current network's prediction is better than the approximate precomputed solution, we will propagate the gradient.

3.3.1   Experiments

They validate the scheme on both synthetic and real data on three applications: tracking multiple targets, graph matching, and TSP. First, mathematical model of tracking multiple targets is linear assignment, which can be formulated as a binary linear program:

$$X^* = \arg\max_X c^T X$$
$$st \quad \begin{aligned} \forall j : \sum_i \chi_{ij} = 1 \\ \forall i : \sum_j \chi_{ij} = 1 \end{aligned} \tag{15}$$

where c are the (linear) coefficients, $X \in \{0, 1\}^{n^2}$ is a binary vector, and $\chi \in R^{n \times n}$ is the same vector reshaped as a square matrix. The constraints ensure the one-to-one correspondences. The network reads the c as input and obtains the hidden state h with full-connected layer. The output is a probability distribution over the number of elements in the second set, obtained by applying a softmax transform. The validation of experiments on tracking multiple targets is better than the label acquired from $JPDA_{10}$ [10]. Second, by using the same network structure they validate the quadratic assignment problem (QAP). Different from linear assignment, the objective of QAP takes on a quadratic form, $X^* = \arg\max_X X^T Q X$, which makes the problem NP-hard. On public Pascal image dataset [19], the matching accuracy is around 75% on eight point pairs. Third, they verify the effectiveness of the scheme on planner TSP with 20 nodes. The objective-based learning gains a better performance than traditional supervised learning with pointer network.

In complex combinatorial problem, especially NP-hard problem, time consuming increases in exponent as dimension does. The proposed objective-based learning improves the supervised learning by incorporating the problem specific objective. This method optimizes the label obtained from suboptimal solution and steers the training towards higher quality solution. Experiments on multiple applications show that the scheme results in very good approximations of the globally optimal solution.

4 Reinforcement Learning for Solving Some Combinatorial Optimization Problems

Supervised learning is training model with a large labeled training dataset of input–output pairs, in order to minimize a loss function that measures the error between the predicted output and labeled real output. It is the most common form of machine learning, but it is often impractical for combinatorial optimization. Supervised learning is undesirable for NP-hard problems of combinatorial optimization because (1) the performance of supervised learning model is tied to the quality of the labels, (2) high-quality labeled data is often expensive, unreliable, or unavailable and may be infeasible for new problem instances, and (3) finding a competitive and general algorithm is more important than replicating the results of another algorithm [2].

By contrast, reinforcement learning systems are trained from their own experience, without supervised data. Trial-and-error search and delayed reward are the two most important distinguishing features of reinforcement learning. Using reinforcement learning to tackle combinatorial optimization problems is feasible, considering that these problems have relatively simple reward mechanisms based on themselves; thus, the reward mechanisms can be even used at test time.

In reinforcement learning, an agent interacts with the environment through a Markov decision process (MDP), learning an optimal strategy through tracking and error methods for sequential decision-making problems. At each time step t of interaction with the environment, the agent receives a state s_t in the state space S and selects an action a_t from the action space A, following a policy $\pi(a_t|s_t)$, which defines the learning agent's way of behaving at a given time, i.e., a mapping from state s_t to actions a_t to be taken. Then the agent receives a scalar reward r_t and transitions to the next state s_{t+1}, for reward function $R(s, a)$ which maps each state–action pair to a scalar number and state transition probability $P(s_{t+1}|s_t, a_t)$, respectively. In an episodic problem, this process continues until the agent reaches a terminal state and then it restarts. The return $R_t = \sum_{k=0}^{\infty} \gamma^k r_{t+k}$ is the discounted, accumulated reward with the discount factor $\gamma \in [0, 1]$. The agent's sole objective is to maximize the expectation return in the long run [20].

Value function [32], which is a fundamental concept in reinforcement learning, is a prediction of the expected return, measuring how good each state or state–action pair is in the long run. The state value function is the total amount of reward an agent can expect to accumulate over the future for following policy π, starting from state s:

$$V_\pi(s) = E[R_t|s_t = s] \tag{16}$$

$V_\pi(s)$ decomposes into the Bellman equation:

$$V_\pi(s) = E[r_t + \gamma V_\pi(s_{t+1})|s_t = s] \tag{17}$$

The action-value function is the expected return for following policy π from selecting action a in state s:

$$Q_\pi(s, a) = E[R_t|s_t = s, a_t = a] \tag{18}$$

It similarly decomposes into the Bellman equation:

$$Q_\pi(s, a) = E[r_t + \gamma Q_\pi(s_{t+1}, a_{t+1})|s_t = s, a_t = a] \tag{19}$$

The Bellman equation shows the relationship between the value function of the current state and the value function of the next state.

Value-based methods, main approaches of reinforcement learning, estimate the optimal value function $Q^*(s, a)$ by traversing all states and actions and then selecting the action corresponding to the maximum value over the long run as the strategy. In model-based environment, i.e., already knowing reward function $R(s, a)$ and state transition probability $P(s_{t+1}|s_t, a_t)$, use dynamic programming methods to evaluate value function. While in model-free environment, Monte Carlo and temporal difference learning are main methods to evaluate value function. Monte Carlo methods require only *experience*—sample sequences of states, actions, and rewards from online or simulated interaction with an environment, then averaging sample returns as evaluation [32]. Temporal difference learning learns value function directly from experience with TD error in bootstrapping way. Q-learning [36] and SARSA as classical learning methods are regarded as temporal difference learning. Q-learning learns action-value function, with the update rule:

$$Q(s, a) \leftarrow Q(s, a) + \alpha[r + \gamma max_{a'} Q(s', a') - Q(s, a)] \tag{20}$$

where $r + \gamma max_{a'} Q(s', a') - Q(s, a)$ is called TD error.

However, classical reinforcement learning methods such as Q-learning can't solve the problems with high-dimensional state and action space. Function approximation is a way for generalization when the state or action spaces are high-dimensional or continuous. Mnih et al. [22] introduce deep Q-network (DQN) which uses deep neural networks to approximate value function, and apply to the Atari game, taking game screen as input and game scores as reward signal of learning, achieving better than human, which is a groundbreaking work of deep reinforcement learning.

Policy descent methods, another main approaches of reinforcement learning, optimize the policy $\pi(a|s; \theta)$ directly and update the parameters θ by gradient ascent on $E[R_t]$. Policy gradient methods are widely applied to continuous state or action spaces. Stochastic policy $\pi_\theta(a|s) = P[a|s; \theta]$ defines probability distribution of actions among total space at state s. Sutton et al. [33] introduce an alternative approach in which the policy is explicitly represented by its own function approximator, independent of the value function, and is updated according to the gradient of expected reward with respect to the policy parameters:

$$\nabla_\theta L(\pi_\theta) = \int_S \rho^\pi(s) \int_A \nabla_\theta \pi_\theta(a \mid s) Q^\pi(s, a) \, da \, ds$$

$$= \mathbb{E}_{s \sim \rho^\pi, a \sim \pi_\theta}[log \pi_\theta(a \mid s) Q^\pi(s, a)]$$

While deterministic policy $a = \mu_\theta(s)$ maps directly state to action. Silver et al. [28] show that the deterministic policy gradient does indeed exist, and furthermore it has a simple model-free form that simply follows the gradient of the action-value function:

$$\nabla_\theta L(\pi_\theta) = \int_S \rho^\mu(s) \nabla_\theta \mu_\theta(s) \nabla_\theta Q^\mu(s, a)|_{a=\mu_\theta(s)} \, ds$$

$$= \mathbb{E}_{s \sim \rho^\mu}[\nabla_\theta \mu_\theta(s) \nabla_\theta Q^\mu(s, a)|_{a=\mu_\theta(s)}]$$

Actor–critic algorithm is a widely used architecture [24]. The actor–critic algorithm consists of two eponymous components. The critic updates action-value function parameters, and the actor updates policy parameters, in the direction suggested by the critic.

Deep neural networks using reinforcement learning have made rapid progress, especially in the game of Go. Go is essentially a game that makes decisions many times, similar to combinatorial optimization problems. Each decision produces a variety of different changes, and the possible changes are millions, like a huge tree search map. The game of Go has vast search space and difficulty in evaluating the position and movement of the board. Humans consider it as a huge challenge for artificial intelligence in very long period of time.

In 2016, Google DeepMind introduced a new program named *AlphaGo* which achieved a 99.8% winning rate against other Go programs, and defeated the human world Go champion by 4 games to 1. This is the first time that a computer program has beaten a professional human chess player in a full-scale Go game, a feat that was thought to take at least 10 years. *AlphaGo* combines Monte Carlo tree search (MCTS) with deep neural networks consisting of value and policy networks, greatly reducing the computation of searching process and improving the accuracy of board estimation. Similar to actor–critic algorithm, policy network selects movement of board as actor, and value network evaluates board positions as critic. These networks are trained by a novel combination of supervised based on data of human expert games, and reinforcement learning from games of self-play [29].

In 2017, Google DeepMind introduced a new program named *AlphaGoZero* based solely on reinforcement learning, without human data, guidance, or domain knowledge beyond game rules, winning 100-0 against the previously published, champion-defeating *AlphaGo*. Other than *AlphaGo*, *AlphaGoZero* doesn't need human data and uses a single neural network to evaluate board positions and select moves, rather than separate policy and value networks. The success *AlphaGoZero* achieved by reinforcement learning inspires people to use reinforcement learning to solve the NP-hard problem of combinatorial optimization, because Go can be seen as a combinatorial optimization problem requiring decision-making at every step.

4.1 Examples

Reinforcement learning that trains a model through trial and error provides a feasible approach for combinatorial optimization problems, considering that these problems have relatively simple reward mechanisms based on themselves.

4.1.1 Traveling Salesman Problem (TSP)

Bello et al. [2] propose neural combinatorial optimization, a framework combining deep neural networks with reinforcement learning to tackle combinatorial optimization problems, focusing on the 2D Euclidean traveling salesman problem (TSP), achieving close to optimal solutions with up to 100 nodes.

Given an input graph $s = \{\mathbf{x}_i\}_{i=1}^n$ of 2D Euclidean TSP with n cities, a tour length defined by a permutation π:

$$L(\pi \mid s) = \left\|\mathbf{x}_{\pi(n)} - \mathbf{x}_{\pi(1)}\right\|_2 + \sum_{i=1}^{n-1} \left\|\mathbf{x}_{\pi(i+1)} - \mathbf{x}_{\pi(i)}\right\|_2$$

It's straightforward to take the negative tour length as reward signal in reinforcement learning. They follow the network structure of [34], using pointer network pointing to a specific position in the input sequence of city coordinates, to learn parameter θ of a stochastic policy $p(\pi \mid s)$ that assigns high probabilities to short tours. The objective function is the expected tour length:

$$J(\theta|s) = \mathbb{E}_{\pi \sim p_\theta(.|s)} L(\pi|s) \tag{21}$$

They use policy gradient methods and stochastic gradient descent to optimize the parameter. The gradient of objective function:

$$\nabla_\theta J(\theta|s) = \mathbb{E}_{\pi \sim p_\theta(.|s)}[(L(\pi|s) - b(s))\nabla_\theta \log p_\theta(\pi|s)] \tag{22}$$

where $b(s)$ denotes a baseline function that estimates the expected tour length to reduce the variance of the gradients.

The first method called RL *pre-training*, an actor–critic algorithm, doesn't require supervision, still requires training data and thus generalization depends on the training data distribution S, described in Algorithm 1. The actor network uses policy gradient method to optimize parameter π of policy and the gradient is approximated with Monte Carlo sampling:

$$\nabla_\theta J(\theta) \approx \frac{1}{B} \sum_{i=1}^{B} (L(\pi_i|s_i) - b(s_i))\nabla_\theta \log p_\theta(\pi_i|s_i) \tag{23}$$

The critic network learns its parameter θ_v by stochastic gradient descent. The loss function is a mean squared error objective between its predictions $b_{\theta_v}(s)$ and the actual tour lengths sampled by the most recent policy:

$$L(\theta_v) = \frac{1}{B} \sum_{i=1}^{B} \left\| b_{\theta_v}(s_i) - L(\pi_i|s_i) \right\|_2^2 \tag{24}$$

At test time, using the learnt policy to inference by greedy decoding that always selects the index corresponding to largest probability, or sampling that samples several candidate solutions, then selects the best one. They show sampling at test time is more effective than greedy decoding in their experiments.

Algorithm 1 Actor–critic training

1: Train (training set S, number of training steps T, batch size B)
2: Initialize pointer network params θ
3: Initialize critic network params θ_v
4: **for** $t = 1$ to T **do**
5:    $s_i \sim$ SAMPLEINPUT(S) for $i \in 1, \ldots, B$
6:    $\pi_i \sim$ SAMPLESOLUTION $(p_\theta(.|s_i))$ for $i \in \{1, \ldots, B\}$
7:    $b_i \leftarrow b_{\theta_v}(s_i)$ for $i \in 1, \ldots, B$
8:    $g_\theta \leftarrow \frac{1}{B} \sum_{i=1}^{B} (L(\pi_i|s_i) - b(s_i)) \nabla_\theta log p_\theta(\pi_i|s_i)$
9:    $\mathbb{L}_v \leftarrow \frac{1}{B} \sum_{i=1}^{B} \|b_i - L(\pi_i)\|_2^2$
10:   $\theta \leftarrow$ ADAM (θ, g_θ)
11:   $\theta_v \leftarrow$ ADAM $(\theta_v, \nabla_{\theta_v}\mathbb{L}_v)$
12: **end for**
13: **return** θ

The second method called active search does not require training data so it's distribution independent, presented in Algorithm 2. The method can start with a random policy, iteratively optimizing parameters of policy on a single test instance with policy gradient, and draw Monte Carlo samples over candidate solutions $\pi_1, \pi_2, \ldots, \pi_B \sim p_\theta(.|s)$ for a single test input. Instead of a parameterized critic network, it resorts to an exponential moving average baseline.

Algorithm 2 Active search

1: TRAIN (ActiveSearch(input s, θ, number of candidates K, B, α))
2: $\pi \leftarrow$ RANDOMSOLUTION()
3: $L_\pi \leftarrow L(\pi|s)$
4: $n \leftarrow \lceil \frac{K}{B} \rceil$
5: **for** $t = 1 \ldots n$ **do**
6:    $\pi_i \sim$ SAMPLESOLUTION $(p_\theta(.|s))$ for $i \in \{1, \ldots, B\}$
7:    $j \leftarrow$ ARGMIN $(L(\pi_1|s) \ldots L(\pi_B|s))$
8:    $L_j \leftarrow L(\pi_j|s)$
9:    **if** $L_j < L_\pi$ **then**
10:      $\pi \leftarrow \pi_j$
11:      $L_\pi \leftarrow L_j$
12:    **end if**
13:    $g_\theta \leftarrow \frac{1}{B} \sum_{i=1}^{B} (L(\pi_i|s) - b) \nabla_\theta log p_\theta(\pi_i|s)$
14:    $\theta \leftarrow$ ADAM (θ, g_θ)
15:    $b \leftarrow \alpha \times b + (1 - \alpha) \times (\frac{1}{B} \sum_{i=1}^{B} b_i)$
16: **end for**
17: **return** π

They conduct experiments for three tasks, 2D Euclidean TSP20, TSP50, and TSP100 which points are drawn uniformly at random in the unit square $[0, 1]^2$, and their results are better than results of pointer network with supervised learning. In their experiments, among five kinds of neural combinatorial optimization, RL pretraining-sampling and RL pretraining-active search are the most competitive methods for test cases.

4.1.2 Maximum Cut (MAXCUT) and Minimum Vertex Cover (MVC)

Maximum Cut (MAXCUT): Given a graph G, find a subset of nodes $S \subseteq V$ such that the weight of the cut-set which edges between S and the complementary subset, $\sum_{(u,v) \in C} w(u, v)$, is maximized.

Minimum Vertex Cover (MVC): Given a graph G, find a subset of nodes $S \subseteq V$ such that every edge is covered, i.e., $(u, v) \in E \Leftrightarrow u \in S$ or $v \in S$, and $|S|$ is minimized.

The learning meta-algorithm [16], using a common formulation to solve graph optimization problems, designs a unique combination of reinforcement learning and graph embedding network which is called structure2vec (S2V) [6] to represent the nodes in the graph. The greedy algorithm constructs a solution by sequentially adding nodes to a partial solution S, based on maximizing evaluation function Q measuring the quality of a node.

An algorithm $A(G, h(\cdot), t(\cdot), c(\cdot))$ consists of a problem instance G, the helper function h used to map set S to a combinatorial structure, the termination criterion t, and the cost function c, and then output a final solution $\widehat{S}$ determined by the evaluation function Q, which is learned by using a set of problem instances:

$$\widehat{S} := A(G, h(\cdot), t(\cdot), c(\cdot)), \text{ where } G \sim \mathbb{D} \tag{25}$$

The partial solution S will be extended as:

$$S := A(S, v^*), \text{ where } v^* := \arg\max_{v \in \overline{S}} Q(h(S), v) \tag{26}$$

For maximum cut (MAXCUT) problem, to satisfy the constraint, the helper function divides V into two sets, S and its complement $\overline{S} = V \setminus S$, and maintains a cut-set $C = (u, v) \in E, u \in S, v \in \overline{S}$. And the cost function is $c(h(S), G) = \sum_{(u,v) \in C} w(u, v)$.

For minimum vertex cover (MVC), $c(h(S), G) = -|S|$.

With function approximation method, Khalil et al. use $\widehat{Q}$ parameterized by Θ instead of Q:

$$\widehat{Q}(h(S), v; \Theta) \approx Q(h(S), v) \tag{27}$$

Structure2vec (S2V), the graph embedding network, computes a p-dimensional feature embedding μ_v for each node $v \in V$ and updates the embeddings synchronously at each iteration:

$$\mu_v^{(t+1)} := F\left(x_v, \{\mu_u^{(t)}\}_{u \in N(v)}, \{w(v, u)\}_{u \in N(v)}; \Theta\right) \tag{28}$$

where x_v is an additional feature on node v: $x_v = 1$ if $v \in S$ and 0 otherwise. $N(v)$ is the set of neighbors of node v in graph G, and F is a generic nonlinear mapping such as a neural network or kernel function. Designing F to update a p-dimensional embedding μ_v:

$$\text{relu}\left(\theta_1 x_v + \theta_2 \sum_{u \in N(v)} \mu_u + \theta_3 \sum_{u \in N(v)} \text{relu}(\theta_4 w(v, u))\right) \tag{29}$$

Using the embeddings from structure2vec to parameterize $\widehat{Q}$:

$$\widehat{Q}(h(S), v; \Theta) = \theta_5^\top \text{relu}[\theta_6 \sum_{u \in V} \mu_u, \theta_7 \mu_v] \tag{30}$$

Using a combination of n-step Q-learning and fitted Q-iteration [26] to learn the parameters in $\widehat{Q}(h(S), v; \Theta)$, as illustrated in Algorithm 3. Using n-step Q-learning, the parameters are trained with stochastic gradient descent on the squared regression loss objective:

$$(y - \widehat{Q}(h(S_t), v_t; \Theta))^2 \tag{31}$$

where

$$y = \sum_{i=0}^{n-1} r(S_{t+i}, v_{t+i}) + \gamma \max_{v'} \widehat{Q}(h(S_{t+n}), v'; \Theta) \tag{32}$$

The fitted Q-iteration approach uses *experience replay* to randomly sample batch from a dataset E consisting of tuples of previous episodes, similarly to deep Q-Network (DQN) [22]. Hence, they call the algorithm *structure2vec deep Q-learning* (S2V-DQN).

Algorithm 3 Q-learning for the greedy algorithm

1: Initialize experience replay memory $\mathbb{M}$ to capacity N
2: **for** episode $e = 1$ to L **do**
3:     Draw graph G from distribution $\mathbb{D}$
4:     Initialize the state to the empty sequence $S_1 = ()$
5:     **for** step $t = 1$ to T **do**
6:         Add v_t to partial solution: $S_{t+1} := (S_t, v_t)$
7:         **if** $t \geq n$ **then**
8:             Add tuple $(S_{t-n}, v_{t-n}, R_{t-n,t}, S_t)$ to $\mathbb{M}$
9:             Sample random batch from $B \sim \mathbb{M}$
10:           Update Θ by SGD
11:         **end if**
12:     **end for**
13: **end for**
14: **return** Θ

From their experiment of two types of graph: Erdos–Renyi (ER) graph and Barabasi–Albert (BA) graph, the performance of S2V-DQN is particularly better than performance of pointer network trained with actor–critic (PN-AC) [2] and other heuristic algorithms. In this experiment, training and testing graphs obey the same distribution and that the sizes of graph are varied in the ranges {15–20, 40–50, 50–100, 100–200, 400–500} for the MVC problem, while for the MAXCUT problem the sizes are {15–20, 40–50, 50–100, 100–200, 200–300}. Particularly, the approximation ratio respect to optimal solution is approximately 1 for the MVC problem. They also show the generalization ability of the proposed model S2V-DQN and show that it can get very low approximation ratio of testing graphs with up to 1200 nodes when training on graphs with 50–100 nodes.

5 Conclusion

Combinatorial optimization problem has developed in many fields and been increasingly addressed theoretically by scholars and practically by programmers. The development and application of deep neural network provides a new method for complicated combinatorial optimization problems. This chapter introduces several models for solving combinatorial optimization problems using deep neural networks, like pointer network, multi-pointer network, and so on. However, traditional supervised learning techniques typically require a large number of high-quality labeled data to learn an accurate model. Therefore, scholars pay more attention to the theory and practice of combining training mechanism based on reinforcement learning with appropriate network model. The present results demonstrate the effectiveness of the proposed framework as compared with manually designed greedy algorithm. For instance, on graphs with 1200 nodes, only using a single GPU, paper [16] find the solution of MVC within 11 s while getting an approximation ratio of 1.0062. In future, with the rapid development of the computer science, we are supposed to design a general frame to solve the optimization problem using machine learning methods.

References

1. Bahdanau, D., Cho, K., Bengio, Y.: Neural machine translation by jointly learning to align and translate (2014). arXiv preprint arXiv:1409.0473
2. Bello, I., Pham, H., Le, Q.V., Norouzi, M., Bengio, S.: Neural combinatorial optimization with reinforcement learning (2016). arXiv preprint arXiv:1611.09940
3. Bing, L.I., Jiang, W.: Chaos optimization method and its application. In: Control Theory & Applications (1997)
4. Bonabeau, E., Dorigo, M., Theraulaz, G.: Swarm Intelligence: From Natural to Artificial Systems. Oxford University Press, Inc., Oxford (1999)
5. Cho, K., Van Merrienboer, B., Gulcehre, C., Bahdanau, D., Bougares, F., Schwenk, H., Bengio, Y.: Learning phrase representations using RNN encoder–decoder for statistical machine translation. In: Proceedings of the 2014 Conference on Empirical Methods in Natural Language Processing, pp. 1724–1734 (2014)

6. Dai, H., Dai, B., Song, L.: Discriminative embeddings of latent variable models for structured data. In: International Conference on Machine Learning, pp. 2702–2711 (2016)
7. Dijkstra, E.W.: A note on two problems in connection with graphs. Numer. Math. **1**(1), 269–271 (1959)
8. Donahue, J., Hendricks, L.A., Guadarrama, S., Rohrbach, M., Venugopalan, S., Darrell, T., Saenko, K.: Long-term recurrent convolutional networks for visual recognition and description. In: IEEE Conference on Computer Vision and Pattern Recognition, pp. 2625–2634 (2015)
9. Dorigo, M., Maniezzo, V., Colorni, A.: Ant system: optimization by a colony of cooperating agents. Syst. Man Cybern. **26**(1), 29–41 (1996)
10. Fortmann, T.E., Bar-Shalom, Y., Scheffe, M.: Multi-target tracking using joint probabilistic data association. In: Conference on Decision and Control Including the Symposium on Adaptive Processes, pp. 807–812. IEEE, Piscataway (1980)
11. Girshick, R., Donahue, J., Darrell, T., Malik, J.: Rich feature hierarchies for accurate object detection and semantic segmentation. In: Proceedings of the IEEE Conference on Computer Vision and Pattern Recognition, pp. 580–587 (2014)
12. Golberg, D.E.: Genetic Algorithms in Search, Optimization, and Machine Learning, vol. 102, p. 36. Addison Wesley, Boston (1989)
13. Gower, J.C., Ross, G.J.S.: Minimum spanning trees and single linkage cluster analysis. Appl. Stat. **18**(1), 54–64 (1969)
14. Graves, A., Mohamed, A.-R., Hinton, G.: Speech recognition with deep recurrent neural networks. In: 2013 IEEE International Conference on Acoustics, Speech and Signal Processing, pp. 6645–6649. IEEE, Piscataway (2013)
15. Hochreiter, S., Schmidhuber, J.: Long short-term memory. Neural Comput. **9**(8), 1735–1780 (1997)
16. Khalil, E., Dai, H., Zhang, Y., Dilkina, B., Song, L.: Learning combinatorial optimization algorithms over graphs. In: Advances in Neural Information Processing Systems, pp. 6348–6358 (2017)
17. Kirkpatrick, S., Gelatt, C.D., Vecchi, M.P.: Optimization by simulated annealing. Science **220**(4598), 671–680 (1983)
18. Krizhevsky, A., Sutskever, I., Hinton, G.E.: ImageNet classification with deep convolutional neural networks. In: International Conference on Neural Information Processing Systems, pp. 1097–1105 (2012)
19. Leordeanu, M., Sukthankar, R., Hebert, M.: Unsupervised learning for graph matching. Int. J. Comput. Vis. **96**(1), 28–45 (2012)
20. Li, Y.: Deep reinforcement learning: an overview (2017). arXiv preprint arXiv:1701.07274
21. Milan, A., Rezatofighi, S.H., Garg, R., Dick, A., Reid, I.: Data-driven approximations to NP-hard problems. In: Thirty-First AAAI Conference on Artificial Intelligence, pp. 1453–1459 (2017)
22. Mnih, V., Kavukcuoglu, K., Silver, D., Graves, A., Antonoglou, I., Wierstra, D., Riedmiller, M.: Playing Atari with deep reinforcement learning. Computer Science (2013). https://arxiv.org/abs/1312.5602
23. Mnih, V., Kavukcuoglu, K., Silver, D., Rusu, A.A., Veness, J., Bellemare, M.G., Graves, A., Riedmiller, M., Fidjeland, A.K., Ostrovski, G., et al.: Human-level control through deep reinforcement learning. Nature **518**(7540), 529 (2015)
24. Nicolas Heess, D.S., Teh, Y.W.: Actor-critic reinforcement learning with energy-based policies. In: European Workshop on Reinforcement Learning (2012)
25. Pineda, F.J.: Generalization of back-propagation to recurrent neural networks. Phys. Rev. Lett. **59**(19), 2229–2232 (1987)
26. Riedmiller, M.: Neural fitted q iteration - first experiences with a data efficient neural reinforcement learning method. In: European Conference on Machine Learning, pp. 317–328 (2005)
27. Schuster, M., Paliwal, K.K.: Bidirectional recurrent neural networks. IEEE Trans. Signal Process. **45**(11), 2673–2681 (1997)

28. Silver, D., Lever, G., Heess, N., Degris, T., Wierstra, D., Riedmiller, M.: Deterministic policy gradient algorithms. In: *ICML* (2014)
29. Silver, D., Huang, A., Maddison, C.J., Guez, A., Sifre, L., van den Driessche, G., Schrittwieser, J., Antonoglou, I., Panneershelvam, V., Lanctot, M.: Mastering the game of Go with deep neural networks and tree search. Nature **529**, 484–489 (2016)
30. Silver, D., Schrittwieser, J., Simonyan, K., Antonoglou, I., Huang, A., Guez, A., Hubert, T., Baker, L., Lai, M., Bolton, A.: Mastering the game of go without human knowledge. Nature **550**(7676), 354–359 (2017)
31. Sutskever, I., Vinyals, O., Le, Q.V.: Sequence to sequence learning with neural networks. In: Advances in Neural Information Processing Systems, pp. 3104–3112 (2014)
32. Sutton, R., Barto, A.: Reinforcement Learning: An Introduction. MIT Press, Cambridge (1998)
33. Sutton, R.S., McAllester, D., Singh, S., Mansour, Y.: Policy gradient methods for reinforcement learning with function approximation. In: Advances in Neural Information Processing Systems, pp. 1057–1066 (2000)
34. Vinyals, O., Fortunato, M., Jaitly, N.: Pointer networks. In: Advances in Neural Information Processing Systems, pp. 2692–2700 (2015)
35. Vinyals, O., Kaiser, Ł., Koo, T., Petrov, S., Sutskever, I., Hinton, G.: Grammar as a foreign language. In: Advances in Neural Information Processing Systems, pp. 2773–2781 (2015)
36. Watkins, C.H., Dayan, P., Christopher, J.: Q-learning. In: Machine Learning, pp. 279–292 (1992)
37. Williams, R.J., Zipser, D.: A learning algorithm for continually running fully recurrent neural networks. Neural Comput. **1**(2), 270–280 (1989)
38. Zhang, M.-L., Zhou, Z.-H.: A review on multi-label learning algorithms. IEEE Trans. Knowl. Data Eng. **26**(8), 1819–1837 (2014)

Modeling Malware Propagation Dynamics and Developing Prevention Methods in Wireless Sensor Networks

Zaobo He, Yaguang Lin, Yi Liang, Xiaoming Wang, Akshita Maradapu Vera Venkata Sai, and Zhipeng Cai

Abstract Modeling malware propagation dynamics and developing prevention methods are very imperative with flourishing and advancement of WSN technologies in a variety of fields, such as smart cities. In the last decade, a lot of effort has been put into designing effective models to characterize the propagation dynamics of malware and developing effective prevention methods, with different focuses such as spatial–temporal model, pulse immunization, trade-off model between prevention cost and network utility, etc. This chapter reviews the state-of-the-art malware modeling and prevention method to present a comprehensive guide on how to choose a more appropriate approach for different applications. First, the application background and definitions of WSNs and malware are introduced, followed by the challenges of modeling malware propagation dynamics and developing prevention methods. Second, the recent advances in modeling and prevention methods are summarized. Third, four recently published papers that focus on spatial–temporal modeling, pulse immunization, and cost-efficiency trade-off are introduced. Finally, this chapter is ended by pointing out some possible future research directions.

Z. He
Department of Computer Science and Software Engineering, Miami University, Oxford, OH, USA
e-mail: hez26@miamioh.edu

Y. Lin · X. Wang
School of Computer Science, Shaanxi Normal University, Shaanxi, Xi'an, China

Y. Liang · A. M. Vera Venkata Sai · Z. Cai (✉)
Department of Computer Science, Georgia State University, Atlanta, GA, USA
e-mail: zcai@gsu.edu

© Springer Nature Switzerland AG 2019
D.-Z. Du et al. (eds.), *Nonlinear Combinatorial Optimization*,
Springer Optimization and Its Applications 147,
https://doi.org/10.1007/978-3-030-16194-1_10

1 Introduction

Wireless sensor networks (WSNs) have been successfully applied to a variety of fields [1–4], such as smart cities [5, 6], environment monitoring[7], and border protection. For all these applications, sensory data is gathered from each sensor through wireless communications and then manipulated. One significant challenge of data gathering and manipulation is to explore how to guarantee data availability, integrity, and reliability, due to limitations of sensor nodes in terms of energy, storage, and computation capacity [8–10]. Furthermore, more challenges rise due to the popularity of mobile wireless sensor networks (MWSNs), which form a dynamically changing network topology. In MWSN, mobility of nodes enables data communication to eliminate the limitation of geographic space [11]. These deficiencies provide great opportunities for malware to inject malicious codes into sensor nodes, which not only compromise data integrity and reliability, but also compromise data availability due to bad influence on network lifetime and channel utilization [12–17].

A lot of effort has been put into designing effective models to characterize the propagation dynamics of malware and developing effective cybersecurity methods[10], with different focuses such as spatial–temporal model[18, 19], pulse immunization[20], trade-off model between prevention cost and network utility[21, 22], and so on. In this chapter, we introduce and summarize the recent advances in propagation modeling and prevention, and present some related works.

In general, the application scenarios of malware propagation modeling and prevention include two parts: static WSNs and MWSNs. Compared with static WSNs, the network topology in MWSNs changes dynamically because of node mobility. Although MWSN technologies are successfully used in smart cities nowadays, cybersecurity issues are emerging increasingly[23, 24]. Furthermore, dynamically changing network topology also poses serious challenges to the implementation of prevention methods.

Meanwhile, prevention methods can be further classified as *immunization* and *quarantine* [25–28]. In immunization, security packages are installed in advance to immunize nodes from malware. Through restoring the security flaws that could be exploited by malware, susceptible nodes could be immunized and their states are transferred to recovered nodes. In quarantine, malware is first removed from the infected nodes, and then these nodes are immunized. Finally, the infected nodes can carry out data collection and manipulation normally, and their states are transferred to recovered nodes.

When dealing with malware propagation modeling and prevention for WSNs, the constraints and limitations of WSNs in terms of energy supply, mobility, communication capacity, and immunization and quarantine costs introduce many challenges. We summarize some of the main challenges as follows.

- **Network Utility and Prevention Efficiency Trade-off.** Installing immunization packets continuously or removing malware at high intensity will seriously occupy the limited communication channels, deplete limited battery energy, which definitely exacerbates communication delay, and shorten the life-cycle of networks. How to conduct propagation modeling and prevention to achieve a trade-off between network utility and prevention efficiency is a significant challenge. The goal is to enable the proposed models and prevention methods to not only prevent further propagation of malware but also to guarantee utility of the network.

– **Prevention Cost and Efficiency Trade-off.** Installing immunization packets and removing malware jams regular network communication and shortens the network lifetime. Furthermore, these immunization packets will occupy expensive memory space and computational capacity of the node, which further exacerbates the performance deduction of data collection and manipulation. Therefore, the cost of immunization and quarantine cannot be ignored, which poses a challenge to achieving the trade-off between prevention cost and efficiency.

– **Prevention Methods with Time Limitation.** It is very important to the decision makers to know whether or not malware in the network will die out under the prevention methods. At the same time, strong prevention methods may lead to the rapid disappearance of the malware, but the cost is considerable. The goal is to develop prevention methods that can guarantee that the malware extincts at the end of an expected time period with the lowest prevention cost.

– **Maximum Immunization Period.** Continuous immunization method is widely applied in large WSNs, which install immunization packets on susceptible nodes continuously. Although continuous immunization largely reduces the chance of propagation of malware, it also depletes network resources continuously. Especially, it jams regular network communication intensively and shortens the network lifetime aggressively. Therefore, there should be a discrete immunization plan that immunizes susceptible nodes at some regular interval. How to find such a proper interval offers another challenge.

– **Spatial–Temporal Dynamic Modeling.** Temporal dynamic modeling is used to predict malware propagation dynamics for some time into the future, but it cannot offer insight on propagation dynamics in geographic space. For preventing malware propagation, expected methods are to either install immunization packets on susceptible nodes or quarantine infected nodes in both spatial–temporal dimensions. How to model the spatial–temporal dynamics of malware propagation in WSNs and develop target immunization and quarantining methods is a significant challenge. The goal is to immunize susceptible nodes prior to the next peak of node infection and quarantine infected nodes to prevent the occurrence of next peak of node infection.

– **Malware Propagation Threshold.** Malware propagation dynamics is inherently determined by network properties, such as node communication range, mobility rate, and packet delivery rate. For example, susceptible nodes are easier to be infected by malware in a network with larger communication, mobility rate, and packet delivery rate, since infected nodes can reach more susceptible neighbors easily. Therefore, the last challenge is to identify a threshold that determines whether the malware continuously propagates or becomes extinct in the future, which also formulates a restrictive relationship between network properties and prevention methods.

The rest of this chapter is organized as follows: In Section 2, the background of malware propagation in WSNs is provided. Then we review and summarize the existing propagation dynamic models and prevention methods, followed by the discussion of four famous modeling and prevention schemes. Finally, we conclude this chapter and point out some possible future research directions in Section 4.

2 Malware Propagation in Wireless Sensor Networks

In this section, we first introduce some background knowledge on malware propagation in WSNs. Subsequently, we review some existing modeling and prevention methods. Finally, we discuss one popular spatial–temporal model that describes the propagation dynamics of malware in WSNs, which is also the best existing model with respect to predicting the propagation dynamics in spatial–temporal dimensions of malware in WSNs [18].

2.1 Introduction of Malware Propagation in WSNs

Recent years have witnessed an overwhelming propagation of malware in WSNs and traditional Internet. The cybersecurity and privacy issues in WSNs are of great importance as the sensory data collected and aggregated by sensor nodes could be very sensitive, and moreover, most of the sensor nodes generally operate in field environment, or even hostile environment [29–31]. For example, although smart cities are becoming ubiquitous in the scientific, social and economic fields, WSNs are the first step in the development of big data collection, aggregation and applications. For those sensors across cities, the collected data could be from smart home, smart grid, and smart government, which pose serious threat on our cities. WSNs are prone to be vulnerable to various attacks, such as eavesdropping, intrusion, or malicious interference with their normal operations, or even malicious hardware damage[32–34]. Malwares are injected into sensor nodes, which can destroy nodes, deplete their energy, block regular communications, or damage the integrity of regular data packets. This terminology refers to a variety of hostile software, including network worms, viruses, Trojan Horses and spyware [35]. Furthermore, when malware is injected into some nodes, they can propagate across the entire network, with the propagation of tampered data.

Existing results showed that the process of malware propagation is closely related to network properties, involving sensor node density, energy consumption (remaining energy), communication range, movement speed, network topology, and work interleaving schedule policy. According to that, the process of malware propagation in WSN has its own unique features, compared with other traditional networks:

1. When malware propagates on a network like the Internet, it injects malicious codes into other hosts by randomly scanning other hosts' IP addresses, whereas malware in a WSN can propagate malicious codes to entire networks by outspreading neighbors, and there are no geographical limitations for malware propagation in MWSNs.
2. Due to the sleep and active intervals, the malware on an active node can propagate to neighbor nodes that are active, but the sleeping neighbors of that active node do not get infected. Moreover, while a node is sleeping, any malware on that node cannot infect other nodes, since there is no communications between this sleeping node and its neighbors.
3. When the energy of a WSN is exhausted, more and more nodes become dead nodes, then the network cannot effectively carry on data collection and aggregation. Meanwhile, malwares also cannot effectively propagate in network, because dead nodes will not participate in the process of propagating malware in a WSN.
4. When malware propagates in MWSN, node movement removes spatial restrictions. Network topology changes enable the malware to propagate to a larger geographic space.
5. When malware propagates on a node with small communication rate, the number of infected neighbors is limited. Only those nodes located in its communication rage can be infected. For a network with high density, malware can easily propagate to a large network area.

In view of the above challenges and properties, existing research works can be broadly divided into two categories as follows: (a) mathematically model the propagation dynamics of malware in WSNs [36–43]; (b) develop prevention methods that can achieve the trade-off among network utility, prevention efficiency, and prevention cost so that the propagation of malware can be prevented while tolerable network utility and prevention cost can also be guaranteed [18, 20–22, 44–49].

2.2 Overview of Modeling and Prevention

In this subsection, we review some recent works on malware propagation modeling and prevention methods in WSNs. Some comprehensive surveys have been conducted in [50–54]. Furthermore, the research advances in malware propagation prevention are summarized in [55–57]. Following this, we summarize new advances in malware propagation modeling and prevention in four aspects: spatial–temporal dynamic model, pulse immunization model, cost-efficient prevention methods, and information propagation dynamics in heterogeneous network.

2.2.1   Spatial–Temporal Modeling of Malware Propagation

In [18], a spatial–temporal dynamic model is proposed that characterizes the process of malware propagation in WSNs, based on epidemic models [58] and diffusion–reaction models [59, 60]. First, the functional relations between network properties and malware propagation are investigated, which play vital roles in determining the general modeling and offering an insight on preventing malware propagation by making some changes to the network properties. Furthermore, epidemic models and diffusion–reaction models are inherently dependent on differential equations, so that it is fundamental to incorporate network properties into models of malware propagation. Second, by analyzing the solution properties of the proposed system (i.e., the stability of the equilibrium solutions [61]), a threshold is obtained which determines whether the malware continually propagates or dies out in a WSN. On the one hand, system solutions indicate the density of susceptible nodes in the future, because they are actually the function of susceptible nodes, infected nodes and recovered nodes at any time t in any geographic space. On the other hand, system solutions offer great insight on carrying out prevention methods since they are also the functions of network properties, as shown in the theorem given in [18]:

Theorem 1 *When there is a positive–equilibrium point and the probability of immunizing the susceptible nodes is greater than the probability of patching the infected nodes, the proposed system finally reaches this equilibrium point, and it indicates that the malware will continuously propagate in the MWSN.*

Since the geographical distribution of infected nodes in the future can be predicted in advance, a targeted immunization strategy can be used to recover the infected nodes in some extensively infected regions.

2.2.2   Pulse Immunization Model of Malware

In [20], a pulse immunization model is proposed that implements prevention methods in every optimized pulse period based on pulse differential equation models. The prevention method that is successfully applied in large WSNs is to install security packets on susceptible sensor nodes, so that those nodes will not be infected by malware in the future. The crucial problem is, when immunization packets should be installed for a WSN. In recent years, many continuous immunization methods are proposed, and they install immunization packets randomly. If immunization packets are installed intensively, network resources will be depleted intensively. Especially, regular network communication will be jammed intensively and network lifetime is also shortened aggressively. However, if immunization packets are installed infrequently, the malware propagation cannot be controlled and more and more susceptible nodes will transform to infected nodes. Therefore, there should be a discrete immunization plan that immunizes susceptible nodes at some regular interval. Furthermore, most of the current prevention methods are designed only for static WSNs without considering the influence of node mobility on malware propagation. Therefore, how to define prevention methods that can be applied to MWSNs is a serious challenge.

Motivated by [18], this work developed a differential equation model to characterize the propagation dynamics of malware in MWSNs based on epidemic model, and then developed an pulse immunization method that carries out the installation of immunization packets at some regular interval. Finally, this work mathematically analyzed the existence and stability of a malware-free solution of the proposed model, and derived the maximal immunization interval T. At every interval T, immunization packets are installed on susceptible nodes to prevent malware. The derived immunization interval guarantees that immunization packets can be installed with minimum frequency, while malware can extinct over time in MWSN.

2.2.3   Cost-Efficient Prevention Methods

In [62], computational models and optimal control strategies are proposed for emotion contagion in the human population in emergencies that implements prevention methods in every optimized pulse period based on Pontryagin's maximum principle [63]. There are two methods for preventing malware propagation in WSNs: immunization and quarantine. In immunization, security packages are installed in advance to immunize nodes from malware. Through restoring the security flaws that could be exploited by malware, susceptible nodes could be immunized and their states are transformed to recovered nodes. In quarantine, malware is first removed from infected nodes, and then these nodes are immunized.

Unfortunately, the costs of carrying out these two methods are usually ignored. Installing immunization packets and removing malware jam regular network communication and shorten network lifetime. Furthermore, these immunization packets will occupy precious memory space and computational capacity of node, which further exacerbates the performance deduction of data collection and manipulation. Therefore, the cost of immunization and quarantine cannot be ignored, which poses a challenge to achieving the trade-off between prevention cost and efficiency. Moreover, immunization and quarantine are considered individually in most existing works [64–68]. Hence, prior works overestimate the efficiency of proposed prevention methods and are costly. In this work, a prevention method is proposed that prevents malware propagation by exploiting both immunization and quarantine methods with limited costs. With this idea, a real-time optimization (RTO) prevention method is proposed based on Pontryagin's maximum principle, which can minimize the malware prevention cost by optimally combining immunization and quarantine, while a rumor can become extinct within an expected time period. With the optimization objective, RTO provides optimized rates for immunization and quarantine in a real-time manner.

2.2.4 Information Propagation Dynamics in Heterogeneous Network [22]

Since information propagation in online social networks (OSNs) has similar dynamics as malware propagation in WSNs, in this subsection, we discuss a recently published cost-efficient strategy for restraining rumor spreading in OSNs. It is a heterogeneous network based epidemic model which tries to achieve a rumor restraining trade-off between cost and efficiency.

This paper takes the influence of network heterogeneity and the cost of prevention methods into account, and proposes a heterogeneous network based epidemic model that incorporates both network heterogeneity influence and two prevention methods. First, by analyzing the existence and stability of equilibrium solutions of the proposed model, the critical conditions that determine whether a rumor continuously propagates or becomes extinct are derived. Second, the cost of the countermeasures, i.e., blocking rumors at influential users and spreading truth to clarify rumors, are incorporated in the propagation model. Finally, based on the Pontryagin's maximum principle, the optimized countermeasures that guarantee that a rumor can become extinct at the end of an expected time period with lowest cost are derived. Both the critical conditions and the optimized countermeasures provide a real-time decision reference to restrain the rumor spreading

In phase one, based on epidemic model, ONS users can be divided into three groups: Susceptible (S) represents the users that are not infected by rumor but are susceptible to it; Infected (I) represents users that are infected and act as rumor spreaders; and Recovered (R) represents the users that are immunized to rumors because of knowing truth. Obviously, one user belongs to only one of the three groups. The acceptance rates for information (rumor and truth) of different users is different because of their different social connectivity. Users can be characterized based on their social connectivity (also called degree) that reflects the social influence of one user. Degree based heterogeneity can effectively characterize information diffusion in scale free networks [4, 6, 13] (i.e., power-law degree distribution). Based on degree, the users in a network can be classified into n groups and the users in one group have same social connectivity. For group i ($i = 1, 2, \ldots, n$), k_i denotes the social connectivity of the individuals in this group. Let $S_{k_i}(t)$, $I_{k_i}(t)$, and $R_{k_i}(t)$ denote the density of susceptible, infected, and recovered individuals in group i at time t, respectively, and $S_{k_i}(t) + I_{k_i}(t) + R_{k_i}(t) = 1$.

Based on the above analysis, this work finally draws the following conclusions.

Theorem 2 *If prevention methods' countermeasures are effective resulting to $r_0 \leq 1$, the rumor will be extinct. Otherwise, if prevention methods' result is $r_0 > 1$, the rumor will continuously propagate and the density of infected users will converge to a stable level.*

In Theorem 2, r_0 is the threshold which determines the existence of the equilibrium solution for the proposed model.

2.3 Reaction–diffusion Modeling of Malware Propagation in MWSN [18]

This work abstracts a WSN as a two dimensional area Ω with area Φ, which is divided into a set of blocks. Each block represents a piece of geographic subarea. The area of block x is denoted by Φ_x. $N(t)$ is used to denote the number of active sensor nodes in Ω at time t. This work uses random walk model to model the behavior of mobile nodes. The mobility behavior of nodes follows random walk model, which indicates nodes randomly roam with an average velocity v in Ω. The node density is denoted by σ, which is defined as $\sigma = N/\Phi$. Then, the number of nodes located in block x is: $N(x, t) = \Phi_x N/\Phi$. Any two neighboring nodes n_i, n_j in Ω, can directly communicate with each other. The number of neighbors of each node is denoted by δ with average value $\pi r^2 \sigma$, assuming nodes are uniformly distributed over Ω, where r is the communication range of sensor node. The packet transmission rate is denoted as ρ, $0 \leq \rho \leq 1$, which means one node sends ρ data packets to its neighbors per unit time.

A node is infected if it is injected with malicious codes by malware. An infected node propagates multiple copies of the malware to its neighbors while transmitting regular data packets. A node is susceptible if it has not yet been infected by malware, but is prone to be infected. A node is recovered if it is installed with immunization packets or is quarantined from the infected nodes. A node is dead if it is unable to conduct data collection and manipulation because of hardware damage or energy exhaustion.

Let $S(x, t)$, $I(x, t)$, and $R(x, t)$ denote the number of susceptible nodes, infected nodes, and recovered nodes in block (x, t) at time t, respectively. For the active nodes in Ω, $S(x, y, t) + I(x, y, t) + R(x, y, t) = N(t)$. In block x, the proportion of susceptible nodes is $S(x, t)/N(x, t) = S(x, t)/(\sigma \Phi_x)$. Meanwhile, the number of susceptible neighbors for each infected node in block x is $\delta S(x, t)/(\sigma \Phi_x) = \pi r^2 S(x, t)/\Phi_x$. Since packet loss generally occurs in WSNs, when an infected node sends ρ data packets to its neighbors per unit time, the probability that one of ρ data packets is at least successfully received by its susceptible neighbors is $1 - (1 - \rho)^\rho$, where ρ is the average probability of one packet that is successfully sent. Thus, the number of susceptible neighbors to which an infected node can successfully send malware is $N_s = \pi r^2 (1 - (1 - \rho)\rho)S(x, t)/\Phi_x$. For simplicity, denote $\lambda = \pi r^2 (1 - (1 - \rho)\rho)/\Phi_x$.

Intuitively, the capability of an infected node to infect a large set of neighbors by sending malware to its neighbors reduces with the increase of infected nodes in network. This effect is called saturation effect [10]. For simplicity, $S(x, t)$, $I(x, t)$, $R(x, t)$, and $D(x, t)$ are represented by S, I, R, and D, respectively. To characterize the saturation effect, a parameter is introduced: $\beta = \frac{I}{1+\alpha I^2}$, where α is an adjusting factor related to the network property. In general, α take value from: $0 \leq \alpha \leq 1$; obviously, $0 \leq \beta < 1$.

Because some nodes will die due to hardware damage and energy exhaustion, new nodes must be added to the network. A is used to represent the rate at which

Fig. 1 Node state transition
relationship, where dots
represent nodes

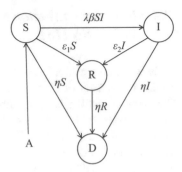

new nodes (susceptible nodes) are added to the network. The death rate of nodes
is denoted by η, $0 < \eta < 1$. At any time, the state transition of sensor nodes can
be described as Figure 1. The weight ς on $F \xrightarrow{\varsigma} H$ means that ς nodes in state
F enter state H, where $F, H \in \{S, I, R, D\}$, $F \neq H$. ε_1 is the probability with
which a susceptible node transforms to a recovered node, and ε_2 is the probability
with which an infected node transforms to a recovered node, $0 < \varepsilon_1 < 1$ and
$0 < \varepsilon_2 < 1$.

Nodes continuously transform to another state, as malware propagates in network. The change rate of the density of nodes in different states with time can
be represented as $\frac{\partial S}{\partial t}$ $\frac{\partial I}{\partial t}$ and $\frac{\partial R}{\partial t}$. The process of malware propagation involves
both spatial and temporal dynamics. Therefore, the spatial–temporal propagation
dynamics of malware can be modeled using reaction–diffusion system:

$$\frac{\partial S}{\partial t} = \gamma(S, I, R) + \mu_S \nabla^2 S,$$

$$\frac{\partial I}{\partial t} = \chi(S, I, R) + \mu_I \nabla^2 I, \tag{1}$$

$$\frac{\partial R}{\partial t} = \gamma(S, I, R) + \mu_R \nabla^2 R,$$

where the left part of each equation is the rate of density change of susceptible,
infected, and recovered nodes in network, respectively. $\gamma(S, I, R)$, $\chi(S, I, R)$ and
$\gamma(S, I, R)$ characterize the state transformation of nodes, which are called reaction
functions (or reaction terms). $\mu_S \nabla^2 S$, $\mu_I \nabla^2 I$ and $\mu_R \nabla^2 R$ are called diffusion
terms, which are used to characterize the influence of node mobility, where μ_S, μ_I
and μ_R are the diffusion coefficients, which reflect the influence of data mobility on
change rate of node density. Assuming all nodes are moving with the same speed,
$\mu_S = \mu_I = \mu_R$. ∇^2 is the Laplace operator, a second-order differential operator
in multivariable calculus [69]. In a two dimensional region, ∇^2 is defined as $\nabla^2 = \partial^2/\partial x^2 + \partial^2/\partial y^2$.

If prevention methods are effective, a malware does not propagate any more.
Otherwise, a malware would continuously propagate. These two cases correspond
to the malware-free equilibrium solution and the positive-equilibrium solution of

System (1), respectively. If a malware-free equilibrium point exists and is stable, the malware in network will finally extinct. If a positive-equilibrium point exists and is stable, the malware will continuously propagate and the density of all nodes will finally reach a steady state. Otherwise, the malware propagation finally goes into an oscillation state. From the non-linear system theory [69], we know that a diffusion system has the same constant equilibrium points with its corresponding non-diffusion system.

Theorem 3 *When there is a positive-equilibrium point and the immunization rate on susceptible nodes is greater than the quarantining probability on infected nodes, the proposed system finally reaches its equilibrium point, and the malware will continuously propagate in the MWSN.*

2.4 Pulse Immunization Model

Consider a MWSN with area A and N nodes. Each node moves in a random direction with speed v in network, which results in uniform distribution of nodes in geographic space. Each node is restricted by limited battery capacity and communication range that can be viewed as a circle with radius r, b represents the probability of a node dying due to hardware damage or energy exhaustion. To keep the active nodes in network steady, new susceptible nodes are continuously placed into network with rate b.

The state transformation among nodes is represented in Figure 2. In Figure 2, ρ is the probability of a node being transformed to a recovered node, and ε is the probability of node being transformed to a dead node. λ is the rate that represents the contacting chance of any two nodes per unit time. Because the total number of acting nodes is steady, then $S(t) + I(t) + R(t) = 1$. The movement process of each node n_j is divided into finite continuous K stages, represented by $t_1, t_2, \ldots t_K$, and $t_{i+1} - t_i = t_{i+2} - t_{i+1} = t$, where $1 \leq i \leq K - 2$. In any stage t_i, node n_j moves or stops in an interleaving way. For an arbitrary node, assuming that the mobility time is T_1 and that the stopping time is T_2, where $T_1 + T_2 = t$. Therefore, a larger T_1 implies a smaller T_2, and vice versa.

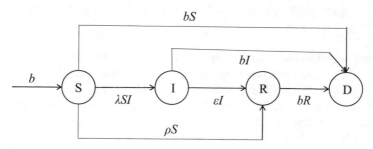

Fig. 2 State transition relationships of nodes

Obviously, the distance that node n_j can move is vT_1 at state t, and the communication region area of node n_j is $\Phi = 2rvT_1 + \pi r^2$. Therefore, the number of neighbors of n_j is $\zeta = \mu\Phi = (2rvT_1 + \pi r^2)N/A$ at one stage. In addition, β is used to represent the rate of n_j to scan neighbors for sending data package. The probability of one infected node to successfully send malware to its neighbors is α. The capability of an infected node to successfully infect its neighbors is defined by $\eta = \alpha\beta$. Therefore, the number of susceptible nodes that can be infected successfully by an infected node at stage t is $\delta = \eta\zeta S(t) = (2rvT_1 + r2)N/A$. Let $\lambda = (2rvT_1 + r^2)N/A$.

The node state transformation part is described as follows, when $t \neq nT$, based on transformation in Figure 2:

$$\frac{dS(t)}{dt} = b - \lambda S(t)I(t) - bS(t);$$

$$\frac{dI(t)}{dt} = \lambda S(t)I(t) - \varepsilon I(t) - bI(t); \tag{2}$$

$$\frac{dR(t)}{dt} = \varepsilon I(t) - bR(t).$$

The pulse immunization part is described as: when $t = nT$, $S(nT^+) = (1 - p)S(nT)$; $I(nT^+) = I(nT)$; $R(nT^+) = R(nT) + pS(nT)$, where n is a positive integer. $[nT, (n+1)T]$ is n-th time intervals between two immunization operation at time nT and $(n+1)T$. Since pulse immunization lets nodes transform states sharply at each time nT, the above proposed system is a pulse differential system. At any time $t \neq nT$, nodes' states are transformed because of malware propagation. We use nT^+ to represent the next immunization point of nT. At any time $t = nT$, a set of nodes' states are transformed because of pulse immunization. As shown in the system, the number of susceptible node is sharply deducted $pS(nT)$, at nT^+ after pulse immunization at nT. Furthermore, the number of infected nodes does not change in the process of pulse immunization, because pulse immunization only transforms susceptible nodes into recovered nodes at each time point $t = nT$.

If there exists a malware-free and stable T-period solution of proposed system, malware will extinct finally. Otherwise, malware will continuously propagate.

Theorem 4 *If the immunization period is T, the T-period zero-equilibrium of proposed System is $S^T(t)$, where $S^T(t) = (1 + (S^* - 1)e^{-b(t-nT)})$ and $S^* = \frac{(1-p)(e^{bT}-1)}{e^{bT}+p-1}$.*

Immunizing susceptible users with the maximum immunization period can minimize the frequency of jamming regular network communication and shortening network lifetime. Then, the maximum immunization period T_{max} is the threshold that determines whether the T-period zero-equilibrium solution is stable. If $T < T_{max}$, i.e., immunization frequency is less than maximum frequency, malware will extinct finally. On the contrary, any immunization period T with $T > T_{max}$ would let malware propagate in the network continuously.

2.5 Emotion Contagion in the Human Population in Emergencies [62]

In this chapter, we discuss a new emotion information diffusion model SLIR, which is extended from the traditional epidemic model and can describe the spatial–temporal diffusion of emotion information more accurately in self-organizing networks.

In SLIR model, the author divides nodes into four categories in an ad hoc network: the susceptible state (S), the latent state (L), the infected state (I), and the recovered state (R). If one node does not receive any information, the node belongs to S-state. If one node receives emotional information, it may be infected by that emotional information. Because of the cognitive psychology of the users, the users who receive the emotional information first enter the thinking and wandering stage, considering the authenticity of the information, and whether or not to believe the information. The node in the state of wandering and thinking belongs to the L-state.

If the L-state nodes believe the emotional information after a period of thinking, the nodes are transformed to I-state. If the nodes lose interest in the received emotional information, or the nodes no longer believe the emotional information, they will not receive the emotional information from then on, and the nodes belong to R state.

When the S-state nodes encounter the I-state nodes, the I-state nodes can transmit emotional information to the S-state nodes through conversation. Then, the S-state nodes are transformed to the I-state nodes with probability $\alpha(t)$, or transformed to the L-state nodes with probability $\beta(t)$, or transformed to the R-state nodes with probability $1 - \alpha(t) - \beta(t)$, at any time t. When the I-state nodes encounter the R-state nodes, the I-state nodes are transformed to the R-state nodes with probability $\gamma(t)$. When the L-state nodes encounter the R-state nodes, the L-state nodes are transformed to the R-state nodes with probability $\delta(t)$. When the L-state nodes encounter the I-state nodes, the L-state nodes are transformed to the I-state nodes with probability $\theta(t)$. The probability of the L-state nodes jump out of thinking and wandering state is ρ. After jumping out of the wandering period, L-state nodes are transformed to the R-state nodes with probability $\sigma(t)$, or transformed to the I-state nodes with probability $1 - \sigma(t)$. The nodes in the I-state may lose interest in emotional information and thus become nodes in R state. Since the nodes in R state may be interested in emotional information again, so it could transform to S-state with probability $\xi(t)$.

In particular, this work proposes a novel method for determining these probability parameters based on the psychological model. This method incorporates these parameters into a time-decay model, which can accurately describe the process of nodes losing interest in information over time. For example, the authors model interest decay as $\alpha(t) = \omega e^{-\psi_1(t-t_0)}$, where t_0 indicates the occurrence time of event. ω and ψ_1 are positive constants. When one event occurs, ω indicates the influence of emotional information involving the event. The ψ_1 can be used to express the attenuation of emotional information's influence over time, and the larger the ψ_1, the faster the attenuation of information influence. The authors use the mean field equations to express the emotional information diffusion model.

To sum up, the advantages of the SLIRS model are that: (1) Based on a systematic discussion of the existing models for characterizing the process of emotion information diffusion, the authors identify some strengths and limitations of the existing models. The concept of latent state is introduced to describe the nodes in thinking or wandering state, which are considered separately from other types of nodes in modeling; (2) The model fully considers the complex interaction between nodes in different states and the transformation process between nodes in different states, and further establishes the computational model of emotion information diffusion dynamics; and (3) According to the attenuated mental model, the authors have different mathematical descriptions of the time varying parameters of node state transition probabilities. The attenuation function is introduced to describe the influence of attenuation of interest on information diffusion more accurately.

2.6 Optimal Control Method for Information Diffusion

In this chapter, the authors introduce an optimal control method for the propagation of panic emotion in the network. Based on the SLIRS model mentioned above, the author further proposes some cooperative control strategies. For the potential loss derived from the propagation of panic emotion information, the authors also consider the cost of preventing the propagation of panic emotion information, based on optimal control theory, with the goal of minimizing the total cost.

In particular, the authors take different control measures for S-state nodes, L-state nodes, and I-state nodes in the SLIRS model. The measure taken by S-state nodes is called vaccination measure, which enables the S-state nodes to hold some information that the panic emotion information is not true, in advance, so that the S-state nodes can be protected from panic emotion information when they receive such information, rather than transforming to R-state nodes. The measure taken by L-state nodes is called quarantine measure, which enables the L-state nodes to know that the panic emotion information they receive is untrue. In this way, L-state nodes will not be infected by emotional information and will not further generate panic emotion, and then turn into R-state nodes. The measure taken by I-state nodes is called treatment measure, which tries to pacify and dispel rumors on those nodes that have been infected by panic emotional information, so that these nodes no longer panic, and then turn into R-state nodes.

Through the implementation of the above three control measures, the panic information in network can be effectively controlled. However, an important issue needs to be considered. Since the cost of implementing the three control measures is different, how to determine the real-time strength of the three control measures, at the minimum total cost to control the propagation of panic information? Moreover, since the damage of panic information to nodes in various states is not feasible, how to minimize the total loss of all nodes in network based on the above three control measures?

In order to solve the above problems, the authors first express the costs of implementing vaccination, quarantine, and treatment control strategies at time t as $v_1^2(t)$, $v_2^2(t)$, and $v_3^2(t)$, respectively, and then express the loss of panic emotion information to L-state nodes and I-state nodes at time t as $pL(t)$ and $zI(t)$, respectively. Finally, the cost and loss can be considered as the total cost of the system. The goal of the author's optimization is to minimize the total cost of the system.

After establishing the optimization problem, the author uses the optimal control theory to analyze the existence of the optimal solution and to solve it. The following conclusions are obtained:

Theorem 5 *There exists only one optimal control pair* $(v_1^*(t),\ v_2^*(t),\ v_3^*(t))$*, such that* $J(v_1^*(t), v_2^*(t), v_3^*(t))\ =\ \min_{v_1(t), v_2(t), v_3(t)} J(v_1(t), v_2(t), v_3(t))$ *subject to SLIRS model with vaccination, quarantine, and treatment control strategies.*

Theorem 6 *For a small time interval* $[0, t]$*, at any time* $t \in [0, t]$ *the bounded solutions of the optimality system are unique.*

To sum up, the advantages of the optimal control method for panic information diffusion are as follows: (1) Three different control strategies (vaccination, quarantine, and treatment) are established for different states of users in the network, which can efficiently control different states of nodes to varying degrees. (2) The authors systematically consider the impact of panic information propagation on users in the network and the cost of implementing control strategies. The two aspects are modeled as the control objectives of the optimization problem. (3) By using the optimal control theory, the existence theory of the optimal solution of the optimization problem is analyzed and the optimal solution is derived.

3 Discrete Models Based on Cellular Automata

Apart from these dynamic models based on epidemic models and differential equations, another type of models based on cellular automata has also been applied in large WSNs successfully. In the following [36, 70–73], cellular automata are introduced to characterize the propagation dynamics of malware in WSNs.

Cellular automata are simple computational models that are also successfully applied in large WSNs, which are finite-state machines that can characterize system dynamics efficiently and effectively. They consist of a finite set of cells that interact with each other via a certain policy, so that at any time, each cell is in one state from among a finite number of possible states. Such set of cells can be viewed as a set of blocks and then each cell represents a piece of geographic space. Nodes located in one cell c_i could be susceptible nodes, infected nodes, or recovered nodes. The topologies that determine the state change of nodes in one cell are malware propagation dynamics and prevention methods. In cellular automata, the state change of nodes in a cell is also determined by its neighbors.

4   Conclusions and Future Work

In this chapter, the techniques of modeling malware propagation and prevention methods in WSNs are discussed. First, the application background and definitions of WSNs and malware are introduced, followed by the challenges of modeling malware propagation dynamics and developing prevention methods. Subsequently, the recent advances in modeling and prevention methods are summarized. Three recently published papers that focus on spatial–temporal modeling, pulse immunization, and cost-efficiency trade-off are then introduced in detail, respectively. Spatial–temporal modeling is applied in MWSNs to characterize malware propagation dynamic from both spatial and temporal dimensions, which can effectively predict the spatial and temporal distribution of malware in future. The result from spatial–temporal analysis offers great opportunity to prevent malware propagation by target immunization and quarantine. Pulse immunization method installs immunization packets on sensors with a maximum period; therefore, network utility can be largely improved. To further achieve the trade-off between network utility and prevention costs, we further introduce a method that derives the optimized immunization and quarantine rate based on Pontryagin's maximum principle; so that malware propagation can stop propagating at the end of given time period, with minimum cost.

In recent years, since the use of WSNs is becoming increasingly wider and deeper, and cybersecurity and data privacy issues are also becoming very serious, there are still many new challenges introduced by different applications involving malware propagation in Internet of Things, smart cities applications. Therefore, the following future research directions exist.

- Most of the existing works on modeling and preventing are deterministic, based on differential equations, cellular automata, or Markov chain. Therefore, they may be unable to simulate the individual dynamics of each node. These works declare that the overall propagation dynamics observed reflect the propagation trends of malware enough.
- Almost all of the existing works on modeling and preventing malware propagation are based on the strong assumption that data movement follows certain data mobility models, such as random direction, random walk, Gauss–Markov model, etc. However, for specific applications, data mobility is very hard to characterize. For example, in smart cities, people taking their smart phones do not follow any models strictly to move around. Therefore, this assumption may not be enough with the wider application of WSN technology in smart cities. Therefore, developing reliable node mobility models is a prerequisite for wide application in smart cities.
- Most of the models based on epidemic models and differential equations ignore network heterogeneity. Especially, almost all works based on epidemic models assume that nodes are infected at the same rate, nodes are uniformly distributed in a geographic space, or have same node mobility speed, etc. These models simply classify all nodes into different groups, such as susceptible nodes, infected

nodes, recovered nodes, or dead nodes. Then there is almost no any heterogeneity for nodes in one group. However, in actual applications, one susceptible node entering the communication range of one infected node is a random event. For example, in smart cities, the route of cars cannot be simulated in 1 day.

References

1. Rashid, B., Rehmani, M.H.: Applications of wireless sensor networks for urban areas: a survey. J. Netw. Comput. Appl. **60**, 192–219 (2016)
2. Kurt, S., Yildiz, H.U., Yigit, M., Tavli, B., Gungor, V.C.: Packet size optimization in wireless sensor networks for smart grid applications. IEEE Trans. Ind. Electron. **64**(3), 2392–2401 (2017)
3. Li, X., Niu, J., Kumari, S., Liao, J., Liang, W., Khan, M.K.: A new authentication protocol for healthcare applications using wireless medical sensor networks with user anonymity. Secur. Commun. Netw. **9**(15), 2643–2655 (2016)
4. Lu, C., Saifullah, A., Li, B., Sha, M., Gonzalez, H., Gunatilaka, D., Wu, C., Nie, L., Chen, Y.: Real-time wireless sensor-actuator networks for industrial cyber-physical systems. Proc. IEEE **104**(5), 1013–1024 (2016)
5. Wu, J., Ota, K., Dong, M., Li, C.: A hierarchical security framework for defending against sophisticated attacks on wireless sensor networks in smart cities. IEEE Access **4**(4), 416–424 (2016)
6. Fadel, E., Gungor, V.C., Nassef, L., Akkari, N., Malik, M.A., Almasri, S., Akyildiz, I.F.: A survey on wireless sensor networks for smart grid. Comput. Commun. **71**, 22–33(2015)
7. Xu, G., Shen, W., Wang, X.: Applications of wireless sensor networks in marine environment monitoring: a survey. Sensors **14**(9), 16932–16954 (2014)
8. Rezvani, M., Ignjatovic, A., Bertino, E., Jha, S.: Secure data aggregation technique for wireless sensor networks in the presence of collusion attacks. IEEE Trans. Dependable Secure Comput. **12**(1), 98–110 (2015)
9. Luo, X., Zhang, D., Yang, L.T., Liu, J., Chang, X., Ning, H.: A kernel machine-based secure data sensing and fusion scheme in wireless sensor networks for the cyber-physical systems. Futur. Gener. Comput. Syst. **61**, 85–96 (2016)
10. Shen, S., Li, H., Han, R., Vasilakos, A.V., Wang, Y., Cao, Q.: Differential game-based strategies for preventing malware propagation in wireless sensor networks. IEEE Trans. Inf. Forensics Secur. **9**(11), 1962–1973 (2014)
11. Liu, B., Zhou, W., Gao, L., Wen, S., Luan, T.H.: Mobility increases the risk of malware propagations in wireless networks. In: Trustcom/BigDataSE/ISPA, 2015 IEEE, vol. 1, pp. 90–95. IEEE, Piscataway (2015)
12. De, P., Liu, Y., Das, S.K.: An epidemic theoretic framework for vulnerability analysis of broadcast protocols in wireless sensor networks. IEEE Trans. Mob. Comput. **8**(3), 413–425 (2009)
13. Haghighi, M.S., Wen, S., Xiang, Y., Quinn, B., Zhou, W.: On the race of worms and patches: Modeling the spread of information in wireless sensor networks. IEEE Trans. Inf. Forensics Secur. **11**(12), 2854–2865 (2016)
14. Illiano, V.P., Lupu, E.C.: Detecting malicious data injections in wireless sensor networks: a survey. ACM Comput. Surv. CSUR **48**(2), 24 (2015)
15. Wang, T., Wu, Q., Wen, S., Cai, Y., Tian, H., Chen, Y., Wang, B.: Propagation modeling and defending of a mobile sensor worm in wireless sensor and actuator networks. Sensors **17**(1), 139 (2017)
16. Osanaiye, O., Alfa, A.S., Hancke, G.P.: A statistical approach to detect jamming attacks in wireless sensor networks. Sensors **18**(6), 1691 (2018)

17. Dâmaso, A., Rosa, N., Maciel, P.: Integrated evaluation of reliability and power consumption of wireless sensor networks. Sensors 17(11), 2547 (2017)
18. Wang, X., He, Z., Zhao, X., Lin, C., Pan, Y., Cai, Z.: Reaction-diffusion modeling of malware propagation in mobile wireless sensor networks. Sci. China Inf. Sci. 56(9), 1–18 (2013)
19. He, Z., Wang, X.: A spatial-temporal model for the malware propagation in MWSNs based on the reaction-diffusion equations. In: International Conference on Web-Age Information Management, pp. 45–56. Springer, Heidelberg (2012)
20. Xiaoming, W., Zaobo, H., Lichen, Z.: A pulse immunization model for inhibiting malware propagation in mobile wireless sensor networks. Chin. J. Electron. 23(CJE-4), 810–815 (2014)
21. He, Z., Cai, Z., Wang, X.: Modeling propagation dynamics and developing optimized counter-measures for rumor spreading in online social networks. In: 2015 IEEE 35th International Conference on Distributed Computing Systems (ICDCS), pp. 205–214. IEEE, Piscataway (2015)
22. He, Z., Cai, Z., Yu, J., Wang, X., Sun, Y., Li, Y.: Cost-efficient strategies for restraining rumor spreading in mobile social networks. IEEE Trans. Veh. Technol. 66(3), 2789–2800 (2017)
23. Sun, B., Osborne, L., Xiao, Y., Guizani, S.: Intrusion detection techniques in mobile ad hoc and wireless sensor networks. IEEE Wirel. Commun. 14(5), 56–63 (2007)
24. Wang, J., Jiang, C., Zhang, K., Quek, T.Q., Ren, Y., Hanzo, L.: Vehicular sensing networks in a smart city: Principles, technologies and applications. IEEE Wirel. Commun. 25(1), 122–132 (2018)
25. Ojha, R.P., Srivastava, P.K., Sanyal, G.: Pre-vaccination and quarantine approach for defense against worms propagation of malicious objects in wireless sensor networks. Int. J. Inf. Syst. Model. Des. IJISMD 9(1), 1–20 (2018)
26. Zhang, Z., Wang, H., Wang, C., Fang, H.: Cluster-based epidemic control through smartphone-based body area networks. IEEE Trans. Parallel Distrib. Syst. 26(3), 681–690 (2015)
27. Mishra, B.K., Tyagi, I.: Defending against malicious threats in wireless sensor network: a mathematical model. Int. J. Inf. Technol. Comput. Sci. IJITCS 6(3), 12 (2014)
28. Nwokoye, C.H., Ozoegwu, G.C., Ejiofor, V.E.: Pre-quarantine approach for defense against propagation of malicious objects in networks. Int. J. Comput. Netw. Inf. Secur. 9(2), 43 (2017)
29. Sookhak, M., Tang, H., He, Y., Yu, F.R.: Security and privacy of smart cities: a survey, research issues and challenges. In: IEEE Communications Surveys and Tutorials (2018)
30. Braun, T., Fung, B.C., Iqbal, F., Shah, B.: Security and privacy challenges in smart cities. Sustain. Cities Soc. 39, 499–507 (2018)
31. Khatoun, R., Zeadally, S.: Cybersecurity and privacy solutions in smart cities. IEEE Commun. Mag. 55(3), 51–59 (2017)
32. Can, O., Sahingoz, O.K.: A survey of intrusion detection systems in wireless sensor networks. In: 2015 6th International Conference on Modeling, Simulation, and Applied Optimization (ICMSAO), pp. 1–6. IEEE, Piscataway (2015)
33. Radhappa, H., Pan, L., Xi Zheng, J., Wen, S.: Practical overview of security issues in wireless sensor network applications. Int. J. Comput. Appl. 40(4), 1–12 (2017)
34. Chen, H., Lou, W.: On protecting end-to-end location privacy against local eavesdropper in wireless sensor networks. Pervasive Mob. Comput. 16, 36–50 (2015)
35. Mallela, S.S., Jonnalagadda, S.K.: Internet security—a brief review. In: Microelectronics, Electromagnetics and Telecommunications, pp. 889–894. Springer, Singapore (2018)
36. del Rey, A.M., Peinado, A.: Mathematical models for malware propagation in wireless sensor networks: an analysis. In: Computer and Network Security Essentials, pp. 299–313. Springer, Cham (2018)
37. Peng, S., Yu, S., Yang, A.: Smartphone malware and its propagation modeling: a survey. IEEE Commun. Surv. Tutorials 16(2), 925–941 (2014)
38. Feng, L., Song, L., Zhao, Q., Wang, H.: Modeling and stability analysis of worm propagation in wireless sensor network. Math. Probl. Eng. 2015 (2015)
39. Zhu, L., Zhao, H.: Dynamical analysis and optimal control for a malware propagation model in an information network. Neurocomputing 149, 1370–1386 (2015)

40. Liu, B., Zhou, W., Gao, L., Zhou, H., Luan, T.H., Wen, S.: Malware propagations in wireless ad hoc networks. IEEE Trans. Dependable Secure Comput. (1), 1–1 (2016)
41. Batista, F.K., del Rey, Á.M., Queiruga-Dios, A.: Malware propagation software for wireless sensor networks. In: International Conference on Practical Applications of Agents and Multi-Agent Systems, pp. 238–241. Springer, Berlin (2017)
42. Lin, Y., Wang, X., Hao, F., Wang, L., Zhang, L., Zhao, R.: An on-demand coverage based self-deployment algorithm for big data perception in mobile sensing networks. Futur. Gener. Comput. Syst. **82**, 220–234 (2018)
43. Lu, J., Wang, X., Zhang, L.: Signal power random fading based interference-aware routing for wireless sensor networks. Wirel. Netw. **20**(7), 1715–1727 (2014)
44. Wang, X., Lin, Y., Zhao, Y., Zhang, L., Liang, J., Cai, Z.: A novel approach for inhibiting misinformation propagation in human mobile opportunistic networks. Peer-to-Peer Netw. Appl. **10**(2), 377–394 (2017)
45. Wang, X., Lin, Y., Zhang, S., Cai, Z.: A social activity and physical contact-based routing algorithm in mobile opportunistic networks for emergency response to sudden disasters. Enterp. Inf. Syst. **11**(5), 597–626 (2017)
46. Lin, Y., Wang, X., Zhang, L., Li, P., Zhang, D., Liu, S.: The impact of node velocity diversity on mobile opportunistic network performance. J. Netw. Comput. Appl. **55**, 47–58 (2015)
47. Wang, X., Lin, Y., Zhang, L., Cai, Z.: A double pulse control strategy for misinformation propagation in human mobile opportunistic networks. In: International Conference on Wireless Algorithms, Systems, and Applications, pp. 571–580. Springer, Heidelberg (2015)
48. Wang, X., Zhang, L., Dou, W., Hu, X.: Fuzzy colored time petri net and termination analysis for fuzzy event-condition-action rules. Inf. Sci. **232**, 225–240 (2013)
49. Zhang, L., Wang, X., Lu, J., Li, P., Cai, Z.: An efficient privacy preserving data aggregation approach for mobile sensing. Secur. Commun. Netw. **9**(16), 3844–3853 (2016)
50. Cheng, S.-M., Ao, W.C., Chen, P.-Y., Chen, K.-C.: On modeling malware propagation in generalized social networks. IEEE Commun. Lett. **15**(1), 25–27 (2011)
51. Suarez-Tangil, G., Tapiador, J.E., Peris-Lopez, P., Ribagorda, A.: Evolution, detection and analysis of malware for smart devices. IEEE Commun. Surv. Tutorials **16**(2), 961–987 (2014)
52. Shen, S., Huang, L., Liu, J., Champion, A.C., Yu, S., Cao, Q.: Reliability evaluation for clustered WSNs under malware propagation. Sensors **16**(6), 855 (2016)
53. Yu, S., Gu, G., Barnawi, A., Guo, S., Stojmenovic, I.: Malware propagation in large-scale networks. IEEE Trans. Knowl. Data Eng. **27**(1), 170–179 (2015)
54. Liu, Y., Dong, M., Ota, K., Liu, A.: ActiveTrust: secure and trustable routing in wireless sensor networks. IEEE Trans. Inf. Forensics Secur. **11**(9), 2013–2027 (2016)
55. Shen, S., Ma, H., Fan, E., Hu, K., Yu, S., Liu, J., Cao, Q.: A non-cooperative non-zero-sum game-based dependability assessment of heterogeneous WSNs with malware diffusion. J. Netw. Comput. Appl. **91**, 26–35 (2017)
56. Liu, L., Ko, R.K., Ren, G., Xu, X.: Malware propagation and prevention model for time-varying community networks within software defined networks. Secur. Commun. Netw. **2017** (2017)
57. Lu, Y., Da Xu, L.: Internet of things (IoT) cybersecurity research: a review of current research topics. IEEE Internet Things J. 13pp. (2018). https://doi.org/10.1109/JIOT.2018.2869847
58. Duan, W., Fan, Z., Zhang, P., Guo, G., Qiu, X.: Mathematical and computational approaches to epidemic modeling: a comprehensive review. Front. Comp. Sci. **9**(5), 806–826 (Oct 2015)
59. Colizza, V., Pastor-Satorras, R., Vespignani, A.: Reaction–diffusion processes and metapopulation models in heterogeneous networks. Nat. Phys. **3**(4), 276 (2007)
60. Doi, M.: Stochastic theory of diffusion-controlled reaction. J. Phys. A Math. Gen. **9**(9), 1479 (1976)
61. Hirsch, M.W., Smale, S., Devaney, R.L.: Differential Equations, Dynamical Systems, and an Introduction to Chaos. Academic, Amsterdam (2012)
62. Wang, X., Zhang, L., Lin, Y., Zhao, Y., Hu, X.: Computational models and optimal control strategies for emotion contagion in the human population in emergencies. Knowl.-Based Syst. **109**, 35–47 (2016)

63. Kopp, R.E.: Pontryagin maximum principle. In: Mathematics in Science and Engineering, vol. 5, pp. 255–279. Elsevier, New York (1962)
64. Mishra, B.K., Srivastava, S.K., Mishra, B.K.: A quarantine model on the spreading behavior of worms in wireless sensor network. Trans. IoT Cloud Comput. 2(1), 1–12 (2014)
65. Li, F., Yang, Y., Wu, J.: CPMC: An efficient proximity malware coping scheme in smartphone-based mobile networks. In: INFOCOM, 2010 Proceedings IEEE, pp. 1–9. IEEE, Piscataway (2010)
66. Tang, S., Mark, B.L.: Analysis of virus spread in wireless sensor networks: an epidemic model. In: 7th International Workshop on Design of Reliable Communication Networks, 2009. DRCN 2009, pp. 86–91. IEEE, Piscataway (2009)
67. Sun, X., Lu, Z., Zhang, X., Salathé, M., Cao, G.: Targeted vaccination based on a wireless sensor system. In: 2015 IEEE International Conference on Pervasive Computing and Communications (PerCom), pp. 215–220. IEEE, Piscataway (2015)
68. Gardner, M.T., Beard, C., Medhi, D.: Using SEIRS epidemic models for IoT botnets attacks. In: Proceedings of DRCN 2017-Design of Reliable Communication Networks; 13th International Conference, pp. 1–8. VDE, Berlin (2017)
69. Perko, L.: Differential Equations and Dynamical Systems, vol. 7. Springer Science and Business Media, New York (2013)
70. Wang, Y., Li, D., Dong, N.: Cellular automata malware propagation model for WSN based on multi-player evolutionary game. IET Netw. 7(3), 129–135 (2017)
71. García, G.G., Ramirez, M.E.L.: Modeling the spatio-temporal dynamics of worm propagation in smartphones based on cellular automata. In: Modelling Symposium (EMS), 2016, European, pp. 196–201. IEEE, Piscataway (2016)
72. Peng, S., Wang, G.: Worm propagation modeling using 2d cellular automata in Bluetooth networks. In: 2011 IEEE 10th International Conference on Trust, Security and Privacy in Computing and Communications (TrustCom), pp. 282–287. IEEE, Piscataway (2011)
73. Zhao, Y., Wang, X., Li, L.: Research on mobile cellular automata model for public sentiment dissemination in opportunistic networks. Appl. Res. Comput. 2 (2015)

Composed Influence Maximization in Social Networks

Smita Ghosh, Jianming Zhu, and Weili Wu

Abstract Influence maximization has been studied extensively in the literature. Its mathematical formulation is a monotone nondecreasing submodular maximization. However, when composed influence is considered, the corresponding problem becomes a nonsubmodular maximization. A composed influence results from a combination of at least two active members in the social network. To study this problem, a hypergraph model is introduced and hence different methodologies are involved.

1 Introduction

The rapid growth of online social network communities such as Facebook and Twitter has intrigued the interest of researchers and scientist all over the world to study and analyze large-scale social structure and behavior. Abundance of rich data is now available from social networks that can be analyzed and studied to understand information flow and social dynamics.

A social network is depicted as a graph with nodes that represent individuals and edges that represent the relationship shared among the nodes in the graph. The influence maximization problem takes in input a graph G(V, E), where V is the set of users and E is the set of (directed/undirected) edges in G. The objective of this problem is to find a set of users with the maximum influence in graph G. The output to this problem is a k-sized seed set, which when initially influenced is expected to give the set of maximum influenced nodes in the graph G. Influence in social networks is propagated through stochastic cascade models. Given a social graph G,

S. Ghosh (✉) · W. Wu
Department of Computer Science, The University of Texas at Dallas, Richardson, TX, USA
e-mail: smita.ghosh1@utdallas.edu; ghosh1@utdallas.edu; weiliwu@utdallas.edu

J. Zhu
School of Engineering Science, University of Chinese Academy of Sciences, Beijing, China
e-mail: jmzhu@ucas.ac.cn

© Springer Nature Switzerland AG 2019
D.-Z. Du et al. (eds.), *Nonlinear Combinatorial Optimization*,
Springer Optimization and Its Applications 147,
https://doi.org/10.1007/978-3-030-16194-1_11

251

a user set $S \subset V$, and a diffusion model M captures the stochastic process for S spreading information on G. The influence spread of S, denoted as $\sigma_{G,M}(S)$, is the expected number of users influenced by S, where $\sigma_{G,M}(.)$ is a non-negative function defined on any subset of users, i.e., $\sigma_{G,M} : 2^V \Rightarrow R_{\geq}0$ [12].

1.1 Composed Influence in Social Networks

Prior work on social network analysis has been focused on traditional graph analysis which assumed interaction between a pair of nodes in the graph where one node influences only on other node (represented by an edge between them). But in reality, the social interaction is not only between pairs of individuals but is most commonly observed among the members of a group. Researchers have studied that the best way of representing these group dynamics is through a different graph model known as hypergraphs. Crowd psychology plays an important role in determining the decisions that an individual makes in their daily life [24]. Influence maximization (IM) problem is a widely researched problem in the area of social network analysis. In this chapter we look into social influence maximization in hypergraphs in social networks. A hypergraph is a variation of a normal graph, in which an edge can join any number of vertices. A hypergraph H is defined as a pair $H = (X, E)$, where X is a set of elements called nodes or vertices, and E is a set of non-empty subsets of X called hyperedges or edges. Therefore, E is a subset of $P(X) \setminus $, where P(X) is the power set of X.

1.2 Submodular and Nonsubmodular Functions

The mathematical formulation of influence maximization is a monotone nondecreasing submodular maximization. A monotonic function, in mathematics [11] (or monotone function) is defined as a function between ordered sets that preserves or reverses the given order. A function is called monotonically increasing (also increasing or nondecreasing), if for all x and y such that $x \leq y$ one has $f(x) \leq f(y)$ so f preserves the order. Similarly a function is called monotonically decreasing (also decreasing or nonincreasing [3]) if, whenever $x \leq y$, then $f(x) \geq f(y)$, so it reverses the order. A submodular set function (also known as a submodular function) is a set function whose value has the property that the difference in the incremental value of the function that a single element makes when added to an input set decreases as the size of the input set increases. Submodular functions have a natural diminishing returns property which makes them suitable for many applications, including approximation algorithms, game theory (as functions modeling user preferences), and electrical networks. Recently, submodular functions have also found immense utility in influence maximization and social network analysis. It also has applications in several real-world problems in machine learning and artificial intelligence, including automatic summarization, multi-document summarization, feature selection, active learning, sensor placement, image collection summarization, and many other domains [13]. The popular problem of influence maximization has been formalized as a monotone nondecreasing submodular maximization. But when we extend the traditional problem of influence maximization to the composed influence maximization problem, the formalization of the corresponding problem becomes nonsubmodular.

1.2.1 Methods for Maximizing Monotone Nondecreasing Set Functions

Research on approximation for maximizing monotone nondecreasing set functions has focused on greedy methods. In [22], the authors have showed that the greedy algorithm can find a solution with value at least $1/1 + \mu$ of the optimum value for a general monotone nondecreasing function. In [17] the authors have worked on the problem of maximizing a monotone submodular functions over the integer lattice. Recently in [1] the authors have analyzed the performance of the greedy algorithm, and also analyzed a discrete semi-gradient based algorithm, for maximizing the sum of a suBmodular and suPermodular (BP) functions (both of which are non-negative monotone nondecreasing) under two types of constraints, either a cardinality constraint or $p \geq 1$ matroid independence constraints.

1.2.2 Maximization of Nonsubmodular Functions

Composed influence in social network analysis is formulated as a nonsubmodular function. Researchers have studied different ways to maximize nonsubmodular functions. One such method is mentioned in [2]. In this paper the authors have investigated the performance of the standard GREEDY algorithm for cardinality constrained maximization of nonsubmodular nondecreasing set functions. There exists strong theoretical guarantees on the performance of GREEDY for maximizing submodular functions, but there are few theoretical analysis on guarantees for nonsubmodular ones. In this paper they prove theoretical guarantees supporting the empirical performance. Their guarantees are characterized by a combination of the (generalized) curvature α and the submodularity ratio γ. They prove that GREEDY has a tight approximation guarantee of $1/\alpha(1 - e^{-}\gamma\alpha)$ for cardinality constrained maximization.

1.3 Diffusion Models

In a diffusion model framework, each user $u \in V$ with a status of either inactive or active. Based on the social graph G, at first, it views the status of a set of chosen users, called seed set $S \subset V$, to be active, while other users in V are inactive.

Then, it considers the diffusion process that the seed users in S can "influence" their neighbors to be active. The newly activated users can further activate their neighbors, and so on. This diffusion process terminates when no new users can be activated [12]. There are various types of diffusion model, namely the independent cascade (IC) model, the linear threshold (LT) model, the triggering (TR) model, time-aware diffusion model, and non-progressive diffusion Model.

1.3.1 Independent Cascade Model

In the IC Model [6], a user v is activated by each of its incoming neighbors independently by introducing an influence probability p_u, v to each edge $e = (u, v)$. Based on the influence probabilities and given a seed set S at time step 0, a diffusion instance of the IC model unfolds in discrete steps. Each active user u in step t will activate each of its outgoing neighbor v that is inactive in step t-1 with probability p_u, v. The activation process can be considered as flipping a coin with head probability p_u, v : if the result is head, then v is activated; otherwise, v stays inactive. The diffusion instance terminates when no more nodes can be activated. The influence spread of seed set S under the IC model is the expected number of activated nodes when S is the initial active node set and the above stochastic activation process is applied.

1.3.2 Linear Threshold Model

In the LT model [7, 16], each edge e = (u,v) E is associated with a weight b_u, v. Let NI(v) be the set of incoming neighbors of user v, and it satisfies that $\Sigma_u \in NI(v)b_u$, $v \leq 1$. Each user v is also associated with a threshold θ_v. The LT model first samples the value of v of each user v uniformly at random from [0,1]. In step 0, it sets the status of users in S as active and others as inactive. Then, it updates the status of each user iteratively. In step t, all users that were active in step t-1 remain active, and any user v that were inactive in step t -1 switches to active if the total weight of its active neighbors in NI (v) is at least θ_v. The diffusion instance terminates when no more user is to be activated. Given multiple instances of the diffusion processes, the influence spread of seed set S under the LT model, i.e., $\sigma(S)$, is the expected number of activated nodes when S is initially activated.

A lot of research has been done since Kempe [9] introduced the concept of influence maximization first in the year 2003. Throughout the years, influence maximization problems have been studied and have wide applications such as viral marketing, network monitoring, rumor control, and social recommendation. However there still exist some challenges while studying this problem. The first challenge is modeling the information diffusion process in a social network which would heavily affect the influence spread of any seed set. The second challenge is that IM problems are theoretically complex in general. It has been proved that obtaining an optimal solution of IM is NP-hard under most of the diffusion models.

The third challenge is due to the stochastic nature of information diffusion, even the evaluation of influence spread of any individual seed set is computationally complex. Active research is going on to determine the stability of IM algorithms. Previous work shows that there is a poor stability of IM algorithms when the input influence probabilities are adversarially noisy [12]. Research is being done to break the boundary of submodularity by modeling influence function with more general functions. Also, a prospective future direction is to consider the concept of group influence.

2   Related Work

Kempe et al. [9] was the first to formulate social influence maximization problem (SIMP) as an optimization problem under the IC model. Inspired by the concept of "word of mouth" in the promotion of new products, the authors provided the first provable approximation guarantees for efficient algorithm. They showed that their designed natural greedy strategy yields $(1 - 1/e - \epsilon)$-approximate solutions for any $\epsilon > 0$ and the obtained solution is provably within 63% of optimal for several classes of models. They backed up their problem formulation with experimental observations showing that their algorithm outperforms node selection heuristics based on the well-studied notions of degree centrality and distance centrality from the field of social network. Inspired by this work, a lot of literature on SIMP has been since developed.

In [10], the authors have presented a general methodology for near optimal sensor placement in a given network. They demonstrate that many realistic outbreak detection objectives (e.g., detection likelihood and population affected) exhibit the property of "submodularity." They exploit submodularity to develop an efficient algorithm that scales to large problems, achieving near optimal placements, while being 700 times faster than a simple greedy algorithm. They also derived online bounds on the quality of the placements obtained by any algorithm. Their algorithms and bounds also handle cases where nodes (sensor locations and blogs) have different costs. In another paper [21], the authors have proposed a new algorithm called community based greedy algorithm for mining top-K influential nodes. The proposed algorithm has two components. The first is, an algorithm for detecting communities in a social network by taking into account information diffusion and second is a dynamic programming algorithm for selecting communities to find influential nodes. They have also provided provable approximation guarantees for their algorithm. Empirical studies on a large real-world mobile social network show that their algorithm is more than an order of magnitudes faster than the state-of-the-art greedy algorithm for finding top-K influential nodes and the error of our approximate algorithm is small.

However most of the existing methods are not fast enough for scaling billions of edges in networks such as those in Facebook, Twitter, and World Wide Web. This problem is targeted by [4], where the authors have developed a novel sketch-based design for influence computation. Their greedy sketch-based influence maximization (SKIM) algorithm scales to graphs with billions of edges, with one to two orders of magnitude speedup over the best greedy methods. It guarantees an approximation ratio, and its quality nearly matches that of exact greedy. They also presented influence oracles, which use linear-time preprocessing to generate a small sketch for each node, allowing the influence of any seed set to be quickly answered from the sketches of its nodes.

Prior studies show that there are two scalable models, namely TIM/TIM+ [18] and IMM [19] with a $(1 - 1/e - \epsilon)$-approximate guarantee for SIMP. In [18] and IMM [19], the authors utilize a novel reverse influence set (RIS) sampling technique introduced in [3]. Both TIM+ and IMM focus on generating a $(1 - 1/e - \epsilon)$-approximate solution by using minimum number of RIS samples. One challenge this method has is it may take a long period of time (even spanning over a number of days) to process a large-scale network with over a billion edges. In [15]; however, the author makes a breakthrough by proposing two novel sampling algorithms, namely SSA and D-SSA. These algorithms were faster than the previously proposed TIM+ and IMM algorithms and also providing the $(1 - 1/e - \epsilon)$-approximate guarantee.

Apart from intensive research being done on SIMP, a lot of work has also been done on group-level influence maximization [23]. Existing work often focuses on the influence of individual nodes, ignoring that infecting different seeds may require different costs. In [23] the authors have investigated the problem of group-level influence maximization with budget constraint. They introduced a statistical method to reveal the influence relationship between the groups, based on which they proposed a propagation model that can dynamically calculate the influence spread scope of seed groups, followed by presenting a greedy algorithm called GLIMB to maximize the influence spread scope with a limited cost budget via the optimization of the seed-group portfolio. Theoretical analysis shows that GLIMB can guarantee an approximation ratio of at least $(1 - 1/\sqrt{e})$. Experimental results on both synthetic and real-world data sets verified the effectiveness and efficiency of their approach. In another paper [8] the authors proposed local information maximization (LIM), considering group impact in terms of local propagation where the influencer(s) of each community has a direct effect on the nodes in the same community. They conducted experiments on synthetic data set and compared the performance of the LIM to various other heuristics.

3   Problem Formulation

In this section we will focus on how group influence can be represented as hypergraphs and introduce the different formulations for social influence maximization in hypergraphs.

Fig. 1 Information diffusion
process in SIMPH with initial
seeds $\{V_1, V_2\}$

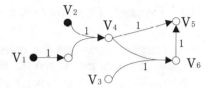

3.1 Information Diffusion in Hypergraphs

Motivated by the crowd influence in social networks, a social influence maximization problem in hypergraph (SIMPH) that aims to maximize the expected number of eventually influenced users under independent cascade(IC) model is proposed in [24]. Given a directed hypergraph $G = (V, E, P)$, where V is a set of nodes (representing users in an online social network (OSN)), E is a set of directed hyperedges, and P is the weight function on hyperedge set E. Hyperedges represent influence propagation directions, including personal and crowd influences. For a hyperedge $e = (H_e, v)$, let H_e denote its head set of nodes and v be the tail node. If H_e contains only one node u, it means e is a normal directed edge and the influence is personal. While H_e contains more than one node, the hyperedge e means there is crowd influence from H_e to v. Let P_e denote the weight of e, representing the influence propagation probability ($0 \le P_e \le 1$). Specifically, P_e is the probability that v is activated by H_e after each node in H_e is activated. The diffusion process of SIMPH is shown in Figure 1.

3.2 Influence Maximization in Hypergraphs

The social influence maximization problem in hypergraph (SIMPH) also considers information diffusion in social network with crowd influence under the IC model. Given a directed hypergraph $G = (V, E, P)$, the objective is to select k initially influenced seed users to maximize the expected number of eventually influenced users:

$$\max \sigma(S) \tag{1}$$

$$s.t.|S| \le k, \tag{2}$$

where S is the initial seed set and $\sigma(S)$ the expected number of eventually influenced nodes.

3.3   Example Objective Function Formulation

Prior literature states that the objective function of influence maximization is submodular under the IC model. However, the objective function in influence maximization problem in hypergraph is not submodular. It is shown that $\sigma(\cdot)$ is not supermodular as well.

Theorem 1 $\sigma(\cdot)$ *is not submodular under IC model.*

Proof Proved by a counterexample. Consider Figure 2. A social network $G = (V, E, P)$ has $V = \{v_1, v_2, v_3, v_4\}$, $E = \{(v_1, v_4), (v_3, v_4), (\{v_1, v_2\}, v_3)\}$, and $\{P_{(v_1,v_4)} = 1, P_{(v_3,v_4)} = 1, P_{(\{v_1,v_2\},v_3)} = 1\}$. Let $A = \emptyset$ and $B = \{v_2\}$, we have $\sigma(A) = 0, \sigma(B) = 1$. Putting v_1 into A and B, we have $\sigma(\{v_1\}) = 2$ and $\sigma(\{v_2, v_1\}) = 4$. Thus,

$$\sigma(A \cup \{v_1\}) - \sigma(A) < \sigma(B \cup \{v_1\}) - \sigma(B).$$

Therefore, $\sigma(\cdot)$ is not submodular.   □

From the proof, it is seen that the reason why $\sigma(\cdot)$ is not submodular is the crowd influence from the newly added node and the existing seed nodes.

Theorem 2 $\sigma(\cdot)$ *is not supermodular under IC model.*

Proof Proved by a counterexample. Consider Figure 2. Let $A = \emptyset$ and $B = \{v_1\}$, we have $\sigma(A) = 0, \sigma(B) = 2$. Putting v_3 into A and B, we have $\sigma(\{v_3\}) = 2$ and $\sigma(\{v_1, v_3\}) = 3$. Thus,

$$\sigma(A \cup \{v_3\}) - \sigma(A) > \sigma(B \cup \{v_3\}) - \sigma(B).$$

Therefore, $\sigma(\cdot)$ is not supermodular.   □

There is no general method to optimize a nonsubmodular function. In [14] the authors proposed a sandwich approximation strategy, which approximates the objective function by looking for its lower bound and upper bound.

In [24], the authors have shown derivations for building the upper bound and the lower bound. Figures 3 and 4 show an instance of the upper bound and lower bound formulation.

Fig. 2 Counterexample

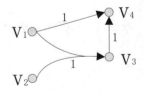

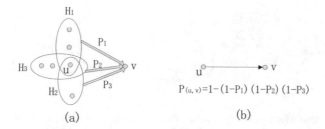

Fig. 3 An example for generation of each pair of nodes for the upper bound. (a) A node pair u and v with three head node sets H_1, H_2, H_3 contain u as shown in hypergraph (a). (b) Shows the generation process for directed edge (u, v) with probability $P(u, v) = 1-(1-P_1)(1-P_2)(1-P_3)$

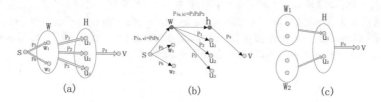

Fig. 4 An example for generation of directed graph for lower bound problem. (a) Sample hypergraph for applying lower bound problem. (b) Directed graph generated from (a). (c) Directed graph for lower bound problem

4 Method for Solving SIMPH

The dynamic-stop-and-stare (D-SSA) algorithm was extended to solve general weighted SIMP. Then a randomized algorithm based on a greedy strategy is designed for solving SIMPH. At the end, a sandwich approximation framework is proposed for analyzing performance of the algorithms.

4.1 RIS Sampling

Given a graph $G = (V, C, E, P, f)$, where $C \subseteq V$ is a candidate seed set. RIS captures the influence landscape of G through generating a set $\mathscr{R}$ of random *weighted reverse reachable(WRR)* sets. Each WRR set R_j is a subset of V and constructed as follows:

Definition 1 (Weighted Reverse Reachable (WRR) Set) Given $G = (V, C, E, P, f)$, a random WRR set R_j is generated from G by (1) selecting a random node $v \in V$; (2) generating a sample graph g from G; (3) returning R_j as the set of nodes that can reach v in g and (4) $w(R_j) = f(v)$.

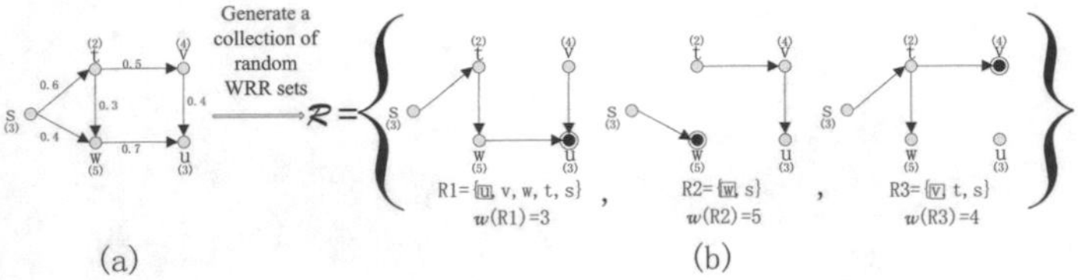

Fig. 5 An example for generating random WRR sets under IC model. R_1, R_2, R_3 with $w(R_1) = 3$, $w(R_2) = 5$, $w(R_1) = 4$ are generated. (**a**) Is the original weighted random graph. (**b**) Contains three WRR sets up to three sample graphs

For a seed set S, denote the coverage number of set S as $Cov_{\mathcal{R}}(S) = \sum_{R_j \in \mathcal{R}} \min\{|S \cap R_j|, 1\}$ and the coverage weight as $WCov_{\mathcal{R}}(S) = \sum_{R_j \in \mathcal{R}} w(R_j) \min\{|S \cap R_j|, 1\}$. $\sigma'(S)$ can be estimated by computing weighted coverage of set S. Figure 5 shows an example of generating a collection of random WRR sets. Suppose seed set $S = \{t\}$, then $Cov_{\mathcal{R}}(S) = 2$ and $WCov_{\mathcal{R}}(S) = 7$.

4.2 Greedy Strategy for SIMPH

4.2.1 Influence Estimation

Given a directed hypergraph $G = (V, E, P)$ with n nodes, $\sigma(S)$ is the expected number of eventually influenced nodes for seed set S. Suppose $g = (V, E')$ is a sample graph of G, let $\sigma_g(S)$ denote the number of eventually influenced nodes. Then $\frac{\sigma_g(S)}{n}$ is random variable distributed in interval $[0, 1]$.

4.2.2 Greedy Algorithm

The nodes in the head set of a hyperedge will try to activate the tail node only when they are all active themselves. The reverse technique in RIS sampling is unsuitable. Then, we design a greedy algorithm, as shown in Algorithm 3. Starting with an empty seed set, the greedy strategy iteratively adds a node that maximizes the marginal gain of $\sigma(S)$, until k nodes are selected.

5 Sandwich Approximation Framework

A lower bound and upper bound was designed so that the sandwich framework [5] can be applied to SIMPH.

For sandwich approximation framework, we can get the following result.

Algorithm 1 D-SSA Algorithm for general weighted SIMP

Require: Graph $G = (V, C, E, P, f)$, $n = |V|$, $0 \leq \epsilon, \delta \leq 1$ and k.
Ensure: An $(1 - 1/e - \epsilon)$-approximation solution $\hat{S}_k$.
1: $\Gamma \leftarrow 4(e - 2)(1 + \epsilon)^2 \ln(2/\delta)(1/\epsilon^2)$
2: $\mathcal{R} \leftarrow$ generate Γ random RR sets by RIS
3: $< \hat{S}_k, \hat{\sigma}'(\hat{S}_k) > \leftarrow$ Weighted Max-Coverage$(\mathcal{R}, k, f(\cdot))$
4: **while** $|\mathcal{R}| \geq (8 + 2\epsilon)n \cdot \frac{\ln(\frac{2}{\delta}) + \ln C_n^k}{\epsilon^2}$ **do**
5: $\quad \mathcal{R}' \leftarrow$ generate Γ random RR sets by RIS
6: $\quad \sigma_c'(\hat{S}_k) \leftarrow WCov_{\mathcal{R}'}(\hat{S}_k) \cdot \sum_{v \in V} f(v) / \sum_{j=1}^{|\mathcal{R}'|} w(R_j)$
7: $\quad \epsilon_1 \leftarrow \hat{\sigma}'(\hat{S}_k)/\sigma_c'(\hat{S}_k) - 1$
8: $\quad$ **if** $(\epsilon_1 \leq \epsilon)$ **then**
9: $\quad\quad \epsilon_2 \leftarrow \frac{\epsilon - \epsilon_1}{2(1 + \epsilon_1)}$, $\epsilon_3 \leftarrow \frac{\epsilon - \epsilon_1}{2(1 - 1/e)}$
10: $\quad\quad \delta_1 \leftarrow e^{-\frac{Cov_{\mathcal{R}}(\hat{S}_k)\epsilon_3^2}{2c(1+\epsilon_1)(1+\epsilon_2)}}$
11: $\quad\quad \delta_2 \leftarrow e^{-\frac{(Cov_{\mathcal{R}'}(\hat{S}_k) - 1)\epsilon_2^2}{2c(1+\epsilon_2)}}$
12: $\quad\quad$ **if** $\delta_1 + \delta_2 \leq \delta$ **then**
13: $\quad\quad\quad$ **return** $\hat{S}_k$
14: $\quad\quad$ **end if**
15: $\quad$ **end if**
16: $\quad \mathcal{R} \leftarrow \mathcal{R} \cup \mathcal{R}'$
17: $\quad < \hat{S}_k, \hat{\sigma}'(\hat{S}_k) > \leftarrow$ Weighted Max-Coverage$(\mathcal{R}, k, f(\cdot))$
18: **end while**
19: **return** $\hat{S}_k$

Proof Let S_L^*, S_U^*, and S^* be the optimal solution to maximizing the lower bound, the upper bound, and the original SIMPH. Then, we have

$$\sigma(S_U) = \frac{\sigma(S_U)}{\sigma_U(S_U)}\sigma_U(S_U) \geq \frac{\sigma(S_U)}{\sigma_U(S_U)}(1 - \frac{1}{e} - \epsilon)\sigma_U(S_U^*)$$

$$\geq \frac{\sigma(S_U)}{\sigma_U(S_U)}(1 - \frac{1}{e} - \epsilon)\sigma_U(S^*) \geq \frac{\sigma(S_U)}{\sigma_U(S_U)}(1 - \frac{1}{e} - \epsilon)\sigma(S^*).$$

and

$$\sigma(S_L) = \sigma_L(S_L) \geq \left(1 - \frac{1}{e} - \epsilon\right)\sigma_L(S_L^*) \geq \frac{\sigma_L(S_L^*)}{\sigma(S^*)}\left(1 - \frac{1}{e} - \epsilon\right)\sigma(S^*)$$

Let $S_{max} = \arg \max_{S_0 \in \{S_L, S_U, S_A\}} \sigma(S_0)$, then

$$\sigma(S_{max}) \geq \max\{\frac{\sigma(S_U)}{\sigma_U(S_U)}, \frac{\sigma_L(S_L^*)}{\sigma(S^*)}\}\left(1 - \frac{1}{e} - \epsilon\right)\sigma(S^*).$$

Since $\forall S_0 \in \{S_L, S_U, S_A\}$, $(1 - \epsilon)\sigma(S_0) \leq \sigma_c(S_0) \leq (1 + \epsilon)\sigma(S_0)$, we have

$$(1 + \epsilon)\sigma(S) \geq \sigma_c(S) \geq \sigma_c(S_{max}) \geq (1 - \epsilon)\sigma(S_{max}).$$

It follows that

$$\sigma(S) \geq \frac{1-\epsilon}{1+\epsilon}\sigma(S_{max}).$$

$$\geq \max\{\frac{\sigma(S_U)}{\sigma_U(S_U)}, \frac{\sigma_L(S_L^*)}{\sigma(S^*)}\}\frac{1-\epsilon}{1+\epsilon}(1 - \frac{1}{e} - \epsilon)\sigma(S^*)$$

$\square$

6 Conclusion

Influence maximization is theoretically formulated as a monotonic submodular function. But when group influence of social networks is represented in the form of a hypergraph, the problem becomes a nonsubmodular function. In this chapter we showed one such method to solve the nonsubmodular formulation of composed influence. In this proposed method the objective function of SIMPH converts to a nonsubmodular function. A method was discussed to transform a function into nonsubmodular as seen in [24] where the authors have modeled the crowd influence in information diffusion process by using a hyperedge. Social influence maximization problem in hypergraph (SIMPH) was formulated to select initially influenced seed users under independent cascade (IC) model to maximize the

---

Algorithm 2 APP-Calculation procedure

Require: a directed hypergraph $G = (V, E, P)$, $n = |V|$, $0 \leq \epsilon, \delta \leq 1$, seed set S.
Ensure: $\sigma_c(S)$ such that $\sigma_c(S) \leq (1 + \epsilon)\sigma(S)$ with at least $(1 - \delta)$-probability
1: $\Upsilon_1 = 1 + 4(1 + \epsilon)(e - 2)\ln(2/\delta)/\epsilon^2$
2: $SumZ = 0$
3: $N = 0$
4: **while** $SumZ \leq \Upsilon_1$ **do**
5:     $g \leftarrow$ generate sample graph of G
6:     $N = N + 1, S_1 = S, S_2 = S$
7:     **while** $S_2 \neq \emptyset$ **do**
8:         $S_1 = S_1 \cup S_2$
9:         $S_2 = \emptyset$
10:        **for** each hyperedge $e = (H_e, v) \in E$ in g and v is inactive **do**
11:           **if** $H_e \subseteq S_1$ **then**
12:             Add v to S_2
13:           **end if**
14:        **end for**
15:     **end while**
16:     $SumZ = SumZ + \frac{|S_1|}{n}$
17: **end while**
18: **return** $\sigma_c(S) = n \cdot \frac{SumZ}{N}$

Algorithm 3 Greedy strategy for SIMPH

Require: a directed hypergraph $G = (V, E, P), k$.
Ensure: a set of seed nodes, S.
1: $S = \emptyset$
2: **for** $i = 1$ to k **do**
3:      $v \leftarrow \arg\max_{v \in V}$ (APP-Calculation$(G, S \cup \{v\})$−APP-Calculation(G, S))
4:      Add v to S
5: **end for**
6: **return** S

Algorithm 4 Sandwich approximation framework

Require: a directed hypergraph $G = (V, E, P), k, \epsilon, \delta$.
Ensure: a set of seed nodes, S.
1: Let S_L be the output seed set of solving the auxiliary problem $G_L = (V \cup V', E_L, P^L)$ for lower bound by D-SSA Algorithm.
2: Let S_U be the output seed set of solving the auxiliary problem $G_U = (V, E_U, P^U)$ for upper bound by D-SSA Algorithm.
3: Let S_A be the output seed set of solving $G = (V, E, P)$ by Greedy Strategy for SIMPH(Algorithm 3).
4: $S = \arg\max_{S_0 \in \{S_L, S_U, S_A\}}$ APP-Calculation(G, S_0)
5: **return** S

---

expected number of eventually influenced users. SIMPH was shown to be NP-hard and the objective function was neither submodular nor supermodular. To read more about optimization problems refer to [20].

References

1. Bai, W., Bilmes, J.A.: Greed is still good: maximizing monotone submodular+ supermodular functions (2018). Preprint. arXiv:1801.07413
2. Bian, A.A., Buhmann, J.M., Krause, A., Tschiatschek, S.: Guarantees for greedy maximization of non-submodular functions with applications (2017). Preprint. arXiv:1703.02100
3. Borgs, C., Brautbar, M., Chayes, J., Lucier, B.: Maximizing social influence in nearly optimal time. In: Proceedings of the Twenty-Fifth Annual ACM-SIAM Symposium on Discrete Algorithms, pp. 946–957. SIAM, Portland (2014)
4. Cohen, E., Delling, D., Pajor, T., Werneck, R.F.: Sketch-based influence maximization and computation: Scaling up with guarantees. In: Proceedings of the 23rd ACM International Conference on Conference on Information and Knowledge Management, pp. 629–638. ACM, New York (2014)
5. Dagum, P., Karp, R., Luby, M., Ross, S.: An optimal algorithm for Monte Carlo estimation. SIAM J. Comput. **29**(5), 1484–1496 (2000)
6. Goldenberg, J., Libai, B., Muller, E.: Talk of the network: a complex systems look at the underlying process of word-of-mouth. Mark. Lett. **12**(3), 211–223 (2001)
7. Granovetter, M.: Threshold models of collective behavior. Am. J. Sociol. **83**(6), 1420–1443 (1978)
8. Ibrahim, R.A., Hefny, H.A., Hassanien, A.E.: Group impact: local influence maximization in social networks. In: International Conference on Advanced Intelligent Systems and Informatics, pp. 447–455. Springer, Cham (2016)

9. Kempe, D., Kleinberg, J., Tardos, É.: Maximizing the spread of influence through a social network. In: Proceedings of the Ninth ACM SIGKDD International Conference on Knowledge Discovery and Data Mining, pp. 137–146. ACM, New York (2003)
10. Leskovec, J., Krause, A., Guestrin, C., Faloutsos, C., VanBriesen, J., Glance, N.: Cost-effective outbreak detection in networks. In: Proceedings of the 13th ACM SIGKDD International Conference on Knowledge Discovery and Data Mining, pp. 420–429. ACM, New York (2007)
11. Leversha, G.: The concise oxford dictionary of mathematics, edited by Christopher Clapham and James Nicholson. pp. 520.£ 9.99 (pbk). 2005. isbn 0 19 860742 3 (oup). Math. Gaz. 90(519), 537–537 (2006)
12. Li, Y., Fan, J., Wang, Y., Tan, K.-L.: Influence maximization on social graphs: a survey. IEEE Trans. Knowl. Data Eng. 30(10), 1852–1872 (2018)
13. Lin, H., Bilmes, J.: A class of submodular functions for document summarization. In: Proceedings of the 49th Annual Meeting of the Association for Computational Linguistics: Human Language Technologies, vol. 1, pp. 510–520. Association for Computational Linguistics, Portland (2011)
14. Lu, W., Chen, W., VS Lakshmanan, L.: From competition to complementarity: comparative influence diffusion and maximization. Proc. VLDB Endowment 9(2), 60–71 (2015)
15. Nguyen, H.T., Thai, M.T., Dinh, T.N.: Stop-and-stare: optimal sampling algorithms for viral marketing in billion-scale networks. In: Proceedings of the 2016 International Conference on Management of Data, pp. 695–710. ACM, New York (2016)
16. Schelling, T.C.: Schelling Micromotives and Macrobehavior (Cloth).Fels Lectures on Public Policy Analysis. W. W. Norton & Company, New York (1978)
17. Soma, T., Yoshida, Y.: Maximizing monotone submodular functions over the integer lattice. In: International Conference on Integer Programming and Combinatorial Optimization, pp. 325–336. Springer, Berlin (2016)
18. Tang, Y., Xiao, X., Shi, Y.: Influence maximization: near-optimal time complexity meets practical efficiency. In: Proceedings of the 2014 ACM SIGMOD International Conference on Management of Data, pp. 75–86. ACM, New York (2014)
19. Tang, Y., Shi, Y., Xiao, X.: Influence maximization in near-linear time: a martingale approach. In: Proceedings of the 2015 ACM SIGMOD International Conference on Management of Data, pp. 1539–1554. ACM, New York (2015)
20. Thai, M.T., Pardalos, P.M.: Handbook of Optimization in Complex Networks: Theory and Applications, vol. 57. Springer Science and Business Media, Berlin (2011)
21. Wang, Y., Cong, G., Song, G., Xie, K.: Community-based greedy algorithm for mining top-k influential nodes in mobile social networks. In: Proceedings of the 16th ACM SIGKDD International Conference on Knowledge Discovery and Data Mining, pp. 1039–1048. ACM, New York (2010)
22. Wang, Z., Moran, B., Wang, X., Pan, Q.: Approximation for maximizing monotone non-decreasing set functions with a greedy method. J. Comb. Optim. 31(1), 29–43 (2016)
23. Yan, Q., Huang, H., Gao, Y., Lu, W., He, Q.: Group-level influence maximization with budget constraint. In: International Conference on Database Systems for Advanced Applications, pp. 625–641. Springer, Berlin (2017)
24. Zhu, J., Zhu, J., Ghosh, S., Wu, W., Yuan, J. Social influence maximization in hypergraph in social networks. In: IEEE Transactions on Network Science and Engineering (2018)

Friending

Shuyang Gu, Hongwei Du, My T. Thai, and Ding-Zhu Du

Abstract The friending is a popular and important operation in online social networks. In this article, we discuss various optimization problems about friending. They can be formulated into nonlinear combinatorial optimization problems.

1 Active Friending

If you have a LinkedIn or Facebook account, then you may frequently receive a message like this "Xuefei Zhang added connections you may know," which reminds you that you may know someone, or someone is your friend's friend. If you open the message, then you may find a link to login your account and from your account, you may find some names who invited you to be their friend, and a list of names whom you may consider to invite for your friends. These activities are called friending.

The active friending is the first optimization problem appeared in the literature [34] about friending. The problem can be described as follows:

Definition 1 (Active Friending) Consider a social network represented as directed graph $G - (V, E)$ with an information diffusion model m. Suppose S is the list of existing friends of a node s and t is a target node that s wants to include in his friend list. Given an integer $r > 0$, the problem is to find a subset R with at most r nodes

S. Gu (✉) · D.-Z. Du
Department of Computer Science, The University of Texas at Dallas, Richardson, TX, USA
e-mail: Shuyang.Gu@utdallas.edu; dzdu@utdallas.edu

H. Du
Department of Computer Science and Technology, Harbin Institute of Technology Shenzhen Graduate School, Shenzhen, China
e-mail: hongwei.du@ieee.org

M. T. Thai
Department of Computer and Information Science and Engineering, University of Florida, Gainesville, FL, USA
e-mail: mythai@cise.ufl.edu

© Springer Nature Switzerland AG 2019
D.-Z. Du et al. (eds.), *Nonlinear Combinatorial Optimization*,
Springer Optimization and Its Applications 147,
https://doi.org/10.1007/978-3-030-16194-1_12

to maximize the success probability $Prob(s, S, R, t)$, i.e., the probability that node t is activated through subgraph induced by $R \cup S \cup \{s, t\}$ when initially set up all nodes in $S \cup \{s\}$ to be active.

There are two popular information diffusion models studied in the literature: the independent cascade (IC) model and the linear threshold (LT) model. They are defined as follows:

The IC Model Each node has two states: active and inactive. Every arc (u, v) is labeled with a probability p_{uv} which means that if u is active and v is inactive, then the event that v accepts the influence of u, i.e., v becomes active because of active u, occurs with probability p_{uv}. Before the process starts, all nodes are inactive. Initially, choose a subset of nodes, called seeds, and activate them. In each of the subsequence steps, every fresh-active node tries to influence its inactive out-neighbors where a node is fresh-active if it becomes active in the step right before current step. If an inactive node v gets influenced by more than one, say k, fresh-active nodes $u_1, u_2, \ldots, u_k$ at the same step, then all k events that u_i influences v successfully are treated as k independent events. This process ends if no fresh-active node is produced.

The LT Model Each node has two states: active and inactive. Every arc (u, v) is labeled with a positive weight w_{uv} such that for any node v, $\Sigma_{u \in N^-(v)} w_{uv} \leq 1$, where $N^-(v) = \{u | (u, v) \in E\}$. Before the process starts, all nodes are inactive. Initially, choose a subset of nodes, called seeds, and activate them; meanwhile each node u chooses a threshold θ_u uniformly and randomly from $[0, 1]$. In each of the subsequence steps, every inactive node v evaluates the total weight of w_{uv} for u over all active in-neighbors. If this total weight is at least θ_v, then v becomes active; otherwise, v keeps inactive. This process ends if no fresh-active node is produced.

The following is proved in [34] by using dynamic programming.

Theorem 1 *For an arborescence directed to t with the IC model, the active friending can be solved in polynomial-time.*

Using this result, they also designed a heuristic by, first, approximating the general network with an in-arborescence with root t. This arborescence is the union of all the most influential paths from each $S \cup \{s\}$ to t where the most influential path from $s' \in S \cup \{s\}$ to t is the shortest path when we consider $-\log p_{uv}$ as the distance from node u to v and p_{uv} is the probability that node v accepts the influence from u in the IC cascade model.

Kempe et al. [13] generalized the LT model and the IC model to the general threshold model and the general cascade model, and proved that every general threshold model is equivalent to a general cascade model, vice versa. For this equivalence, the LT model is equivalent to a general cascade model, called the mutually exclusive cascade (MC) model. The MC model can be defined in the same way as that of the IC model, except that when k fresh-active nodes $u_1, u_2, \ldots, u_k$ try to influence an inactive node v at the same step, this is considered as that k mutually exclusive events occur. In the equivalence relation between the LT model and the MC model, $w_{uv} = p_{uv}$. The MC model (of course, the LT model, too) has an important property. Consider a social network with four nodes u, v, x, y and three arcs $(u, x), (v, x), (x, y)$ in the MC model. Suppose u and v are seeds. Then the probability that y becomes active is

$$(p_{ux} + p_{vx})p_{xy} = p_{ux}p_{xy} + p_{vx}p_{xy},$$

that is, this probability is the sum of the probability that y accepts the influence of u through the path from u to y and the probability that y accepts the influence of v through the path from v to y. In general, this property can be stated in the following lemma:

Lemma 1 *In the LT model, a node v accepts the influence from a seed set S with probability equal to*

$$\sum_{P \in \mathcal{P}} Prob(P),$$

where $\mathcal{P}$ is the set of paths from S to v and $Prob(P)$ is the probability that v accepts the influence of a seed in S along path P.

This property makes that the problem in the LT model sometimes is easier than that in the IC model. For example, the influence maximization in-arborescence directed to the root is polynomial-time solvable in the LT model [27], however NP-hard in the IC model [18]. (This result was first conjectured in [2] and then proved in [18].)

With this special property of the LT model, Yuan et al. [35] proved the following result about $Prob(s, S, R, t)$.

Theorem 2 *$Prob(s, S, R, t)$ is a monotone nondecreasing, supermodular function with respect to R for social network G in the linear threshold model, that is, for any $R' \in R$, $Prob(s, S, R', t) \leq Prob(s, S, R, t)$, and for any R and R',*

$$Prob(s, S, R, t) + Prob(s, S, R', t) \leq Prob(s, S, R \cup R', t)$$
$$+ Prob(s, S, R \cap R', t).$$

Proof It is easy to see the property of monotone nondecreasing. We next show the supermodularity. Before doing so, let us first recall a special property proved in [27] that the linear threshold model is equivalent to the mutually exclusive cascade model in which when k fresh-active nodes influence an inactive node, this event is considered as a composed event of k mutually exclusive events. This property yields that in the linear threshold model, $Prob(s, S, R, t)$ is equal to the sum of accepting probabilities each of which is the probability that t accepts the invitation from a node $s' \in S \cup \{s\}$ along a path p to t where p is over all paths from a node in $S \cup \{s\}$ to t and with all nodes in R. Let $P(R)$ denote the set of all such paths p.

Now, we compare $P(R) \cup P(R')$ with $P(R \cup R')$ and $P(R \cap R')$. Clearly, both $P(R)$ and $P(R')$ are subsets of $P(R \cup R')$. Moreover, if a path p appears in both $P(R)$ and $P(R')$, then p must appear in $P(R \cap R')$. Therefore,

$$Prob(s, S, R, t) + Prob(s, S, R', t) \leq Prob(s, S, R \cup R', t)$$
$$+ Prob(s, S, R \cap R', t).$$

By Theorem 2, the active friending with the LT model can be formulated into the following problem:

$$\max \ Prob(s, S, R, t)$$
$$\text{subject to } |R| \leq r,$$

that is, a monotone supermodular maximization with size constraint. This formulation suggests that the discrete Lagrangian method [21] is suitable to solve the active friending problem for the LT model. The greedy algorithm in [1] can also be used. However, the estimation of the curvature is a trouble, which may be done possibly only for some special networks, such as power-law graphs.

Next, we move our attention to the IC model. Let P be the set of all paths from $\{s\} \cup S$ to t. Denote by $Prob(R; P)$ the probability that the randomized subgraph induced by $R \cup S \cup \{s, t\}$ containing all paths in P. Denote $P_i = \sum_{|P|=i, P \subseteq \mathcal{P}} Prob(R; P)$. By the inclusive–exclusive formula,

$$Prob(s, S, R, t) = P_1 - P_2 + P_3 - P_4 + \cdots + (-1)^{|\mathcal{P}|} P_{|\mathcal{P}|}.$$

By an argument similar to that in the proof of Theorem 2, we can show the following result.

Lemma 2 *$Prob(R; P)$ is monotone nondecreasing supermodular with respect to R.*

Proof It is clear that $Prob(R; P)$ is monotone nondecreasing. Next, we show the supermodularity. Consider two node subsets R_1 and R_2. Note that the randomized subgraph induced by $(R_1 \cup R_2) \cup S \cup \{s, t\}$ contains those paths contained by the randomized subgraph induced by $R_j \cup S \cup \{s, t\}$ for $j = 1, 2$. In addition, it also contains those paths contained by union of these two randomized subgraphs. Therefore,

$$Prob(R_1; P) + Prob(R_2; P) \leq Prob(R_1 \cup R_2; P) + Prob(R_1 \cap R_2; P),$$

that is, $Prob(R; P)$ is supermodular.

By the above lemma, the following holds.

Theorem 3 *In the IC model, $Prob(s, S, R, t)$ can be represented as a difference of two nonnegative monotone nondecreasing supermodular functions, i.e.,*

$$Prob(s, S, R, t) = (P_1 + P_3 + \cdots) - (P_2 + P_4 + \cdots).$$

By Theorem 3, we may employ the sandwich method [7, 17, 25, 28], the submodular–supermodular method [20], the modular–modular method [11], and the iterated sandwich method [31] to solve the active friending problem for the IC model.

2 Target Friending

The second optimization problem on friending is the target friending described as follows:

Definition 2 (Target Friending) Consider a social network represented as directed graph $G = (V, E)$ with an information diffusion model m. Suppose S is the list of existing friends of a node s and t is a target node that s wants to include in his friend list. Given an integer $0 < \rho < 1$, the problem is to find a minimum node subset R such that $Prob(s, S, R, t) \geq \rho$.

By Theorem 2, the target friending for the LT model is a supermodular cover problem as follows:

$$\min \ |R|$$
$$\text{subject to} \ \ Prob(s, S, R, t) \geq \rho.$$

The target friending for the IC model is a generalization of the well-known submodular cover problem [30], the same as above except that $Prob(s, S, R, t)$ is a nonsubmodular and nonsupermodular function in the cover constraint. It is an interesting research subject to see how to generalize the approximation analysis for the submodular cover problem. In fact, there are so many different proofs for the same theorem regarding the approximation performance ratio of a greedy algorithm for the submodular cover [9, 26, 30]. None of them is able to give a generalization for the above nonsubmodular cover problem so far.

3 Group Friending

The group friending was first studied in [6]. They consider a romantic scenario as follows: A boy found an attractive girl. However, they do not really know each other. The boy worries that he may get rejected if he asks her directly. Hence, he wants to influence her friends at the first stage. Thus, her friends form target group for friending. The objective in this problem is the expected number of her friends who become his friends after the friending process. This problem has no much difference from active friending.

Definition 3 (Active Group Friending) Consider a social network represented as directed graph $G = (V, E)$ with an information diffusion model m. Suppose S is the list of existing friends of a node s and T is a set of target nodes that s wants to include in his friend list. Given an integer $r > 0$, the problem is to find a subset R with at most r nodes to maximize the expected number of active nodes in T, which are activated through subgraph induced by $R \cup S \cup \{s, t\}$ when initially set up all nodes in $S \cup \{s\}$ to be active.

The mathematical formulations are similar, respectively, to that of active friending in the LT model and the IC model.

Shen et al.[22] proposed another formulation based on quite different scenario. Suppose we want to organize a social activity with at least p persons, in order to make new friendship between members in a big social organization. Two factors are very important for us, existing friendship between members and potential friendship between members. To evaluate the success of the activity, we may give each potential friendship a positive weight in (0, 1] and a measure of making new friends which is the ratio between the total weight and group size.

Definition 4 (Hop-Bounded Group Friending) Consider a heterogeneous social graph $G = (V, E, R)$ with edge weight $w : R \rightarrow (0, 1]$, where V is the set of nodes, E is the set of friend edges, and R is the set of potential friend edges. Given a hop constraint h and a group size constraint p, find a subset of at least p nodes, H, such that every pair of nodes u and v is within distance h in graph with node set V and edge set E and $\sigma(H)$ reaches the maximum, where $\sigma(H) = w(H)/|H|$ and $w(H)$ is the total weight of potential friend edges in the subgraph induced by H.

This problem has been proved to be NP-hard and has no polynomial-time approximation with a performance ratio $\rho < 1$ unless NP=P [22].

Finding a cohesive group from a social network with existing friend edges is an important research topic in the literature. However, before [22], all efforts are based on existing friendship [10, 19, 23, 24, 29, 32, 33, 36] and no "friending" is involved. In order to have "friending" involved, the potential friend edges are employed in the hop-bounded group friending. How to know the potential friend edges? The link prediction methods are used [8, 12, 14–16]. They analyze the features, the similarity, and/or the interactive patterns to make recommendation for potential friendship.

In the community expansion [3–5], each community consists of all customers for a certain business which always wants to expanse their service. Therefore, a different type of "friending" problems is raised. They can all be formulated into nonlinear combinatorial optimization problems.

References

1. Bai, W., Bilmes, J.A.: Greed is still good: maximizing monotone submodular+ supermodular functions (2018). Preprint. arXiv:1801.07413
2. Bharathi, S., Kempe, D., Salek, M.: Competitive influence maximization in social networks. In: International Workshop on Web and Internet Economics, pp. 306–311. Springer, Berlin (2007)

3. Bi, Y., Wu, W., Wang, A., Fan, L.: Community expansion model based on charged system theory. In: International Computing and Combinatorics Conference, pp. 780–790. Springer, Berlin (2013)
4. Bi, Y., Wu, W., Wang, L.: Community expansion in social network. In: International Conference on Database Systems for Advanced Applications, pp. 41–55. Springer, Berlin (2013)
5. Bi, Y., Wu, W., Zhu, Y., Fan, L., Wang, A.: A nature-inspired influence propagation model for the community expansion problem. J. Comb. Optim. **28**(3), 513–528 (2014)
6. Chen, H., Xu, W., Zhai, X., Bi, Y., Wang, A., Du, D.-Z.: How could a boy influence a girl? In: 2014 10th International Conference on Mobile Ad-hoc and Sensor Networks (MSN), pp. 279–287. IEEE, Piscataway (2014)
7. Chen, W., Lin, T., Tan, Z., Zhao, M., Zhou, X.: Robust influence maximization. In: Proceedings of the 22nd ACM SIGKDD International Conference on Knowledge Discovery and Data Mining, pp. 795–804. ACM, New York (2016)
8. Clauset, A., Moore, C., Newman, M.E.J.: Hierarchical structure and the prediction of missing links in networks. Nature **453**(7191), 98 (2008)
9. Du, D.-Z., Ko, K.-I., Hu, X.: Design and Analysis of Approximation Algorithms, vol. 62. Springer Science & Business Media, New York (2011)
10. Feige, U., Peleg, D., Kortsarz, G.: The dense k-subgraph problem. Algorithmica **29**(3), 410–421 (2001)
11. Iyer, R., Bilmes, J.: Algorithms for approximate minimization of the difference between submodular functions, with applications (2012). Preprint. arXiv:1207.0560
12. Kashima, H., Abe, N.: A parameterized probabilistic model of network evolution for supervised link prediction. In: Sixth International Conference on Data Mining, 2006. ICDM'06, pp. 340–349. IEEE, Washington (2006)
13. Kempe, D., Kleinberg, J., Tardos, É: Maximizing the spread of influence through a social network. In: Proceedings of the Ninth ACM SIGKDD International Conference on Knowledge Discovery and Data Mining, pp. 137–146. ACM, New York (2003)
14. Kunegis, J., Lommatzsch, A.: Learning spectral graph transformations for link prediction. In: Proceedings of the 26th Annual International Conference on Machine Learning, pp. 561–568. ACM, New York (2009)
15. Leung, C.W., Lim, E.-P., Lo, D., Weng, J.: Mining interesting link formation rules in social networks. In: Proceedings of the 19th ACM International Conference on Information and Knowledge Management, pp. 209–218. ACM, New York (2010)
16. Liben-Nowell, D., Kleinberg, J.: The link-prediction problem for social networks. J. Am. Soc. Inf. Sci. Technol. **58**(7), 1019 1031 (2007)
17. Lu, W., Chen, W., Lakshmanan, L.V.S.: From competition to complementarity: comparative influence diffusion and maximization. Proc. VLDB Endow. **9**(2), 60–71 (2015)
18. Lu, Z., Zhang, Z., Wu, W.: Solution of Bharathi–Kempe–Salek conjecture for influence maximization on arborescence. J. Comb. Optim. **33**(2), 803–808 (2017)
19. Mokken, R.J.: Cliques, clubs and clans. Qual. Quant. **13**(2), 161–173 (1979)
20. Narasimhan, M., Bilmes, J.: A submodular-supermodular procedure with applications to discriminative structure learning. In: Proceedings of the Twenty-First Conference on Uncertainty in Artificial Intelligence, pp. 404–412. AUAI Press, Arlington (2005)
21. Shang, Y., Wah, B.W.: A discrete Lagrangian-based global-search method for solving satisfiability problems. J. Glob. Optim. **12**(1), 61–99 (1998)
22. Shen, C.-Y., Yang, D.-N., Lee, W.-C., Chen, M.-S.: Maximizing friend-making likelihood for social activity organization. In: Pacific-Asia Conference on Knowledge Discovery and Data Mining, pp. 3–15. Springer, Cham (2015)
23. Shuai, H.-H., Yang, D.-N., Yu, P.S., Chen, M.-S.: Willingness optimization for social group activity. Proc. VLDB Endow. **7**(4), 253–264 (2013)
24. Surian, D., Liu, N., Lo, D., Tong, H., Lim, E.-P., Faloutsos, C.: Recommending people in developers' collaboration network. In: 2011 18th Working Conference on Reverse Engineering (WCRE), pp. 379–388. IEEE, Washington (2011)

25. Tong, A., Du, D.-Z., Wu, W.: On misinformation containment in online social networks. In: Advances in Neural Information Processing Systems, pp. 339–349 (2018)

26. Wan, P.-J., Du, D.-Z., Pardalos, P., Wu, W.: Greedy approximations for minimum submodular cover with submodular cost. Comput. Optim. Appl. **45**(2), 463–474 (2010)

27. Wang, A., Wu, W., Cui, L.: On Bharathi–Kempe–Salek conjecture for influence maximization on arborescence. J. Comb. Optim. **31**(4), 1678–1684 (2016)

28. Wang, Z., Yang, Y., Pei, J., Chu, L., Chen, E.: Activity maximization by effective information diffusion in social networks. IEEE Trans. Knowl. Data Eng. **29**(11), 2374–2387 (2017)

29. Wasserman, S., Faust, K.: Social Network Analysis: Methods and Applications, vol. 8. Cambridge University Press, Cambridge (1994)

30. Wolsey, L.A.: An analysis of the greedy algorithm for the submodular set covering problem. Combinatorica **2**(4), 385–393 (1982)

31. Wu, W., Zhang, Z., Du, D.-Z.: Set function optimization. J. Oper. Res. Soc. China 1–11, (2018)

32. Yang, D.-N., Chen, Y.-L., Lee, W.-C., Chen, M.-S.: On social-temporal group query with acquaintance constraint. Proc. VLDB Endow. **4**(6), 397–408 (2011)

33. Yang, D.-N., Shen, C.-Y., Lee, W.-C., Chen, M.-S.: On socio-spatial group query for location-based social networks. In: Proceedings of the 18th ACM SIGKDD International Conference on Knowledge Discovery and Data Mining, pp. 949–957. ACM, New York (2012)

34. Yang, D.-N., Hung, H.-J., Lee, W.-C., Chen, W.: Maximizing acceptance probability for active friending in online social networks. In: Proceedings of the 19th ACM SIGKDD International Conference on Knowledge Discovery and Data Mining, pp. 713–721. ACM, New York (2013)

35. Yuan, J., Wu, W., Li, Y., Du, D.: Active friending in online social networks. In: Proceedings of the Fourth IEEE/ACM International Conference on Big Data Computing, Applications and Technologies, pp. 139–148. ACM, New York (2017)

36. Zhu, Q., Hu, H., Xu, C., Xu, J., Lee, W.-C.: Geo-social group queries with minimum acquaintance constraint (2014). Preprint. arXiv:1406.7367

Optimization on Content Spread in Social Network Studies

Yi Li, Ruidong Yan, and Weili Wu

Abstract With the rapid growth of online social networks, people change the way of generating, sharing, and spreading various social contents. The contagiousness of social content is highly depending on the size of of seed nodes and connectivity of the network. In this study, we propose the optimization problems of information content diffusion over social networks. The content here can be either useful information such as news, innovation ideas, and marketing purpose content or negative content such as misinformation and malicious rumors. We show that the optimization problem on information diffusion has been discussed in previous researches from different aspects using different approaches. In our study, we formulate two optimization problems—content spread maximization and misinformation minimization—which are both NP-hard and non-submodular. To tackle the difficulty of these problems we sandwich approximation which has data-dependent guarantees.

1 Introduction

In the past decade, social networks have gained popularity at a rapid pace and become an integral part of our lives. Online social network sites such as Facebook, Twitter, LinkedIn, Instagram, etc. not only help us keep in touch with friends and families but also keep abreast of emerging contents and share daily activities. These social network sites have become significant platforms for users to generate, share, and spread a large amount of social content. They provide access to a vast source of information on an unprecedented scale. Statistic shows that Facebook Messenger

Y. Li (✉) · W. Wu
Department of Computer Science, The University of Texas at Dallas, Richardson, TX, USA
e-mail: yi.li@utdallas.edu; weiliwu@utdallas.edu

R. Yan
School of Information, Renmin University of China, Beijing, China
e-mail: yanruidong@ruc.edu.cn

© Springer Nature Switzerland AG 2019
D.-Z. Du et al. (eds.), *Nonlinear Combinatorial Optimization*,
Springer Optimization and Its Applications 147,
https://doi.org/10.1007/978-3-030-16194-1_13

273

and WhatsApp handle 60 billion messages a day. A 2011 study by AOL/Nielsen showed that 27 million pieces of content were shared every day, and 3.2 billion images are shared each day [18]. Billions of active social network users are engaging in spreading content such as photos, videos, comments, news, or even rumors and misinformation over social networks. For the positive information such as innovation ideas, or useful information we usually hope to maximize through social network, while for the negative contents such as fake or inaccurate information it should be limited or contained. The background and motivations of these two problems will be discussed separately in the following sections.

1.1 Positive Content Maximization

The extent to which a social network spreads content is a key metric that impacts both user engagement and network revenues. The more the novel content spread, the more the useful information users end up discovering, and the more the value users derive from being part of the social network. Also from the perspective of social networks, higher content spread helps users engagement which in turn leads to improved user retention and audience growth [4]. Therefore, it is very important to explore the maximization problem of positive content disseminate across the entire social graph.

In social networks, users recursively share contents with their neighbors that will be expected to quickly reach and influence a large number of audience. But in some cases content spread efficiency is not what we expected. There is a research that shows that a piece of content such as a photo spread on Flickr usually only influences the users within two hops then burnout quickly [3]. In addition, even though the breadth and depth of information dissemination are somehow related to the selection of initial seeds, sometimes the seed users are predetermined. For example, if a beauty company wants to use viral marketing to promote their products with minimum startup cost they usually send free samples to predetermined users such as well-known beauty bloggers and celebrities. Due to the limited cost or companies' preference, we hope to find a way to boost content spread with fixed seed users in advance.

There is a very straight forward way to think about this content spread maximization problem. We need to simply increase the connectivity of social networks. Some social network sites such as Facebook, Twitter, etc. already have the friend recommendation function to help people make possible connections. But in this way, the possible links are usually based on common friends, interests, communities, and other personal related features. While in some cases, personal related information are considered as privacy data that cannot be accessed easily. And the recommendations based on friends of friends or interests similarity can be significantly large which can have diverse content spread characteristics. So simply considering recommendation connections based on a number of mutual friends or common interests may not maximize content spread in the social network [4]. We will show some existing works on maximizing content spread in Section 2 and then propose our formulation and analysis for this positive content maximization problem.

1.2 Negative Content Minimization

Even though social networks bring us the convenience of spreading information, negative contents such as misinformation and malicious rumors will also diffuse widely and quickly. With widespread of negative contents, social network would lose its reliability and cause panic over community and society.

One of the most valuable characteristics of social networks is its capability for user generated contents circulating rapidly through the network. But when it comes to misinformation this valuable characteristic will make things even worse. For example, when the devastating wildfires happened in California in October 2017, the officers not only need to help with evacuating residents and searching for missing persons they still had to take time to deal with fake news [13]. Although the misinformation was shot down by the officers and was debunked by some government websites afterwards, the original story was shared 60,000 times and similar stories were shared 75,000 times on Facebook in a very short time.

Another major aspect of online social networks is its openness to everyone. They enable not only organizations and government agencies to publish information and news, but also our ordinary people to post from own perspectives and experience [24]. Because of the openness, anyone could share any content without validating it. Therefore, taking effective strategies to minimize the negative influence from misinformation should be very crucial to social networks. Or it may cause catastrophic effects in the physical world in a short period.

Existing works have explored the negative content spread problem from different perspectives. Some previous works show that by removing nodes in decreasing order of outdegree and blocking edges can be effective for minimizing the negative information [9, 14, 21, 23]. There are also some other works trying to find minimal set of protectors to limit the diffusion of misinformation [7] or introduce a positive cascade competing against the negative content [20]. We will expand the details of these existing works in Section 2 and discuss our proposed problem in Section 3.

2 Related Works

In this section, we will show some previous researches on the optimization of content spread over social networks. We will also discuss in two parts—the existing works focus on maximizing the positive content and the previous works explore to minimize the misinformation and malicious rumors on social networks.

2.1 Classic Influence Diffusion Problems

There have been abundant studies on various models and computational methods for maximizing and minimizing influence. Kempe et al. [8] first formulate the influence maximization problem which asks to find a set S of k nodes so that the expected influence spread is maximized under a predetermined influence propagation model. The problem is NP-hard under both IC and LT models. In [5], Chen et al. show that to compute the expected influence spread for a given set is #P-hard. But it can be formulated as a submodular and monotone function of S for both IC and LT models which can use a simple greedy algorithm [8] to guarantee the results.

There are also some existing studies that have explored negative influence minimization problem. Nguyen et al. [15] study a set of problems named node protectors, which aims to find the smallest set of highly influential nodes whose decontamination with good information helps to contain the viral spread of misinformation. Kimura et al. [9] proposed a link blocking method to minimize the expected contamination area of the network.

In this work, we think of the influence optimization problems in a different perspective. Instead of choosing the initial seed set we aim to add edges to maximize or minimize the content spread. We focus on the diffusion process from the seed nodes to the nodes with high probability to influence other nodes and low probability to be activated by seed nodes. In our settings, the seed nodes are predetermined.

2.2 Optimizations on Content Spread

2.2.1 Boosting Content Spread

Vineet Chaoji et al. [4] formulate the problem of boosting content spread on social network by adding up to k connections per user such that the probabilistic propagation of content in the social network is maximized. Since the content maximizing problem is NP-hard and the content spread function is not submodular they construct a more restricted variant that is submodular and devise an approximation algorithm that computes an edge set which satisfies constraints. But their content spread function under IC and RMPP model has a few limitations. First, computing the spread of specific content C with any given seed set is #P-hard which leads to substantial computation time for running expensive simulations. Second, the restrictions on information propagation may not reflect the real flow on the network. Additionally their model assumes that a predefined number of new links should be added for each user in the network, thus leading to all the users in the network to accept the same number of recommended connections, a case which not necessarily reflects the power law property of real world social network.

The authors of [19] also raise the question of changing the structure of networks-to add or remove edges from a network to speed up a dissemination. The problem boils down to the eigenvalue optimization problem. They propose an algorithm to optimize the key graph parameter such as leading eigenvalue of the graph adjacency matrix which controls the information dissemination process in their models.

In [1, 16, 17], the authors define the link injection problem which is aiming at boosting overall diffusion of information over the social networks, unlike other link prediction methods which do not consider the optimization of information cascades as an explicit objective. They propose the method that the injected links are being predicted in a collaborative-filtering fashion, based on factorizing the adjacency matrix that represents the structure of the social network and controls the number of injected links to avoid an aggressive injection scheme that may compromise the experience of users. Then they perform the link injection by attaching links to users according to their scores.

Lin et al. [11] propose a k-boosting problem which aims to find k users who are initially uninfluenced and increase their probability to be influenced. It is different from influence maximization problem because boosted users are initially uninfluenced. Their work consider the content spread maximization problem from initial users' perspective.

2.2.2 Negative Content Minimization

The problem of minimizing the negative influence of rumors and misinformation although is an important research topic but gets less attention compared to influence maximization. There are mainly two types of strategies that include blocking influential users and clarifying rumor by spreading truths.

Wang et al. [21] consider the situation that when negative information such as a rumor emerges in the social network and part of users have already adopted it, how to minimize the size of ultimately contained users. They propose a greedy method which efficiently finds a good approximate solution to discover and block k uninfected users to minimize the negative content diffusion. In [22], authors study the problem of minimizing the misinformation spread via changing the connectivity of social network.

Comin et al. [6] analyze three spreading schemes and then propose an effective methodology for the identification of the source nodes. If the source nodes are detected, then using any method to block them could achieve our goal of minimizing the negative content. Their method is based on the calculation of the centrality (degree, betweenness, closeness, and eigenvector) of the nodes on sampled network. Similar to [6], Kitsak et al. [10] study the problem of identifying the most efficient "spreaders" in a network which is very useful for optimizing the information spread problem. They find that the most efficient spreaders are those located with the core of the network as identified by the k-shell decomposition and that when multiple spreaders are considered simultaneously the distance between them becomes the crucial parameter that determines the extent of the spreading.

Budak et al. [2] study the notion of competing campaigns in a social network and address the problem of influence limitation where a "bad" campaign starts propagating from a certain node in the network and use the notion of limiting campaigns to counteract the effect of misinformation. The problem can be summarized as identifying a subset of individuals that need to be convinced to adopt the competing (or "good") campaign so as to minimize the number of people that adopt the "bad" campaign at the end of both propagation processes. This problem is proved to be NP-hard but they use a greedy algorithm to achieve approximation grantee due to the submodularity of objective function. Our work differs from the above because we focus on the manipulation of edges and we consider the network structure changing after each edge is removed.

3 Problem Description and Formulation

In this section, we will discuss how we formulate our problems with detailed explanation of objective functions.

3.1 Information Diffusion Model

Before discussing the proposed problems and formulations, we first want to show how a piece of content will spread over the whole networks. So we briefly introduce the information diffusion model: *independent cascade* (IC) model [8]. Given a social network $G = (V, E, p)$, where V is the node set (users) and $E \subseteq V \times V$ is the edge set (the relationships between users). $e_{vu} \in E$ denotes an arbitrary edge and p_{vu} of edge e_{vu} denotes the probability that node v can successfully activate node u. We call a node *active* if it accepts information from other nodes, *inactive* otherwise. Influence propagation process unfolds in discrete time steps. The initial seed set is S_0, let S_T denotes the *active* nodes in time step T, and each node v in S_T has single chance to activate each *inactive* neighbor u through its out-edge with probability p_{vu} at time $T + 1$. Repeat this process until no more new nodes can be activated. Note that a node can only switch from *inactive* to *active*, but not in reverse direction.

3.2 Content Spread Maximization

In this section, we show how we formulate the content spread function from a marginal increment perspective. Our formulation based on the classical independent cascade (IC) model is discussed in the last part.

For the given acyclic directed social network $G(V, E, P)$, we denote p_i as the probability with which node i shares content independently with each of its neighbors and $q_{v,S}^{cE}$ is the spread of a content $c \in C$ contained at $v \in V$ under the topology of E (which means only the edges in E can be used in the propagation of content c) with seed set S, that is, every node in S contains content c and $q_S^{cE} = (\cdots, q_{v,S}^{cE}, \cdots)^T$ is the content spread vector. Then we need to find out how to calculate the marginal gain $\Delta_{e_{st}} q_{v,S}^{cE}$ of content spread c at node v when an edge $e_{st} \in X$ from a candidate set is added to current topology of E. We give the following theorem.

Theorem 1 *The marginal gain $\Delta_{e_{st}} q_{v,S}^{cE}$ of content spread of c at node v when an edge $e_{st} \in X$ from a candidate set is added to current topology of E is calculated recursively as follows:*

$$\Delta_{e_{st}} q_{t,S}^{cE} = (1 - q_{t,S}^{cE}) p_s q_{s,S}^{cE}.$$

And for any $v \in N^{out}(t)$, where $N^{out}(t)$ is the out-neighbor set of vertex t, we have

$$\Delta_{e_{st}} q_{v,S}^{cE} = \frac{1 - q_{v,S}^{cE}}{1 - p_t q_{t,S}^{cE}} p_t \Delta_{e_{st}} q_{t,S}^{cE}. \tag{1}$$

In addition, for other vertex $v \in V$ that can be reachable from vertex t, we can update the marginal gain similarly according to the topology order in recursive manner. We have $\Delta_{e_{st}} q_{v,S}^{cE} = 0$, for the vertex which is unreachable from vertex t during this process.

During the process of updating marginal spread, if there are paths from vertex t reaching to different in-neighbor nodes of node w, the marginal gain of spread of w should be updated according to Equation (1) multi-times. But the overall marginal gain of content spread for w is independent of the updating orders.

In fact, suppose there exist two paths from t to both node u and v and $w \in N_E^{out}(u) \cap N_E^{out}(v)$. We first consider update from u to w, a marginal gain of spread $\Delta_{e_{st}}^u q_{w,S}^{cE} = \frac{1 - q_{w,S}^{cE}}{1 - p_u q_{u,S}^{cE}} p_u \Delta_{e_{st}} q_{u,S}^{cE}$ is obtained. Then considering update from v to w, another marginal gain of spread $\Delta_{e_{st}}^{u+v} q_{w,S}^{cE} = \frac{1 - (q_{w,S}^{cE} + \Delta_{e_{st}}^u q_{w,S}^{cE})}{1 - p_v q_{v,S}^{cE}} p_v \Delta_{e_{st}} q_{v,S}^{cE}$. Thus the overall marginal gain of spread of w is $\Delta_{e_{st}} q_{w,S}^{cE} = \Delta_{e_{st}}^u q_{w,S}^{cE} + \Delta_{e_{st}}^{u+v} q_{w,S}^{cE}$ which is also equal to $\Delta_{e_{st}}^v q_{w,S}^{cE} + \Delta_{e_{st}}^{v+u} q_{w,S}^{cE}$.

From Theorem 1 and the note above, the objective function of content spread in the marginal gain form can be expressed as

$$f(X) = \sum_{c \in C} \sum_{v \in V} \left(q_{v,S}^{cE} + \sum_{e_{st} \in X} \Delta_{e_{st}} q_{v,S}^{c(E \cup X^{st})} \right), \tag{2}$$

where X^{st} denotes the edge set that has already been added into the network before edge e_{st}. This definition is consistent with the content propagation process and there is no loss during content spread process. Using $f(X)$ as the objective function to be maximized we give our content spread maximization problem (GSMP) as follows.

Definition 1 (GSMP) Given a directed acyclic graph $G = (V, E, P)$, a constant K, and content set C with given initial seed sets S_c for each $c \in C$, find an edge set $X \subseteq \overline{X} = \{e_{ij} : i, j \in V, i \in N_j, j \in N_i\}$, where N_i is the candidate node set of i to be connected such that: (1) At most K edges from X, and (2) $f(X)$ is maximum.

3.3 Negative Content Minimization

Here we define an opposite problem called negative content minimization problem. Given a directed acyclic social network $G = (V, E, p)$, where V represents users and E represents relationships between users, a diffusion model $\mathcal{M}$, a candidate edge set $E' \subseteq E$, and the predetermined seed set S_c for each negative content $c \in C$ such as rumors and gossips, etc. Further, each node v has the following parameters: (1) Let p_{vu} denote the probability that v independently shares content with its neighbor u; (2) Let $\theta_E^c(v, S_c, \mathcal{M})$ denote the probability that seed set S_c successfully activates v on topology E under information diffusion model $\mathcal{M}$. We omit the parameter $\mathcal{M}$ if the context is clear, i.e., $\theta_E^c(v, S_c)$.

The goal of this problem is to identify K edges denoted by $\mathscr{E}$ from candidate edge set E'. Then we remove $\mathscr{E}$ from original graph G such that the negative content spread is minimized. We define *negative content spread minimization* problem as follows.

Definition 2 (Negative Content Spread Minimization (NCSM)) Given a directed acyclic social network $G = (V, E)$, an diffusion model $\mathcal{M}$, a blocking candidate edge set E', and the predetermined seed set S_c for each negative content $c \in C$, NCSM finds a K edges set $\mathscr{E}$ from candidate edge set E' such that the negative content spread $L(\mathscr{E}) = \sum_{c \in C} \sum_{v \in V} \theta_{E \setminus \mathscr{E}}^c(v, S_c)$ is minimized, namely it is equivalent to seek

$$\mathscr{E}^* = \arg \min_{\mathscr{E} \subseteq E', |\mathscr{E}|=K} \sum_{c \in C} \sum_{v \in V} \theta_{E \setminus \mathscr{E}}^c(v, S_c), \tag{3}$$

where $\theta_{E \setminus \mathscr{E}}^c(v, S_c)$ denotes the probability that the seed set S_c activates v successfully on topology $E \setminus \mathscr{E}$.

Let $\theta_{E \setminus \{e_{st}\}}^c(v, S_c)$ denote the probability that the seed set S_c activates v successfully when the edge $(s, t) = e_{st} \in E'$ is removed from E. We focus on the marginal decrement when an edge is removed from network. Then we have the following formula to calculate the marginal decrement $\Delta_{e_{st}} \theta_{E \setminus \{e_{st}\}}^c(v, S_c)$ when edge e_{st} is removed where $v \in V$ and $c \in C$ (see Figure 1).

When an edge e_{st} is removed from current topology E, we consider the following two steps. We first update the marginal decrement of the t. Now we update the content spread of each node in the newly formed network structures and then calculate the marginal decrement of t's neighbor. We recursively use these two steps to update the probability of each node when the edge e_{st} is removed from current

Fig. 1 An instance for
NCSM problem

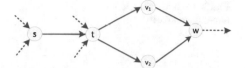

topology until no more nodes can be updated. In particular, if there exist multiple paths from t to w (e.g., $t \to v_1 \to w$ and $t \to v_2 \to w$, see Figure 1), we show that the marginal decrement of $w \in N^{out}_{E\setminus\{e_{st}\}}(v_1) \bigcap N^{out}_{E\setminus\{e_{st}\}}(v_2)$ is independent of updating order. Here we omit the proof.

Based on the above discussion, the objective function (3) can be rewritten with respect to marginal decrement, i.e.,

$$L(\mathscr{E}) = \sum_{c \in C} \sum_{v \in V} (\theta^c_E(v, S_c) - \Delta_{\mathscr{E}} \theta^c_{E\setminus\mathscr{E}}(v, S_c)), \qquad (4)$$

where $\mathscr{E} \in E'$ denotes the edges set removed from network and $|\mathscr{E}| = K$. The item $\sum_{c \in C} \sum_{v \in V} \theta^c_E(v, S_c)$ is fixed with the given initial network G. So minimize function (4) is equivalent to maximize total marginal decrement. Thus we focus on total marginal decrement caused by removing $\mathscr{E}$. Our final objective function is shown as follows:

$$f(\mathscr{E}) = \sum_{c \in C} \sum_{v \in V} \Delta_{\mathscr{E}} \theta^c_{E\setminus\mathscr{E}}(v, S_c). \qquad (5)$$

4 Problem Analysis

4.1 NP Hardness

Theorem 2 *The content optimization problems (content spread maximization problem and negative content spread minimization problem) are NP-hard under IC model.*

Proof Follows from the reduction of the set cover problem to the content spread maximization problem. The CSMP is NP-hard. NCSM can be proved NP-hard from the reduction of knapsack problem. Details omitted due to space constraints.

4.2 Submodularity

Theorem 3 *The objective functions of CSMP and NCSM are both non-submodular.*

Fig. 2 Counterexample for
non-submodularity

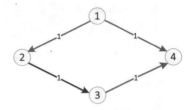

We will show the proof of non-submodularity of NCSM problem by giving a counterexample. The proof of CSMP is similar so we will omit it here.

Proof Submodular functions have a natural diminishing returns property. If E is a finite set, a submodular function is a set function $F : 2^E \rightarrow \Re$, where 2^E denotes the power set of E, which satisfies: for every $A \subseteq B \subseteq E$ and $e \in E \backslash B$, $F(A \cup \{e\}) - F(A) \geq F(B \cup \{e\}) - F(B)$. We prove that function (3) is not submodular by the counterexample in Figure 2.

Suppose that only node 1 (seed) has a piece of negative content c, and each node shares content with its neighbors with probability of 1. Let $A = \{e_{14}\}$, $B = \{e_{14}, e_{34}\}$, and $e = \{e_{23}\}$. Note that $A \subseteq B \subseteq E$ and $e \in E \backslash B$. $L(A) = 4$, $L(A \cup \{e\}) = 2$, $L(B) = 3$, and $L(B \cup \{e\}) = 2$. Thus $L(A \cup \{e\}) - L(A) < L(B \cup \{e\}) - L(B)$ indicates that function L is not submodular.

4.3 Methods

Since the proposed problems lack submodularity, we cannot achieve $(1 - 1/e)$-approximation to the optimal solution. So we adopt a sandwich approximation strategy [12] that leads to a data-dependent approximation factor. Since the original content spread function $f(X)$ is non-submodular, we need to obtain both submodular lower $\underline{f}(X)$ and upper bounds $\overline{f}(X)$. Therefore the sandwich framework can be applied. The sandwich approximation strategy works as follows. First, a solution to the original problem with any strategy is found. Then, an approximate solution to the submodular lower-bound and the submodular upper-bound is found, respectively. At last, the solution that has the best result for the original problem is returned.

5 Conclusion

We introduce two problems of optimizations on content spread over social network. For the positive content such as innovation ideas or product promotion contents we hope to boost the diffusion, while for the negative content of malicious rumors and misinformation we need to contain. In the proposed problems, we focus on the connectivity of network structures by adding and removing edge set from the network to maximize and minimize the content spread from a marginal increment/decrement perspective. We show that both problems are NP-hard and non-submodular. So we need to derive sandwich framework and marginal increment based algorithm to give a data-dependent approximation factor guaranteed solution.

References

1. Antaris, S., Rafailidis, D., Nanopoulos, A.: Link injection for boosting information spread in social networks. Soc. Netw. Anal. Min. **4**(1), 236 (2014)
2. Budak, C., Agrawal, D., Abbadi, A.E.: Limiting the spread of misinformation in social networks. In: Proceedings of the 20th international conference on World wide web, pp. 665–674. ACM, New York (2011)
3. Cha, M., Mislove, A., Gummadi, K.P.: A measurement-driven analysis of information propagation in the Flickr social network. In: Proceedings of the 18th International Conference on World Wide Web, pp. 721–730. ACM, New York (2009)
4. Chaoji, V., Ranu, S., Rastogi, R., Bhatt, R.: Recommendations to boost content spread in social networks. In: Proceedings of the 21st international conference on World Wide Web, pp. 529–538. ACM, New York (2012)
5. Chen, W., Wang, C., Wang, Y.: Scalable influence maximization for prevalent viral marketing in large-scale social networks. In: Proceedings of the 16th ACM SIGKDD International Conference on Knowledge Discovery and Data Mining, pp. 1029–1038. ACM, New York (2010)
6. Comin, C.H., da Fontoura Costa, L.: Identifying the starting point of a spreading process in complex networks. Phy. Rev. E **84**(5), 056105 (2011)
7. Fan, L., Lu, Z., Wu, W., Thuraisingham, B., Ma, H., Bi, Y.: Least cost rumor blocking in social networks. In: 2013 IEEE 33rd International Conference on Distributed Computing Systems (ICDCS), pp. 540–549. IEEE, Piscataway (2013)
8. Kempe, D., Kleinberg, J., Tardos, É.: Maximizing the spread of influence through a social network. In: Proceedings of the Ninth ACM SIGKDD International Conference on Knowledge Discovery and Data Mining, pp. 137–146. ACM, New York (2003)
9. Kimura, M., Saito, K., Motoda, H.: Minimizing the spread of contamination by blocking links in a network. In: AAAI, vol. 8, pp. 1175–1180 (2008)
10. Kitsak, M., Gallos, L.K., Havlin, S., Liljeros, F., Muchnik, L., Stanley, H.E., Makse, H.A.: Identification of influential spreaders in complex networks. Nat. Phys. **6**(11), 888 (2010)
11. Lin, Y., Chen, W., Lui, J.C.S.: Boosting information spread: An algorithmic approach. In: 2017 IEEE 33rd International Conference on Data Engineering (ICDE), pp. 883–894. IEEE, Piscataway (2017)
12. Lu, W., Chen, W., Lakshmanan, L.V.S.: From competition to complementarity: comparative influence diffusion and maximization. Proc. VLDB Endowment **9**(2), 60–71 (2015)
13. Moffitt, M.: The Worst Rumors About the Wine Country Fire. https://www.sfgate.com/bayarea/article/The-worst-rumors-about-the-Wine-Country-fires-12270530.php (October 2017). Accessed 17 Nov 2018
14. Newman, M.E.J., Forrest, S., Balthrop, J.: Email networks and the spread of computer viruses. Phys. Rev. E **66**(3), 035101 (2002)
15. Nguyen, N.P., Yan, G., Thai, M.T., Eidenbenz, S.: Containment of misinformation spread in online social networks. In: Proceedings of the 4th Annual ACM Web Science Conference, pp. 213–222. ACM, New York (2012)
16. Rafailidis, D., Nanopoulos, A.: Crossing the boundaries of communities via limited link injection for information diffusion in social networks. In: Proceedings of the 24th International Conference on World Wide Web, pp. 97–98. ACM, New York (2015)

17. Rafailidis, D., Nanopoulos, A., Constantinou, E.: "With a little help from new friends": boosting information cascades in social networks based on link injection. J. Syst. Softw. **98**, 1–8 (2014)
18. Smith, K.: 121 amazing social media statistics and facts. https://www.brandwatch.com/blog/amazing-social-media-statistics-and-facts/ (October 2018). Accessed 17 Nov 2018
19. Tong, H., Aditya Prakash, B., Eliassi-Rad, T., Faloutsos, M., Faloutsos, C.: Gelling, and melting, large graphs by edge manipulation. In: Proceedings of the 21st ACM International Conference on Information and Knowledge Management, pp. 245–254. ACM, New York (2012)
20. Tong, G., Wu, W., Guo, L., Li, D., Liu, C., Liu, B., Du, D.-Z.: An efficient randomized algorithm for rumor blocking in online social networks. In: IEEE Transactions on Network Science and Engineering (2017)
21. Wang, S., Zhao, X., Chen, Y., Li, Z., Zhang, K., Xia, J.: Negative influence minimizing by blocking nodes in social networks. In: AAAI (Late-Breaking Developments), pp. 134–136 (2013)
22. Yan, R., Li, Y., Wu, W., Li, D., Wang, Y.: Rumor blocking through online link deletion on social networks. ACM Trans. Knowl. Discov. Data **13**(2), 16 (2019)
23. Yao, Q., Zhou, C., Xiang, L., Cao, Y., Guo, L.: Minimizing the negative influence by blocking links in social networks. In: International Conference on Trustworthy Computing and Services, pp. 65–73. Springer, Berlin (2014)
24. Zubiaga, A., Liakata, M., Procter, R., Hoi, G.W.S., Tolmie, P.: Analysing how people orient to and spread rumours in social media by looking at conversational threads. PLoS One **11**(3), e0150989 (2016)

Interaction-Aware Influence Maximization in Social Networks

Shuyang Gu, Chuangen Gao, and Weili Wu

Abstract Influence maximization problem is among the most important topics in the area of social networking, it has attracted a lot of research work. Recently, the influence maximization problem has been extended to practical scenarios. In this chapter, we present one cutting-edge problem named interaction-aware influence maximization, which involves nonsubmodular optimization.

1 Introduction

With the advancements in information science in the last two decades, online social networks find important applications in viral marketing, under this circumstance, influence maximization becomes a very popular research direction, which could be described as the problem of finding a small set of most influential nodes in a social network so that the number of influenced nodes under certain diffusion model in the network is maximized.

Kempe et al. [10] first formulate it as the influence maximization problem: a social network is modeled as a graph with vertices representing individuals and edges representing relationship between two individuals. Influence is propagated in the network according to a stochastic cascade model. One of the most popular cascade models is independent cascade (IC) model: each edge (u, v) in the graph is associated with a propagation probability $p(u, v)$, which is the probability that node u independently activates node v at step $t + 1$ if u is activated at step t. Given a social network graph, the IC model, and a number k, the influence maximization problem is to find k nodes in the graph (referred to as seeds) such that under the influence cascade model, the expected number of nodes activated by the k seeds

S. Gu (✉) · W. Wu
Department of Computer Science, The University of Texas at Dallas, Richardson, TX, USA
e-mail: Shuyang.Gu@utdallas.edu; weiliwu@utdallas.edu

C. Gao
School of Computer Science and Technology, Shandong University, Jinan, China

© Springer Nature Switzerland AG 2019
D.-Z. Du et al. (eds.), *Nonlinear Combinatorial Optimization*,
Springer Optimization and Its Applications 147,
https://doi.org/10.1007/978-3-030-16194-1_14

(referred to as the influence spread) is the largest possible. Kempe et al. [10] prove that the optimization problem is NP-hard, and they present a greedy algorithm which guarantees that the influence spread is within $(1 - 1/e - \epsilon)$ of the optimal influence spread, where e is the base of natural logarithm, and ϵ depends on the accuracy of their Monte Carlo estimate of the influence spread given a seed set.

A large amount of efforts have been made in this research topic since Kempe et al. [10] first defined the problem and obtained plentiful results in many ways. Most of the works focus on maximization of the spread of influence, which considers the number of users influenced by viral marketing or "word-of-mouth" effect in online social networks. These works are based on the assumption that the number of influenced users determines the profit of product. However, some types of products earn profit in a continuous way besides the sales of product itself. The revenue model of online games is a good example, the sales of game is just one source of a game company's profit, another important part of revenue depends on the participation and interaction of players who have already bought the game.

The interactions among users contribute to game profit in several ways. First, the interactive users play games in an online manner, which will attract more in-game advertising. In-game advertising allows advertisers to pay to have their name or products featured in games, in 2017, $109 billion dollars was spent on in-game advertising. Second, the virtual goods transactions in games depend on players' interactions. In 2009, the sale of virtual goods brought in $1 billion dollars.

We analyze such revenue model and define a novel problem of interaction-aware influence maximization. Since the first part of revenue, sales of game, depends on the spread of influence, the objective is same as the classical influence maximization. The second part of revenue hinges on the interactions among users. We use interaction profit to represent such revenues related to the strength of interactions among players. We then define an interaction-aware profit maximization problem, which is how to select a seed set to maximize both the number of influenced users and the interaction profits among active nodes.

For traditional influence maximization problem, since its submodularity, the greedy algorithm can achieve a guaranteed approximation with $1 - 1/e$. But unfortunately, interaction-aware influence maximization problem is not submodular; thus, the greedy strategy can't be directly applied to our problem to get a guaranteed approximate solution. To solve this problem, we propose a new method called decomposition strategy in which we decompose our objective function as a difference of two submodular functions. And based on the decomposition we replace them with the modular functions which are upper or lower bound of them to address the nonsubmodularity part of problem and design an iterated sandwich algorithm.

2 Related Work

Influence maximization was first described as an algorithm problem by Domingos and Richardson [7, 15], they model the problem using Markov random fields and propose heuristic solutions. Kempe et al. [10] formulated the influence maximization problem from the view of combinatorial optimization and showed that the problem is NP-hard under both the IC and LT models, and they proposed a simple greedy algorithm with an approximation ratio of $1 - 1/e$. However a drawback of their work is the scalability of the greedy algorithm. Since then a number of efficient heuristic algorithms haven been proposed in many works to address the issue, one direction is to improve the greedy algorithm, and the other is to propose effective heuristics. In [11], Leskovec et al. present a "lazy-forward" optimization in selecting new seeds, which exploits submodularity and greatly reduces the number of evaluations on the influenced nodes, the main idea is that the marginal gain of a node in the current iteration cannot be better than its marginal gain in the previous iterations.

In [3], Chen et al. improved the greedy algorithm by combining with the CELF optimization proposed in [11], and they also proposed a degree discount heuristics under the independent cascade model. The main idea of degree discount heuristics is when selecting a node based on its degree, and the degree does not include the neighbors that are already activated. In[4], they show that computing influence spread in the independent cascade model is #P-hard, and they propose that a heuristic algorithm uses local arborescence structures of each node to approximate the influence propagation. The heuristic algorithm restricts computations on the local influence regions of nodes. Moreover, by tuning the size of local influence regions, this heuristic is able to achieve tunable tradeoff between efficiency (in terms of running time) and effectiveness (in term of influence spread). In [8], Goyal et al. introduce CELF++ that further optimizes CELF by exploiting submodularity, the advantage of the algorithm CELF++ is that it avoids unnecessary re computations of marginal gains incurred by CELF.

The influence maximization problem has also been extended to practical scenarios in recent works. Chen et al. [6] studied the topic-aware influence maximization problem which considers user interests. In real-world social networks, users have their own interests (topics) and are more likely to be influenced by their friends with similar topics. To address this problem, they study topic-aware influence maximization, that is, given a topic-aware influence maximization (TIM) query, find k seeds from a social network such that the topic-aware influence spread of the k seeds is maximized.

In [12], a keyword-based targeted influence maximization is proposed, where users who are relevant to a given advertisement are targeted. In [9], the problem of privacy reserved influence maximization in both cyber-physical and online social networks is studied, and they propose a model that merges both GPS data and relationship data from social network. Bharathi et al. [2] studied the game of innovation diffusion with multiple competing innovations, for example, multiple companies market competing products using viral marketing. In [5], Chen et al. propose an extension to the independent cascade model that incorporates the emergence and propagation of negative opinions, and the new model has a quality factor to model the natural behavior of people turning negative to a product due to product defects.

Most of the works only consider the number of activated users or the nodes of social graphs but a few works consider interactions among users in viral marketing. The interaction activities between users are first processed by [17]. They consider a specific problem of how one can stimulate the discussion about a topic in a social network as much as possible within a budget. They model the problem as *activity maximization*. Given a propagation network, which records user interaction activity strength along each edge, the goal is to find an optimal set of seed users under a given budget, such that starting information propagation from the seed users leads to the maximum sum of activity strengths among the influenced users. They show that the activity maximization problem is NP-hard under IC model and LT model. The objective function of the problem is proved neither submodular nor supermodular.

Activity maximization does not include maximizing the influence spread in the meantime and only count activity strength of the directly connected users. We propose a different problem-interaction-aware influence maximization, which takes both parts into consideration, in the following section we will go through the formulations of these two problems and then we will discuss a method to solve interaction-aware influence maximization.

3   Problem Formulations

In this section, let us introduce the different formulations on influence maximization problems that consider activity/interactions among users.

3.1   Activity Maximization

This problem was first processed by [17]. Consider a social network represented by a directed graph $G = (V, E)$, together with an information diffusion model m. In this model, each node has two states, active and inactive. Initially, all nodes are in inactive state. The influence diffusion consists of discrete steps. At beginning, a set of nodes are activated. Nodes in this set are called *seeds*. At each subsequent step, every inactive node v evaluates its status and decides whether it should be activated or not, based on the rule in the model m. The process ends at a step in which no more inactive node becomes active.

Let S denote the set of seeds and $I_m(S)$ the set of active nodes at the end of diffusion process. Suppose that for each pair of active nodes $u, v \in I_m(S)$, if (u, v) is an edge of G, i.e., $(u, v) \in E$, then an activity profit $A(u, v)$ will be generated where

$A : E \rightarrow R_+$ is a nonnegative activity profit function. The activity maximization is the following problem:

$$\max \alpha(S) = \sum_{(u,v) \in E : u,v \in I_m(S)} A(u, v) \tag{1}$$

$$\text{subject to} \quad |S| \le k$$

$$S \subseteq V.$$

This problem has been proved to be NP-hard in [17]. There are also counterexamples in [17], which show that $\alpha(S)$ is neither submodular nor supermodular. However, Wang et al. [17] introduced two monotone nondecreasing submodular set functions $\beta : 2^V \rightarrow R_+$ and $\gamma : 2^V \rightarrow R_+$ such that for any $S \in 2^V$, $\beta(S) \ge \alpha(S) \ge \gamma(S)$. These two set functions are defined as follows:

$$\beta(S) = \sum_{(u,v) \in E : u \in I_m(S)} A(u, v)$$

and

$$\gamma(S) = \sum_{s \in S} \sum_{(u,v) \in E : u,v \in I_m(\{s\})} A(u, v).$$

By a theorem of Nemhauser and Wolsey [14], there is a greedy algorithm which is able to find $(1 - e^{-1})$-approximation solutions for the following two problems:

$$\max \beta(S) \tag{2}$$

$$\text{subject to} \quad |S| \le k,$$

$$S \subseteq V$$

$$\max \gamma(S) \tag{3}$$

$$\text{subject to} \quad |S| \le k,$$

$$S \subseteq V.$$

Let S_β and S_γ be $(1 - e^{-1})$-approximation solutions for problems 2 and 3, respectively. Let S_α be a feasible solution for problem 1. Choosing the best one from S_α, S_β, and S_γ, we would obtain a data-dependent approximation solution for problem (α), i.e., the data-dependent approximation solution is

$$S_{data} = \text{argmax}_{S \in \{S_\alpha, S_\beta, S_\gamma\}} \alpha(S).$$

3.2   Interaction-Aware Influence Maximization

The goal of interaction-aware influence maximization is to find a set of initial users to maximize the total profit related to both the number of the influenced nodes and the interaction among the influenced nodes.

Again the social network is represented as directed graph $G = (V, E)$ to represent a social network, where V is the set of users and E is the set of social relations between users. Each edge $(u, v) \in E$ is assigned with a probability p_{uv} so that when u is active, v is activated by u with probability p_{uv}. And the benefit related to the interaction between nodes is represented by a nonnegative function $b : V \times V \to \mathbb{R}_{\geq 0}$, in which $b(u, v) = b(v, u)$ for the unordered pair $\{u, v\}$ of node u and v. Note that for each $\{u, v\}$, we only compute once the benefit between them, i.e., $b(u, v)$ or $b(v, u)$ instead of $b(u, v) + b(v, u)$.

Consider a moment in the propagation process under IC model, when node u has just become active, and it attempts to activate its neighbor v, succeeding with probability $p_{u,v}$. We can view the outcome of this random event as being determined by flipping a coin of bias $p_{u,v}$. With all the coins flipped in advance, the edges in G for which the coin flip indicated a successful n activation are declared to be live; the remaining edges are declared to be blocked [10]. We use g to represent the outcome of this process which is called a live graph of G since it consists of all edges declared to be live. We denote as $g \sim D$, where D is the distribution of g. For any seed set S, denote by $I_g(S)$ the set of all active nodes at the end of the cascade process in live graph g. Its cardinality is represented by $|I_g(S)|$.

The total expected benefit would be defined as

$$f(S) = \mathbb{E}_{g \sim D}[\alpha \cdot |I_g(S)| + \beta \cdot \sum_{\{u,v\} \subseteq I_g(S)} b(u, v)]$$

$$= \sum_g Prob[g] \cdot (\alpha \cdot |I_g(S)| + \beta \cdot \sum_{\{u,v\} \subseteq I_g(S)} b(u, v)).$$

The benefit consists of two parts, the first part denoted as $\alpha \cdot I_g(S)$ is related to the number of nodes that are finally activated, and the second part $\beta \cdot \sum_{\{u,v\} \subseteq I_g(S)} b(u, v)$ is related to the strength of interaction between the active nodes. The parameters α, β are used to balance the weight of the two parts of the profits, and $\{u, v\} \subseteq I(S)$ denotes the all unordered pair in the set $I(S)$. Note that for each unordered pair $\{u, v\}$, since $b(u, v) = b(v, u)$, we only compute once the benefit between them. The expectation is respected to g.

The interaction-aware influence maximization is the following problem: given a social network $G = (V, E)$, a propagation probability p_{uv} for each edge (u, v) under the IC model, a benefit function $b : V \times V \to \mathbb{R}_{\geq 0}$, and a positive integer k, find a set S of k seeds to maximize the expected profit through influence propagation:

Fig. 1 Counter example

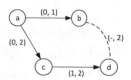

$$\max f(S)$$

$$s.t. |S| \le k.$$

This problem can be proved NP-hard by showing that a special case of interaction-aware influence maximization problem is NP-hard, where $\alpha = 0$, since a problem being NP-hard in a special case implies NP-hardness in general case. The seeds size equals k. Then the problem is transferred to seek k seeds that maximize the benefit between activated nodes. Now we prove by reducing from the set cover problem, which is NP-complete [1]. Given a ground set $U = \{u_1, u_2, \ldots, u_n\}$ and a collection of sets $\{S_1, S_2, \ldots, S_m\}$ whose union equals the ground set, the set cover problem is to decide if there exist k sets in S so that the union equals U. Given an instance of the set cover problem, we construct a corresponding graph with $m + 2n$ nodes as follows. For each set S_i we create one node p_i, and for each element u_j we create two nodes q_j and q'_j. If the S_i contains the element u_j, then we create two edges (p_i, q_j) and (p_i, q'_j). Note that each edge is live which means the probability is 1. Now we design the benefit function over pairs of nodes. For the pairs $\{q_j, q'_j\}$, the benefit equals to 1, and the other pairs equal to 0. Then the set cover problem is equivalent to deciding if there is a set S of k nodes such that the benefit of S equals to n. The NP-hardness follows immediately.

There are also counterexamples which show that $f(S)$ is neither submodular nor supermodular. We prove by the counter example shown in Figure 1. The first element in the tuple tied on each edge represents the propagation probability, and the second one denotes the benefit between its two end nodes. For pairs $\{u, v\}$ between which there is no edge set $b(u, v) = 0$ except pair $\{b, d\}$. In Figure 1, $(0, 1)$ on edge (a, b) means propagation probability $p_{ab} = 0$ and $b(a, b) = 1$, then we have $f(\{a\}) = 1 + 0 = 1$, $f(\{a, b\}) = 2 + 1 = 3$, $f(\{a, d\}) = 2 + 0 = 2$, and $f(\{a, b, d\}) = 3 + 3 = 6$. Thus, $f(\{a, d\}) - f(\{a\}) < f(\{a, b, d\}) - f(\{a, b\})$, which implies $f(S)$ is not submodular. Also, we have $f(\{c\}) = 2 + 2 = 4$, $f(\{d, c\}) = 2 + 2 = 4$, and $f(\{d\}) = 1$. Thus, $f(\{c\}) - f(\emptyset) > f(\{d, c\}) - f(\{c\})$ which implies $f(S)$ is not supermodular.

4 A Method for Interaction-Aware Influence Maximization

We have the following theoretical result leading us to a new method to solve our nonsubmodular problem.

Theorem 1 *For any set function* $f : 2^X \to R$ *and any set* $Y \subset X$, *there are two modular/submodular/supermodular functions* $m_f^u : 2^X \to R$ *and* $m_f^l : 2^X \to R$ *such that* $m_f^u(X) \geq f(X) \geq m_f^l(X)$ *and* $m_f^u(Y) = f(Y) = m_f^l(Y)$.

We can apply the theorem as long as we have a decomposition of the objective function into two submodular functions. This decomposition sometimes can be obtained trivially from the set function structure (or problem structure). However, in general, it is conjectured to be NP-hard [13]. In our case, it is not trivial, but we successfully found a decomposition with special technique and moreover, we made the obtained submodular functions computationally possible.

The following shows how we decompose our objective function $f(S)$ as the difference of $f_1(S)$ and $f_2(S)$ both of which are submodular proved as following, i.e., $f(S) = f_1(S) - f_2(S)$.

Given a seed set S and a live graph g, we define the $B_1(S)$ as benefit between activated users $I_g(S)$ and all users V, and define $B_2(S)$ as the benefit among all activated users $I_g(S)$ plus the benefit between the activated users $I_g(S)$ and the non-activated users $V \setminus I(S)$, which are formulated as follows:

$$B_1(S) = \sum_{u \in I_g(S)} \sum_{v \in V} b(u, v)$$

$$= \sum_{\{u,v\} \subseteq I_g(S)} 2 \cdot b(u, v) + \sum_{u \in I_g(S)} \sum_{v \in V \setminus I_g(S)} b(u, v)$$

$$B_2(S) = \sum_{\{u,v\} \subseteq I_g(S)} b(u, v) + \sum_{u \in I_g(S)} \sum_{v \in V \setminus I_g(S)} b(u, v).$$

Thus we have

$$B(S) = B_1(S) - B_2(S)$$

$$= \sum_{\{u,v\} \subseteq I_g(S)} b(u, v).$$

And given a seed set S, we define the following functions:

$$f_1(S) = \mathbb{E}_{g \sim D}[\alpha \cdot |I_g(S)| + \beta \cdot B_1(S)]$$
$$f_2(S) = \mathbb{E}_{g \sim D}[\beta \cdot B_2(S).]$$

Then we have

$$f(S) = \mathbb{E}_{g \sim D}[\alpha \cdot |I_g(S)| + \beta \cdot \sum_{\{u,v\} \subseteq I(S)} b(u, v)]$$

$$= \mathbb{E}_{g \sim D}[\alpha \cdot |I_g(S)| + \beta \cdot (B_1(S) - B_2(S))]$$

$$= \mathbb{E}_{g \sim D}[\alpha \cdot |I_g(S)| + \beta \cdot B_1(S)] - \mathbb{E}_{g \sim D}[\beta \cdot B_2(S)]$$
$$= f_1(S) - f_2(S).$$

Thus $f(S)$ is decomposed as a difference between function f_1 and f_2, and both of them are submodular. According to Theorem 1 and our decomposed submodular functions, we can design iterated sandwich algorithm to solve interaction-aware influence maximization problem. The main idea of our algorithm is to find the upper bound function and lower bound function based on current seed set, then solve the three functions: the upper bound function, the lower found function, and the objective function, and then we choose the best solution from those three solutions, this best solution is then the seed set for generation of upper and lower bound functions in next iteration. The procedure iterates until converged.

5 Conclusion

When the influence maximization problem is converted to capture the interactions among users as we discussed above, the objective functions are transformed to nonsubmodular functions, one of the possible method to solve such function is discussed in this chapter. The key is DS function maximization where a DS function is a difference of two submodular function. Some fundamental theoretical problems are still open. For more information on social networks, please refer to [16].

References

1. Alon, N., Awerbuch, B., Azar, Y.: The online set cover problem. In: Proceedings of the Thirty-Fifth Annual ACM Symposium on Theory of Computing, pp.100–105. ACM, New York (2003)
2. Bharathi, S., Kempe, D., Salek, M.: Competitive influence maximization in social networks. In: International Workshop on Web and Internet Economics, pp. 306–311. Springer, Berlin (2007)
3. Chen, W., Wang, Y., Yang, S.: Efficient influence maximization in social networks. In: Proceedings of the 15th ACM SIGKDD International Conference on Knowledge Discovery and Data Mining, pp. 199–208. ACM, New York (2009)
4. Chen, W., Wang, C., Wang, Y.: Scalable influence maximization for prevalent viral marketing in large-scale social networks. In: Proceedings of the 16th ACM SIGKDD International Conference on Knowledge Discovery and Data Mining, pp. 1029–1038. ACM, New York (2010)
5. Chen, W., Collins, A., Cummings, R., Ke, T., Liu, Z., Rincon, D., Sun, X., Wang, Y., Wei, W., Yuan, Y.: Influence maximization in social networks when negative opinions may emerge and propagate. In: Proceedings of the 2011 SIAM International Conference on Data Mining, pp. 379–390. SIAM, Philadelphia (2011)
6. Chen, S., Fan, J., Li, G., Feng, J., Tan, K.-L., Tang, J.: Online topic-aware influence maximization. Proc. VLDB Endowment 8(6), 666–677 (2015)

7. Domingos, P., Richardson, M.: Mining the network value of customers. In: Proceedings of the Seventh ACM SIGKDD International Conference on Knowledge Discovery and Data Mining, pp. 57–66. ACM, New York (2001)
8. Goyal, A., Lu, W., Lakshmanan, L.V.S.: Celf++: optimizing the greedy algorithm for influence maximization in social networks. In: Proceedings of the 20th International Conference Companion on World Wide Web, pp. 47–48. ACM, New York (2011)
9. Han, M., Li, J., Cai, Z., Han, Q.: Privacy reserved influence maximization in gps-enabled cyber-physical and online social networks. In: 2016 IEEE International Conferences on Big Data and Cloud Computing (BDCloud), Social Computing and Networking (SocialCom), Sustainable Computing and Communications (SustainCom)(BDCloud-SocialCom-SustainCom), pp. 284–292. IEEE, Piscataway (2016)
10. Kempe, D., Kleinberg, J., Tardos, É.: Maximizing the spread of influence through a social network. In: Proceedings of the Ninth ACM SIGKDD International Conference on Knowledge Discovery and Data Mining, pp. 137–146. ACM, New York (2003)
11. Leskovec, J., Krause, A., Guestrin, C., Faloutsos, C., VanBriesen, J., Glance, N.: Cost-effective outbreak detection in networks. In: Proceedings of the 13th ACM SIGKDD International Conference on Knowledge Discovery and Data Mining, pp. 420–429. ACM, New York (2007)
12. Li, Y., Zhang, D., Tan, K.-L.: Real-time targeted influence maximization for online advertisements. Proc. VLDB Endowment 8(10), 1070–1081 (2015)
13. Narasimhan, M., Bilmes, J.A.: A submodular-supermodular procedure with applications to discriminative structure learning (2012). Preprint. arXiv:1207.1404
14. Nemhauser, G.L., Wolsey, L.A., Fisher, M.L.: An analysis of approximations for maximizing submodular set functions—I. Math. Program. 14(1), 265–294 (1978)
15. Richardson, M., Domingos, P.: Mining knowledge-sharing sites for viral marketing. In: Proceedings of the Eighth ACM SIGKDD International Conference on Knowledge Discovery and Data Mining, pp. 61–70. ACM, New York (2002)
16. Thai, M.T., Pardalos, P.M.: Handbook of Optimization in Complex Networks. Springer, New York (2012)
17. Wang, Z., Yang, Y., Pei, J., Chu, L., Chen, E.: Activity maximization by effective information diffusion in social networks. IEEE Trans. Knowl. Data Eng. 29(11), 2374–2387 (2017)

Multi-Document Extractive Summarization as a Non-linear Combinatorial Optimization Problem

Meghana N. Satpute, Luobing Dong, Weili Wu, and Ding-Zhu Du

Abstract Multi-document summarization deals with finding the core theme presented in multiple documents. This can be done by selecting the important information from the text in the multiple documents. Extractive summarization selects and extracts such sentences which represent the gist of the documents. In this paper, we have surveyed how research in multi-document summarization has evolved from simple sentence-based techniques like sentence position to complex neural network based supervised learning techniques. In recent years, more and more supervised learning methods are proposed to tackle this problem along with some unsupervised approaches described in LSA (Deerwester et al. J Am Soc Inf Sci 41(6): 391–407, 1990) and TextRank (Mihalcea et al. Textrank: Bringing order into text. In: Proceedings of the 2004 conference on empirical methods in natural language processing, 2004). In this chapter, we have proposed an alternative unsupervised method where the problem of multi-document summarization can be viewed as a non-linear combinatorial optimization problem. We have formulated the problem and discussed possible solution to this problem.

1 Introduction

Automatic multi-document summarization is a process of creating shorter version of given text from different but related documents in such a way that it retains the important information the documents are meant to convey. With the advent of Internet, vast amount of data became accessible to everyone. People are better equipped to gain knowledge and make decisions. From online shopping to reading

M. N. Satpute (✉) · W. Wu · D.-Z. Du
Department of Computer Science, The University of Texas at Dallas, Richardson, TX, USA
e-mail: mns086000@utdallas.edu; weiliwu@utdallas.edu; dzdu@utdallas.edu

L. Dong
School of Telecommunications Engineering, Xidian University, Xian, China
e-mail: lbdong@xidian.edu.cn

© Springer Nature Switzerland AG 2019
D.-Z. Du et al. (eds.), *Nonlinear Combinatorial Optimization*,
Springer Optimization and Its Applications 147,
https://doi.org/10.1007/978-3-030-16194-1_15

books, people can read reviews and summaries. But due to time constraint, it is not possible to read every webpage or document available. Thus, people are more inclined to read summaries, i.e., summary of book, summary of news articles, etc. Hence, multi-document summarization research is gaining momentum due to its practical usefulness in day to day life.

There are two main approaches of automatic summarization techniques: extractive and abstractive. Extraction-based summarizers extract individual sentences from the given text (multiple documents). Depending on some criteria, these sentences are deemed important by summarizer. The final summary is composed by using these extracted sentences. Thus, the sentences in the summary come directly from the given text. Abstraction-based summarizers select important sentences or paragraphs from the given text but the final summary is composed by generating new sentences using the selected sentences. Thus, the sentences in summary are often different from sentences in given text.

Initial research in automatic summarization has emphasized on extracting summary from single document. Later on, as more and more text becomes available online, reader wanted to gather information from different but related documents. Hence, the research has diverted more towards multi-document summarization.

Multi-document summarization is a complex problem. A good summarizer is expected to cover all important information from the text from different documents, avoid redundancy, and produce coherent sentences as summary. To train the summarizer, large annotated data is needed but often not available. Even the available data does not cover different document styles, i.e., email data and social network data. Furthermore, there is no consensus among researchers about which sentences are needed to be in summary and which are not, making it harder problem to solve.

In this paper, we study how summarization techniques have evolved from simple heuristic-based techniques to applying complex neural network based learning mechanisms. Lin and Blimes [16] first noticed submodularity of natural language processing (NLP) problems and proposed that these problems can be solved as optimization problems. Extractive multi-document summarization is essentially selecting subset of sentences from a set of related documents based on some constraints. We discuss this problem as a combinatorial optimization problem and formulate the problem.

Section 2 describes background and approaches of summarization problem in early days. Section 3 describes the survey of how summarization techniques have been evolved for summarizing texts. Section 4 proposes a new way to look at extracting summary problem as non-linear combinatorial optimization problem and depicts possible formulation of this problem as an optimization problem. Section 5 concludes the contributions of this work.

2 Background

Automatic summarization efforts were started by researchers in late 1950s when they wanted to have condensed version of scientific and research papers which covers the important content.

2.1 Heuristic-Based Methods

Even though initial summarization methods were not as complex as today's summarization techniques, they were efficient and good enough for summarization needs for that time. These methods were mainly based on some rules or heuristics about how to decide which sentences are important. Once this decision is made, the important sentences are extracted as summary. In the method proposed by Luhn et al. [18], first the stop-words are removed from the sentences since even if they occur frequently, they do not add much meaning to summary. For all remaining words, their frequency is calculated and frequent words are deemed important. Sentences having many frequent words are considered summary-worthy and included in summary. In the work by Baxendale et al. [2], sentence position is given more importance. First and last sentences in a paragraph are extracted as significant sentences and included in summary. It is also observed that this assumption is true in the data set of scientific papers for which the summary is desired. Another such observation was that, in case of news articles, first two sentences of a paragraph are more significant than remaining sentences [26]. Edmundson et al. [7] proposed four components to weigh sentences, instead of just word frequency like previous research. He experimented with different weights for the presence of high-frequency keywords, pragmatic words, title and heading words, and sentence location. This research indicated that considering several linguistic features while deciding extract-worthy sentences offers better results.

3 Multi-Document Summarization Approaches

In multi-document summarization problem, the information comes from multiple documents which are related and often complement each other. While deciding the sentences to be selected, we need to make sure that they are coherent, not redundant and cover all important content. Various approaches are used in the research of multi-document summarization. Some approaches are extension of the work done for single document summarization and some are newly evolved approaches.

3.1 Statistical Approaches

The paper by Gambir and Gupta [12] has described the process of automatic extractive summarization using the following block diagram in Figure 1. Automatic summarization process begins with collecting the different but related documents from different sources. These documents are then pre-processed, i.e., removal of stop-words, stemming, etc. Then linguistic and statistical features are extracted from the documents. Based on the occurrence of features in a sentence, each sentence is scored using score function. Sentences with high score are extracted as a summary.

In [10] by Ferreira and others, an unsupervised system is built based on statistical and linguistic features in the text. They proposed a clustering algorithm to ensure coherence and reduce redundancy when multiple statements of the same meaning are present in the documents. In [5] Latent Semantic Analysis (LSA) is used to index and find topics by creating vectors of a documents based on the semantics in the text. Considering features among sentences such as statistical similarity, semantic similarity, coreference, and discourse relations, text is converted into a graph model. Main sentences are identified by using TextRank algorithm [21]. Based on similarity among the sentences, clusters of sentences are formed. Finally, main sentences from the clusters are selected to form summary.

In research by Ko et al. [15], a hybrid method is proposed which makes use of contextual and statistical information in the given text. In this method, two consecutive sentences are merged and bigram pseudo sentences are formed. Several statistical features are combined to score the pseudo sentence, such as how far the sentence is from the title, the location of sentence, score of a sentence based on aggregation similarity (which is the sum of similarities with all other sentences), term frequency of terms in the sentence, and term frequency based query (where high-frequency terms are used to query the document to find important sentence). After the extraction of high score bigram pseudo sentences, the sentences are fragmented to original sentences and summary is generated. They achieved performance gain due to combination of several important features for deciding which sentences need to be extracted.

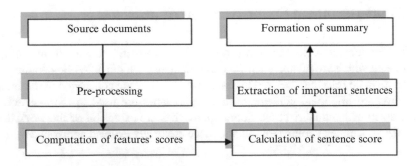

Fig. 1 Extractive summarization using statistical approach [12]

Yeh et al. [31] used different kinds of statistical and contextual features, including sentence position in the paragraph (first sentence in paragraph introduces paragraph and last sentence in paragraph summarizes paragraph), positive or negative keywords (positive keywords are most likely included in summary, while negative keywords are omitted), and centrality of sentence (similarity of the sentence with the other sentences in text. If sentence is too similar with other sentences, then it means it reflects central theme, resemblance with the title of the document. The summary is generated by using linear weighted combination of these features to obtain score function. Genetic algorithm is implemented to find optimal weights of the features while extracting the summary.

3.2 Topic Based Approaches

Topic signatures extract topic-related sentences as summary through two steps, i.e., topic recognition and interpretation. These two steps are considered as two basic steps in a typical automated text summary system. Lin and Hovy [17] proposed the idea of topic signatures. Each topic signature is represented as the terms related to the topic and weight of the term to that topic. One example mentioned in their paper is of topic *restaurant visit* which can be inferred by terms such as menu, waiter, order, etc. It is observed that often the topic words co-occur and hence their co-occurrence suggests that they belong to same topic. The sentences are scored based on their relevance to the topic signatures and high scored sentences are included in summary.

Harabagiu and Lacatusu [13] used two novel topic representations based on topic themes. Then based on their topics, the documents are classified as relevant or non-relevant to the pertaining topic. Sentences are ranked based on their score. In this paper, they considered relation between sentences and within the sentences. They used shallow semantic information from the text on top of lexical information. Extraction of summary sentences is done in different ways based on topic signature, sentence score, weights on topic relevant terms, etc.

3.3 Graph-Based Approaches

These methods converted text into graph by using vertices to represent sentences or concepts and edges to represent the semantic relatedness between two sentences or concepts. Graph-based summarization became effective summarization technology due to capturing contextual information among concepts.

In [20], Mani and Bleodorn described a graphical model which captures concepts shown by words, proper nouns, and phrases and then designate those as vertices. Edges represent the semantic relations between vertices. Figure 2 from [20] depicts three possible relations between concepts. Adj links are shown between adjacent

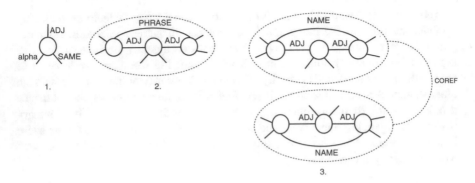

Fig. 2 Possible semantic relations between concepts [20]

concepts in the text. Name links can show person or entities. Phrase links can tie concepts together in a phrase. Coref links tie a concept to another whenever there is coreference. Alpha links are used when two concepts point to the same meaning, i.e., "President" in one sentence can be of same meaning as "Mr. Donald Trump" in another. Sentence selection for summary is based on the coverage of same vertices in the common lists and different lists. Sentences are selected greedily based on the average activated weight of the covered words.

In LexPageRank system by Erkan and Radev [8], sentences are depicted as vertices and link between vertices exists if the cosine similarity between two sentences exceeds predefined threshold. Sentence clusters are formed on the basis of sentence similarity. They hypothesize that the sentence which is more similar to other sentences contain main theme and hence central. The degree of each vertex is calculated. Each link or edge between vertices represents a vote. They used PageRank [24] algorithm to calculate vote of each link. Most voted sentences are included in summary. Their architecture also took care of sentence subsumption. When one sentence subsumes information from another sentence and possesses some additional information, then it is included in summary and another sentence is omitted from summary.

3.4 Machine Learning Based Approaches

As the research progressed, machine learning based methods caught the attention of scientific research. Machine learning methods enable computer to summarize documents by learning from the original documents and "understanding" the potential semantics. For example, methods using classifiers such as Naive Bayesian, support vector machine (SVM) [9], recurrent neural network (RNN) [23], neural convolution network (NCN) attention [30], and recursive neural network [4] have shown significant performance gains.

Machine learning based methods fall into one of the following categories: supervised, semi-supervised, or unsupervised. Supervised learning relies on large data sets to learn features and then using those features, it can classify the test data. Availability of big data set which covers all possible variations is a bottleneck in research of NLP tasks, as it is very time consuming to annotate the data. Unsupervised learning methods learn from the available target data for which the NLP task is to be performed. Classification task is done by learning features from that data itself. Semi-supervised approach relies on some seed examples provided by the user. From these examples patterns are learned and classification task is done.

Fattah et al. [9] experimented summarization problem using many machine learning classifiers such as Naive Bayesian, maximum entropy, support vector machine, decision trees, neural networks, mathematical regression, etc. The method considers summary generation task as a classification problem and each sentence is either included in summary or excluded. Fattah et al. [9] employed a hybrid model for summary generation task which constitutes the following three classifiers: maximum entropy, Naive Bayes, and support vector machine. Several features are taken into account to train the classifiers. For example, words similarity between sentences and between paragraphs, score using term frequency of document, key phrases, position of sentence, occurrence of not-needed information, text format, etc. These features are provided to the three classifiers in training phase. In testing phase, features are extracted and sentences are ranked by feature weights learned in the training phase. Using hybrid model of three classifiers, final summary is created.

Cao et al. [4] presented recursive neural networks (R2N2) to score sentences for extractive multi-document summarization. Each sentence is first converted into a parse tree. Information from different parts of sentence is gathered and fed to R2N2. Some features used in this process are term frequency, inverse document frequency, sentence length, named entity, position of sentence, etc. Sentence relevance is evaluated and sentence rank is given by a hierarchical regression process. On the basis of information from word level to sentence level, features are learned by recursive neural networks apart from the given features. Important sentences for summary are selected based on their ranking score. They employed greedy algorithm and integer linear programming (ILP) for selecting the sentences to be part of summary.

Nallapati et al. [23] presented SummaRuNNer, a model based on recurrent neural network (RNN) for extractive multi-document summarization. SummaRuNNer is trained using reference summaries. Summarization is considered as sequential binary classification problem where each sentence is classified as summary sentence or non-summary sentence. They used two-layer bidirectional RNN where one layer operates at word level, while another layer operates at sentence level. Using words and word embedding, hidden states are generated. They use greedy approximation to create labels from given summaries. The entire document is modeled as follows:

$$d = tanh\left(W_d * (1/N_d)\sum_{j=1}^{N_d}[h_j^f, h_j^b] + b\right), \tag{1}$$

where states starting with h are the hidden states corresponding to the sentence of the forward and backward sentence-level RNNs, respectively, and N_d is the number of sentences in the document. Features such as novelty of a sentence, position, and content are considered in their study. These features are learned rather than hand crafting and providing it to system.

3.5 Optimization Based Approaches

Lin and Blimes [17] were first to notice the submodularity in summarization. They modeled the problem as a knapsack constraint of selecting subset of sentences S from the sentences of whole document set V under the constraint of length of summary. They devised it as a maximization function of quality of summary shown in the following formula, where c_i is cost for adding sentence s_i in summary and b is budget constraint

$$S' \in \arg \max_{S \subseteq V} F(S)$$

$$s.t. \sum_{i \subseteq S} c_i \geq b. \tag{2}$$

Although it is NP-hard problem, it can be solved using greedy algorithm. It becomes too computationally expensive for real-world applications [17].

Shigematsu and Kobayashi [29] used differential evolution approach to overcome the problem of computational complexity of optimization function for summarization. First they used LDA [3] to detect topics in the text. Sentences are ranked based on the topical information each sentence possesses. Resulting number of summary sentences will depend upon the length constraint over summary sentences. This method has reduced the calculation time to generate summary, to great extent but precision is worse than the method with an explicit solution technique using greedy algorithm [29].

Galanis and others used combination of support vector regression (SVR) along with integer linear programming (ILP) [11]. They used features such as sentence position, named entities, Levenshtein distance, word overlap, and content word frequency. These features are given to SVR to score each sentence from the document set. Instead of using the sentence scores directly to formulate summary, they first normalize the sentence scores. These scores are multiplied by the length of the sentence to take care of problem of the method picking short sentences. Importance of summary is calculated by adding the normalized scores of sentences. The importance of summary is maximized. While forming final summary, the number of distinct bigrams it can cover is also maximized. The underlying assumption was that the more the number of bigrams, the summary covers the less redundant it is. Like previous method by Shigematsu [29], length of summary sentences is considered the constraint over which ILP is done.

4 Our Approach to Multi-Document Summarization as a Non-linear Combinatorial Optimization Problem

In summarization, a group of sentences is selected from a bigger group of sentences. Whenever a subset of elements is to be selected from a set of elements based on some constraint, then we can formulate that problem as submodular or supermodular optimization problem.

The maximization or minimization of a set function can be formulated as combinatorial optimization problem. They are widely used in many areas of computer science and applied mathematics [6]. Minimization or maximization problems are defined on the set of subsets of a given base set S. In combinatorial optimization, submodular/supermodular functions have a role somewhat similar to that played by convex/concave functions in continuous optimization [1]. Researchers also proved that some existing extractive summarization methods can be viewed as a problem of submodular function maximization [16], such as maximum marginal relevance (**MMR**).

Given a finite set S, we use 2^S to denote the power set of S. A set function $f : 2^S \rightarrow \mathbb{R}$ is submodular if it satisfies one of the following equivalent conditions:

- For every $A, B \subseteq S$ with $A \subseteq B$ and every $a \in S - B$ we have that $f(A \cup \{a\}) - f(\{a\}) \geq f(B \cup \{a\}) - f(\{a\})$.
- For every $A, B \subseteq S$, we have that $f(A) + f(B) \geq f(A \cup B) + f(A \cap B)$.
- For every $A \subseteq S$ and $a_1, a_2 \in S - A$, we have that $f(A \cup \{a_1\}) + f(A \cup \{a_2\}) \geq f(A \cup \{a_1, a_2\}) + f(A)$.

A set function f is monotonically increasing, if for every $A \subseteq B$ we have $f(A) \leq f(B)$. A function f is supermodular if and only if $-f$ is submodular. A set function f is monotonically increasing, if for every $A \subseteq B$ we have $f(A) \leq f(B)$.

When a single element is added to an input set, as the size of the input set increases, the difference in the incremental value of a submodular function decreases. For a combinatorial optimization problem, a greedy algorithm can be designed if the objective function is submodular. Greedy algorithm can give an approximate solution in polynomial time with an approximation guaranteed to be within $\frac{e-1}{e} \approx 0.63$ of the optimal solution [22].

4.1 Diversity in Summary

Main objective of summary is to obtain maximum information from a given document set in short version in such a way that it captures the gist of the document set. In order to capture more information from document, it is necessary to cover diverse topics from the documents.

Diversity is a central theme in ecology. The diversity concept was first used by ecologists to measure the number of different species in community quantitatively.

Ecological communities with many species are more diverse than ecological communities with fewer species. Ecologists tried to sample with high diversity to increase the probability of finding small species [19]. Ecologists have proposed many methods to measure the diversity of species in these decades, such as the Shannon index or the Simpson index [28]. A diversity index called as quadratic diversity (Q) is proposed by Rao [27]. $Q = \sum_{i=1}^{S} \sum_{j=1}^{S} d_{ij} p_i p_j$ quadratic diversity incorporates both species relative abundances ($p_i p_j$) and a measure of the pairwise distances between species (d_{ij}).

The main theme of summarization is to get as much content from the documents as possible in a compact manner. Thus we want to ensure that the summary gathers information about all topics from the documents. If document set has sentences that convey the same meaning, then these are redundant sentences and must be omitted from summary. If there is sentence limit on summary and redundant sentences exist in the summary, then some other information, which should be part of summary, is missed from the summary. Summarization diversity controls this problem of redundancy and gives the reader insight into different distinct topics covered in the text.

Shannon entropy was originally proposed to quantify the amount of information in a signal or event. For a discrete random variable X with possible states $x_1, x_2, \cdots x_n$, its Shannon entropy is defined as Formula (3). In Formula (3), $p(x_i) = Pr(X = x_i)$ is the probability of x_i

$$H(X) = \sum_{i=1}^{n} p(x_i) \log_2 \left(\frac{1}{p(x_i)} \right). \tag{3}$$

Intuitively, when there is only one possible state of X, then $H(X)$ becomes 0. $H(x)$ increases as number of possible states of X increase denoting more diversity.

4.2 Problem Formulation

In this paper, we address the problem of how to extract a summary from a set of documents that covers as much real content of all subtopics as possible. We first identify all subtopics from the document set and then summarize the documents.

After extracting subtopics, assume that we get a subtopic set C. In this subsection, we propose a new formulation for a method to extract a small and limited set of sentences from set C which can be representative of the entire document set, which is also the goal of the summarization.

We cannot give equal weightage to all subtopics because some subtopics constitute a very few number of sentences, while other subtopics might have many sentences written about them. Thus the subtopics having small number of sentences do not contribute much to the document set and hence can be deleted. For simplicity, we use the symbol C' to denote the subtopic set after the small sized subtopics are deleted.

In this step, for the process of calculation of semantic distance, single sentence is taken into consideration as the basic unit of calculation. C' can be represented as a subtopics covering set of sentences $C' = \{st_1, st_2, \ldots, st_q\}$, where each sentence st_i belongs to a subtopic s_k and it also belongs to a paragraph set c_k. If $st_i \in p_{ef}$ and $p_{ef} \triangleq s_k$, then $st_i \in c_k$.

We use notation $SDS(st_i, st_j)$ to represent the semantic distance between two sentences st_i and st_j. $SDSS(st_i, A)$ $(A \subseteq C')$ is used to denote the semantic distance between a sentence st_i to a sentence set A. It is defined as the following Formula (4):

$$SDSS(st_i, A) = \min_{st_j \in A} SDS(st_i, st_j). \tag{4}$$

The semantic distance between the two subsets of C' is represented as a function $SDTS : (2^{C'}, 2^{C'}) \to \mathbb{R}$. $SDTS$ can be defined as Formula (5)

$$SDTS(A, B) = \sum_{st_j \in (B-A)} SDSS(st_j, A) \qquad A \subseteq C', B \subseteq C'. \tag{5}$$

The set of summary sentences is a size limited subset I of C' which is a set of sentences from document set D. We need to find I such that the similarity between sets I and D is as high as possible, which intuitively means the semantic distance between I and C' is as small as possible. At the same time, we also want subtopic diversity, the more the subtopics that I can cover, the better. Shannon entropy can be used to measure the diversity (Formula (6))

$$HD(I) = \sum_{i=1}^{q} \frac{|I_k|}{|I|} \log_2 \left(\frac{|I|}{|I_k|} \right) \qquad I_k = \{st_i | st_i \in C', st_i \in c_k\}. \tag{6}$$

Therefore, there are two targets: minimizing the distance between I and C' (Formula (7)) and maximizing the subtopic diversity (Formula (8)). At the same time, we have a constraint that the sentence number of the final summary is less than some constant b ($|I| \leq b$). The summary subtopic diversity $HD(I)$ is known to be submodular and monotone increasing [25]. Interestingly, $SDTS(I, C')$ is monotone decreasing and supermodular.

$$\min_{I \subseteq C'} SDTS(I, C') \tag{7}$$

$$\max_{I \subseteq C'} HD(I). \tag{8}$$

We know that the minimizing value of the supermodular function $SDTS(I, C')$ maximizes value of $-SDTS(I, C')$ which is submodular and the maximization of a submodular function with cardinality constraint is NP-hard [14]. Fortunately, $HD(I) - \gamma SDTS(I, C')$ is submodular, and we can formulate our objective function as Formula (9) which can be approximately solved

$$\arg\max_{I \subseteq C'} HD(I) - \gamma SDTS(I, C')$$

$$s.t. \qquad |I| \leq b,$$

(9)

where γ is a parameter which can be adjusted experimentally.

5 Conclusion

Through this work, we have surveyed different approaches in multi-document summarization and proposed it as a combinatorial optimization problem. The proposed formulation can extract meaningful summary from multiple documents. Using the sentence distances and subtopics as backbones we have formulated the problem as submodular combinatorial optimization problem of minimizing distance between summary and document set and maximizing subtopic diversity in the summary.

References

1. Bach, F., et al.: Learning with submodular functions: a convex optimization perspective. Found. Trends® Mach. Learn. **6**(2-3), 145–373 (2013)
2. Baxendale, P.B.: Machine-made index for technical literature: an experiment. IBM J. Res. Dev. **2**(4), 354–361 (October 1958)
3. Blei, D.M., Ng, A.Y., Jordan, M.I., Lafferty, J.: Latent Dirichlet allocation. J. Mach. Learn. Res. **3**, 2003 (2003)
4. Cao, Z., Wei, F., Dong, L., Li, S., Zhou, M.: Ranking with recursive neural networks and its application to multi-document summarization. In: Proceedings of the Twenty-Ninth AAAI Conference on Artificial Intelligence, AAAI'15, pp. 2153–2159. AAAI, Palo Alto (2015)
5. Deerwester, S., Dumais, S.T., Furnas, G.W., Landauer, T.K., Harshman, R.: Indexing by latent semantic analysis. J. Am. Soc. Inf. Sci. **41**(6), 391–407 (1990)
6. Dong, L., Guo, Q., Wu, W.: Speech corpora subset selection based on time-continuous utterances features. J. Comb. Optim. 1–12 (2018). https://doi.org/10.1007/s10878-018-0350-2
7. Edmundson, H.P.: New methods in automatic extracting. J. ACM **16**(2), 264–285 (1969)
8. Erkan, G., Radev, D.R.: LexPageRank: prestige in multi-document text summarization. In: Proceedings of the 2004 Conference on Empirical Methods in Natural Language Processing (2004)
9. Fattah, M.A.: A hybrid machine learning model for multi-document summarization. Appl. Intell. **40**(4), 592–600 (2014)

10. Ferreira, R., Cabral, L.D.S., çalves de Freitas, F.L.G., Lins, R.D., de França Pereira e Silva, G., Simske, S.J., Favaro, L.: A multi-document summarization system based on statistics and linguistic treatment. Expert Syst. Appl. **41**(13), 5780–5787 (2014)
11. Galanis, D., Lampouras, G., Androutsopoulos, I.: Extractive multi-document summarization with integer linear programming and support vector regression. In Proceedings of COLING, pp. 911–926. IIT, Bombay (2012)
12. Gambhir, M., Gupta, V.: Recent automatic text summarization techniques: a survey. Artif. Intell. Rev. **47**, 1–66 (2016)
13. Harabagiu, S.M., Lacatusu, V.F.: Using topic themes for multi-document summarization. ACM Trans. Inf. Syst. **28**(3), 13:1–13:47 (2010)
14. Iyer, R.K., Bilmes, J.A.: Submodular optimization with submodular cover and submodular knapsack constraints. In: Advances in Neural Information Processing Systems, pp. 2436–2444 (2013)
15. Ko, Y., Seo, J.: An effective sentence-extraction technique using contextual information and statistical approaches for text summarization. Pattern Recogn. Lett. **29**(9), 1366–1371 (2008)
16. Lin, H., Bilmes, J.: A class of submodular functions for document summarization. In: Proceedings of the 49th Annual Meeting of the Association for Computational Linguistics: Human Language Technologies, vol. 1, pp. 510–520. Association for Computational Linguistics, Stroudsburg (2011)
17. Lin, C.-Y., Hovy, E.: The automated acquisition of topic signatures for text summarization. In: Proceedings of the 18th Conference on Computational Linguistics - Volume 1, COLING '00, pp. 495–501. Association for Computational Linguistics, Stroudsburg (2000)
18. Luhn, H.P.: The automatic creation of literature abstracts. IBM J. Res. Dev. **2**(2), 159–165 (1958)
19. Magurran, A.E.: Why diversity? In: Ecological diversity and its measurement, pp. 1–5. Springer, Dordrecht (1988)
20. Mani, I., Bloedorn, E.: Multi-document summarization by graph search and matching. In: Proceedings of the Fourteenth National Conference on Artificial Intelligence and Ninth Conference on Innovative Applications of Artificial Intelligence, AAAI'97/IAAI'97, pp. 622–628. AAAI, Cambridge (1997)
21. Mihalcea, R., Tarau, P.: Textrank: bringing order into text. In: Proceedings of the 2004 Conference on Empirical Methods in Natural Language Processing (2004)
22. Minoux, M.: Accelerated greedy algorithms for maximizing submodular set functions. In: Optimization Techniques, pp. 234–243. Springer, Berlin (1978)
23. Nallapati, R., Zhai, F., Zhou, B.: Summarunner: a recurrent neural network based sequence model for extractive summarization of documents. In. AAAI, pp. 3075–3081. AAAI, Cambridge (2017)
24. Page, L., Brin, S., Motwani, R., Winograd, T.: The PageRank citation ranking: bringing order to the web. Technical Report 1999-66, Stanford InfoLab, November 1999. Previous number = SIDL-WP-1999-0120
25. Polyanskiy, Y.: Lecture notes, chapter 1: Information measures: entropy and divergence (January 2016)
26. Radev, D.: [Artificial Intelligence - All in one]. (2016, April 5). Summarization Techniques (NLP) University of Michigan [Video file]. Retrieved from https://www.youtube.com/watch?v=N5N-HCUE3G4
27. Rao, C.R.: Diversity and dissimilarity coefficients: a unified approach. Theor. Popul. Biol. **21**(1), 24–43 (1982)
28. Ricotta, C., Szeidl, L.: Towards a unifying approach to diversity measures: bridging the gap between the Shannon entropy and Rao's quadratic index. Theor. Popul. Biol. **70**(3), 237–243 (2006)
29. Shigematsu, H., Kobayashi, I.: Topic-based multi-document summarization using differential evolution for combinatorial optimization of sentences. In: Proceedings of the 28th Pacific Asia Conference on Language, Information and Computing (2014)

30. Yasunaga, M., Zhang, R., Meelu, K., Pareek, A., Srinivasan, K., Radev, D.: Graph-based neural multi-document summarization. In: Proceedings of the 21st Conference on Computational Natural Language Learning (CoNLL 2017), pp. 452–462. Association for Computational Linguistics, Vancouver (2017)
31. Yeh, J.-Y., Ke, H.-R., Yang, W.-P., Meng, I.-H.: Text summarization using a trainable summarizer and latent semantic analysis. Inf. Process. Manag. **41**(1), 75–95 (2005). An Asian Digital Libraries Perspective

Viral Marketing for Complementary Products

Jianxiong Guo and Weili Wu

Abstract When you purchase a product in internet, a recommendation may appear "the customer who buys A may also buy B." When you purchase both products A and B, another recommendation may appear "the customer who buys A and B may also buy C." Products A, B, and C are complementary. Both recommendations are established based on historical data and statistical analysis. Here, complementary products are those that tend to be purchased together, for example, iPhone and its accessories, computer and monitor, etc. In this article, we would like to address this issue in viral marketing.

1 Who Buys A May also Like to Buy B

First, we consider the complementary relationship between two products, that is, there is a certain probability that buying product A would make the customer to buy product B.

Consider a (directed) social network $G = (V, E)$ with independent cascade model for information diffusion, that is, each arc (u, v) is associated with a number p_{uv} which is the probability that v accepts influence of u. Consider k products $g_1, g_2, \ldots, g_k$ with prices $c_1, c_2, \ldots, c_k$, respectively. Suppose that a customer u who buys product g_i would also buy product g_j with probability p_{uij}. We study the following problem.

Research Problem 1 (Viral Marketing for Complementary Products) *Given a budget B, find a set of customers for giving free samples within the budget B to maximize the expected total sales of the product.*

For this problem, at each step of information diffusion process, there are two types of influences to each customer v for each product g_j:

J. Guo (✉) · W. Wu
Department of Computer Science, The University of Texas at Dallas, Richardson, TX, USA
e-mail: jianxiong.guo@utdallas.edu; weiliwu@utdallas.edu

© Springer Nature Switzerland AG 2019
D.-Z. Du et al. (eds.), *Nonlinear Combinatorial Optimization*,
Springer Optimization and Its Applications 147,
https://doi.org/10.1007/978-3-030-16194-1_16

(1) The influence from a customer u who purchases product g_j and has the probability p_{uv} for success of the influence.
(2) The influence from a product g_i because the customer v purchased g_i at a previous step. The success of this influence has probability p_{vij}.

Theorem 1 *Suppose above influences at each customer are independent and influence can be successful only at fresh (i.e., at the step that the influence just occurs). Then the problem of viral marketing for complementary products is a special case of following influence maximization problem in weighted case: Consider a social network $G = (V, E)$ with independent cascade model. For each node u, assign a cost $c(u)$ and suppose that it costs $c(u)$ to choose u as a seed and obtain benefit $c(u)$ if u becomes active. Given a budget B, find a set of customers for giving free samples within the budget B to maximize the expected total cost of active nodes.*

Proof Let G be the social network in the problem of viral marketing for complementary products. For each product g_i, make a copy G^i of G. Denote by u^i the copy of node u. Consider $G^1 \cup G^2 \cup \cdots \cup G^k$. For every two nodes u^i and u^j, add an arc (u^i, u^j) associated with a probability $p_{ij}(u)$. The problem of viral marketing for complementary products would be equivalent to the influence maximization on constructed network with cost function $c(u^i) = c_i$. $\qquad \square$

By Theorem 1, the viral marketing for complementary products can be formulated as a knapsack-constrained submodular maximization as follows:

$$\max \quad f(A)$$
$$\text{subject to} \quad \sum_{x \in A} c(x) \leq B$$
$$A \subseteq V,$$

where f is a monotone (nondecreasing) submodular function on 2^V with $f(\emptyset) = 0$.

2 Knapsack-Constrained Submodular Maximization

There are three algorithms [7, 10, 13] in the literature for knapsack-constrained submodular maximization. The first two are motivated from two algorithms for the well-known knapsack problem as follows:

$$\max \quad c_1 x_1 + c_2 x_2 + \cdots + c_n x_n$$
$$\text{subject to} \quad s_1 x_1 + s_2 x_2 + \cdots + s_n x_n \leq B$$
$$x_1, x_2, \ldots, x_n \in \{0, 1\}.$$

Assume $s_i \leq B$. Following is its 2-approximation.

2-Approximation Algorithm for Knapsack

Sort $\frac{c_1}{s_1} \geq \frac{c_2}{s_2} \geq \cdots \geq \frac{c_n}{s_n}$.

Choose k such that $\sum_{i=1}^{k} \leq B < \sum_{i=1}^{k+1}$.

output $c_G = \max(\sum_{i=1}^{k}, c_{k+1})$.

Next, we see a similar algorithm for knapsack-constrained submodular maximization. Here denote $\Delta_v f(A) = f(A \cup \{v\}) - f(A)$.

Algorithm 1 Knapsack-constrained submodular maximization

1: $A \leftarrow \emptyset$
2: **while** $c(A) < B$ **do**
3:    choose $v \in V \setminus A$ to maximize $\frac{\Delta_v f(A)}{c(v)}$
4:    set $A \leftarrow A \cup \{v\}$
5: **end while**
6: **return** $\operatorname{argmax}(f(A \setminus \{v\}), f(\{v\}))$

Theorem 2 ([10]) *Algorithm 1 is a* $(1 - e^{-1})/2$-*approximation for knapsack-constrained submodular maximization.*

Proof Let $v_1, v_2, \ldots, v_{k+1}$ be generated by Algorithm 1 and denote $A_i = \{v_1, \ldots, v_i\}$. Then

$$v_{i+1} = \operatorname{argmax}_{v \in V \setminus A_i} \frac{\Delta_v f(A_i)}{c(v)}$$

and $c(A_k) < B \leq c(A_{k+1})$. Suppose optimal $A^* = \{u_1, u_2, \ldots, u_h\}$.

$$f(A^*) \leq f(A_i \cup A^*)$$

$$= f(A_i) + \Delta_{u_1} f(A_i) + \Delta_{u_2} f(A_i \cup \{u_1\}) + \cdots + \Delta_{u_h}$$
$$f(A_i \cup \{u_1, \ldots, u_{h-1}\})$$

$$\leq f(A_i) + \delta_{u_1}(A_i) + \Delta_{u_2} f(A_i) + \cdots + \Delta_{u_h} f(A_i)$$

$$\leq f(A_i) + \frac{c(u_1)}{c(v_{i+1})} \Delta_{v_{i+1}} f(A_i) + \frac{c(u_2)}{c(v_{i+1})} \Delta_{v_{i+1}}$$

$$f(A_i) + \cdots + \frac{c(u_h)}{c(v_{i+1})} \Delta_{v_{i+1}} f(A_i)$$

$$= f(A_i) + \frac{c(A^*)}{c(v_{i+1})} (f(A_{i+1}) - f(A_i)).$$

Denote $\alpha_i = f(A^*) - f(A_i)$. Then $\alpha_i \leq \frac{c(A^*)}{c(v_{i+1})}(\alpha_i - \alpha_{i+1})$. This,

$$\alpha_{i+1} \le \left(1 - \frac{c(v_{i+1})}{c(A^*)}\right)\alpha_I \le e^{-c(v_{i+1})/B}\alpha_i$$

since $1 + x \le e^x$. Hence

$$\alpha_{k+1} \le e^{-c(A_{k+1})/B}\alpha_0 \le e^{-1}f(A^*),$$

that is, $f(A^*) - f(A_{k+1}) \le e^{-1}f(A^*)$. Hence, $f(A_{k+1}) \ge (1 - e^{-1})f(A^*)$. Since $f(A_k) + f(\{v_{k+1}\}) \ge f(A_{k+1}) + f(\emptyset) \ge f(A_{k+1})$, we have $\max(f(A_k), f(\{v_{k+1}\})) \ge (1 - e^{-1})/2 \cdot f(A^*)$. $\qquad\square$

This is a PTAS for the knapsack problem.

PTAS for Knapsack

Let $c + G$ be obtained from 2-approximation algorithm for knapsack problem.

Set $a = c_G \cdot \frac{\varepsilon}{1+\varepsilon}$.

Suppose that for $1 \le i \le m$, $c_i \le a$ and for $m + 1 \le i \le n$, $c_i > a$.

Sort $\frac{c_1}{s_1} \ge \frac{c_2}{s_2} \ge \cdots \ge \frac{c_m}{s_m}$.

For each $I \subseteq \{m + 1, \ldots, n\}$ with $|I| \le 2(1 + \varepsilon)/\varepsilon$,

    if $\sum_{i \in I} s_i > B$,

        then define $c(I) = 0$

        else choose maximum $k \le m$ such that $\sum_{i=1}^{k} s_i \le B - \sum_{i \in I} s_i$

                and define $c(I) = \sum_{i \in I} c_i + \sum_{i=1}^{k} c_i$;

    output $I_{output} = \mathrm{argmax}_I c(I)$.

A similar algorithm for knapsack-constrained submodular maximization is as follows.

Algorithm 2 Knapsack-constrained submodular maximization

1: **for** every $I \subseteq V$ with $|I| \le d$ **do**
2:     $A \leftarrow I$
3:     $T \leftarrow V \setminus I$
4:     **while** $T \ne \emptyset$ **do**
5:         choose $v \in T$ to maximize $\frac{\Delta_v f(A)}{c(v)}$
6:         set $T \leftarrow T \setminus \{v\}$
7:         **if** $c(A \cup \{v\}) \le B$ **then**
8:             $A \leftarrow A \cup \{v\}$
9:         **end if**
10:    **end while**
11: **end for**
12: **return** $A_G = \mathrm{argmax}_I f(A(I))$

Theorem 3 ([7]) *For $d \ge 3$, Algorithm 2 is a $(1 - e^{-1})$-approximation for knapsack-constrained submodular maximization.*

Proof Suppose optimal solution $A^* = \{u_1, u_2, \ldots, u_h\}$ in ordering

$$u_i = \text{argmax}_{u \in A^*}(f(\{u_1, \ldots, u_{i-1}, u\}) - f(\{u_1, \ldots, u_{i-1}\})).$$

Let $I = \{u_1, u_2, \ldots, u_d\}$. Define $g(X) = f(X \cup I) - f(I)$. Then computation of $A(I)$ can be seen as a greedy algorithm applying to monotone submodular function $g(\cdot)$. Suppose this computation produces $A \setminus I = \{v_1, \ldots, v_t\}$. If $A^* \setminus I = A \setminus I$, then $A_G = A(I) = A^*$ and hence the theorem holds. Therefore, we may assume $A^* \setminus I \neq A \setminus I$. In this case, there exists element in $A^* \setminus I$, but not in $A \setminus I$. Let $v_{t+1} \in A^* \setminus I$ be the first one which violates the knapsack constraint, that is,

$$c(I \cup \{v_1, \ldots, v_t\}) \leq B \text{ and } c(I \cup \{v_1, \ldots, v_t, v_{t+1}\}) > B.$$

By the proof of Theorem 2, we would obtain

$$g(\{v_1, v_2, \ldots, v_t, v_{t+1}\}) \geq (1 - e^{-1})g(A^* \setminus I).$$

Thus,

$$f(I \cup \{v_1, v_2, \ldots, v_t, v_{t+1}\}) \geq (1 - e^{-1})f(A^*) + e^{-1}f(I).$$

Hence,

$$f(A(I)) \geq f(I \cup \{v_1, v_2, \ldots, v_t\})$$
$$\geq (1 - e^{-1})f(A^*) + e^{-1}f(I) - \Delta_{v_{t+1}}f(I \cup \{v_1, v_2, \ldots, v_t\}).$$

Note that

$$\Delta_{v_{t+1}}f(I \cup \{v_1, v_2, \ldots, v_t\}) \leq \Delta_{v_{t+1}}f(I) \leq \Delta_{u_{d+1}}f(I) \geq \cdots \geq \Delta_{u_1}f(\emptyset)$$
$$= f(\{u_1\}).$$

Thus,

$$\Delta_{v_{t+1}}f(I \cup \{v_1, v_2, \ldots, v_t\}) \leq \frac{1}{d} \cdot (\Delta_{u_d}f(\{u_1, \ldots, u_{d-1}\}) + \cdots + \Delta_{u_1}f(\emptyset)) = \frac{1}{d}f(I).$$

Therefore,

$$f(A(I)) \geq f(I \cup \{v_1, v_2, \ldots, v_t\}$$
$$\geq (1 - e^{-1})f(A^*) + e^{-1}f(I) - \frac{1}{d}f(I)$$
$$\geq (1 - e^{-1})f(A^*) \quad \text{for } d \geq 3.$$

$\square$

Algorithm 1 runs faster, but its performance is worse than Algorithm 2. An improvement is given in [13], which gives a $(1-e^{-2})/2$-approximation with running time comparable with Algorithm 1.

There exist more approximation algorithms in the literature for the knapsack problem and its variations. *Can we get motivation from them to design better approximations for the knapsack-constrained submodular maximization and its variations?*

3   Who Buys A and B May also Like to Buy C

In general, the viral marketing for complementary product would induce a maximization problem on monotone nonsubmodular function with a knapsack constraint, especially, when we consider influence of two or more products on one. This case is much similar to the composed influence studied in [14].

For monotone nonsubmodular maximization, there are several choices of methodology, such as super modular degree [2–4, 6], sandwich methods [1, 8, 11], and algorithms based on DS decomposition (submodular–supermodular algorithm [9] modular–modular algorithm [5], and iterated sandwich method [12]). Among them, the DS decomposition is more challenging.

References

1. Chen, W., Lin, T., Tan, Z., Zhao, M., Zhou, X.: Robust influence maximization. In: Proceedings of the 22nd ACM SIGKDD International Conference on Knowledge Discovery and Data Mining, KDD '16, pp. 795–804. ACM, New York (2016)
2. Feige, U., Izsak, R.: Welfare maximization and the supermodular degree. In: Proceedings of the 4th Conference on Innovations in Theoretical Computer Science, ITCS '13, pp. 247–256. ACM, New York (2013)
3. Feldman, M., Izsak, R.: Building a good team: secretary problems and the supermodular degree. In: Proceedings of the Twenty-Eighth Annual ACM-SIAM Symposium on Discrete Algorithms. Society for Industrial and Applied Mathematics, pp. 1651–1670 (2017)
4. Feldman, M., Izsak, R.: Constrained monotone function maximization and the supermodular degree. CoRR, abs/1407.6328 (2014)
5. Iyer, R.K., Bilmes, J.A.: Algorithms for approximate minimization of the difference between submodular functions, with applications. CoRR, abs/1207.0560 (2012)
6. Izsak, R.: Working Together: Committee Selection and the Supermodular Degree, vol. 11, pp. 103–115 (2017)
7. Leskovec, J., Krause, A., Guestrin, C., Faloutsos, C., Faloutsos, C., VanBriesen, J., Glance, N.: Cost-effective outbreak detection in networks. In: Proceedings of the 13th ACM SIGKDD International Conference on Knowledge Discovery and Data Mining, KDD '07, pp. 420–429. ACM, New York (2007)
8. Lu, W., Chen, W., Lakshmanan, L.V.S.: From competition to complementarity: comparative influence diffusion and maximization. Proc. VLDB Endow. **9**(2), 60–71 (2015)
9. Narasimhan, M., Bilmes, J.A.: A submodular-supermodular procedure with applications to discriminative structure learning. CoRR, abs/1207.1404 (2012)

10. Sviridenko, M.: A note on maximizing a submodular set function subject to a knapsack constraint. Oper. Res. Lett. **32**(1), 41–43 (2004)
11. Wang, Z., Yang, Y., Pei, J., Chu, L., Chen, E.: Activity maximization by effective information diffusion in social networks. IEEE Trans. Knowl. Data Eng. **29**(11), 2374–2387 (Nov 2017)
12. Wu, W.L., Zhang, Z., Du, D.Z.: Set function optimization. J. Oper. Res. Soc. China **12** (2018)
13. Zhang, H., Zhang, H., Kuhnle, A., Thai, M.T.: Profit maximization for multiple products in online social networks. In: IEEE INFOCOM 2016 - The 35th Annual IEEE International Conference on Computer Communications, April, pp. 1–9 (2016)
14. Zhu, J., Zhu, J., Ghosh, S., Wu, W., Yuan, J.: Social influence maximization in hypergraph in social networks. In: IEEE Transactions on Network Science and Engineering, p. 1-1 (2018)

Printed in the United States
By Bookmasters